Charles L [illegible] Jr.
San Francisco August 1980

RAILWAY MECHANICAL ENGINEERING

A CENTURY OF PROGRESS

Car and Locomotive Design

ASME RAIL TRANSPORTATION DIVISION

THE AMERICAN SOCIETY OF MECHANICAL ENGINEERS
United Engineering Center • 345 East 47th Street • New York, N.Y. 10017

Library of Congress Catalog Card Number 79-54377

Foreword

The commemoration of the 100th anniversary of ASME gives recognition to the contributions made by mechanical engineers to industry and society over the past century. It also looks ahead to continued pushing back of the frontiers of science to provide needed and useful systems, products and services to the general public.

Transportation of goods, raw materials, agricultural products as well as people became a pressing need very early in the history of the dispersed population of North America. The 9 million sq miles (approximately 23 million sq km) of this continent was not blessed with interregional water routes to easily connect east and west. Overland transportation early assumed major importance. The railway first supplemented and later largely supplanted canal and river boat services because it could easily go over and through both mountains and deserts. The railway provided shorter, more direct routes and faster transportation service for large masses of raw materials and finished products.

Mechanical engineering as a separate and distinct discipline evolved in the middle third of the 19th century. The early push of mechanical engineering was the development and application of steam power to general industry and, in particular, to railroading. In the late 1800's and early 1900's mechanical engineers developed the internal combustion engine, which became the mainstay of the North American transportation system. This type of power became the major competitor to railroading beginning about 60 years ago. However, the railroad borrowed the internal combustion principle and applied it to motive power to enable the North American railroads to handle far more freight today than was transported during the peak years of World Wars I and II.

Railway mechanical engineers in North America developed rolling stock and train handling proceedures for transport of far bigger and heavier freight tonnages per train than anywhere else in the world. These railway mechanicals were ingeniously adaptive. While many of the initial improvements were of British or European origin, North American mechanical engineers put these in place and to work in grand style. This adaptive work and the improvements beyond are the hallmark of engineers on the west side of the Atlantic.

Breakthroughs originating in North America were essential for true high capacity interchange railroading. These include the quick action automatic air brake, the automatic car coupler, the friction draft gear, and the track circuit essential for modern signalling systems.

Virtually all of these developments were interdisciplinary in their practical application to reliable and usable products.

While largely overshadowed by recent developments in other more publicly noticed fields, railway mechanical engineers continue to make major contributions. The Rail Transportation Division feels the past contributions and prelude to future progress should be recorded. Thus we present this brief history of the major and landmark developments in the railroad and rail transit ("mass transit") field.

The History and Technology Committee of Rail Transportation Division wishes to thank all of the generous contributions provided by well over two dozen individuals.

D. G. Blaine, Chairman RR-10
William H. Chidley
Paul H. Baker
Norman E. Bateson
Carl E. Tack
J. P. Van Overveen

OFFICIAL MEMBERS OF RAIL TRANSPORTATION DIVISION COMMITTEE RR-10 HISTORY AND TECHNOLOGY, AS CONSTITUTED IN 1976

D. G. Blaine, *Chairman*
J. P. Van Overveen, *Vice Chairman*
Paul H. Baker
Norman E. Bateson
Robert Byrne
William H. Chidley
Rene H. Brodeur
Bernard J. Eck
Sergei G. Guins
Gus D. Holabeck

LIST OF REVIEWERS

Norman E. Bateson
Retired Vice President, Engineering and Research
Pullman Standard

Robert Byrne
Chief Mechanical Officer
Southern Pacific Transportation Co.

William H. Chidley
Chairman, Committee on Railroad Materials
American Iron & Steel Institute

Bernard J. Eck
Director, Product Development
Griffin Wheel Company

Sergei G. Guins
Retired Director of Mechanical Research
Chesapeake & Ohio — Baltimore & Ohio Railroad System

Mark F. Hengel
Mechanical Engineer
Missouri Pacific Lines

Leonard A. McLean
General Superintendent, Equipment Engineering
Seaboard Coast Line Railroad

Andreii A. Potter
Dean Emeritus, School of Engineering
Purdue University

Robert W. Radford
Chief Mechanical and Electrical Engineer
Canadian National Railways

Charles F. Roselius
Vice President, Engineering
Ellcon National, Inc.

Carl E. Tack
President
Trans Dyne Inc.
Retired Vice President, Engineering
American Steel Foundries

J. P. Van Overeen
Mechanical Engineer
Bay Area Rapid Transit District

AUTHORS AND CONTRIBUTORS
(Listed in order of appearance)

David G. Blaine
Senior Marketing Representative
Westinghouse Air Brake Division, WABCO

E. Thomas Harley
Chief Mechanical Officer
Trailer Train Company

Paul H. Baker
Manager, Locomotive Projects
General Electric Company

R. A. Rinehart
Locomotive Department
General Electric Company

J. C. Caldwell
Locomotive Department
General Electric Company

M. A. Sleigh
Locomotive Department
General Electric Company

A. C. Bieber
Locomotive Department
General Electric Company

*A. Norman Addie**
Assistant Chief Engineer, Development, Electro Motive Division
General Motors Corporation

Charles W. Parker
Chief Mechanical Engineer
Canadian Pacific Consulting Services

Kelly J. Arrey
Chief of Car Design
Canadian National Railways

William G. Blevins
Senior Mechanical Engineer — Motive Power
Canadian National Railways

William P. Manos
Consultant, Formerly Vice President, Research and Development
Pullman Standard, Inc.

William Van der Sluys
Retired Vice President, Passenger & Transit Engineering
Pullman Standard, Inc.

*Deceased May 1979.

John H. White, Jr.
Curator, Transportation Division
The Smithsonian Institute

F. T. Burgess
Patent Counsel, GSI Engineering
General Steel Industries

Keith L. Jackson
Vice President and General Manager, GSI Engineering
General Steel Industries

Oliver M. Chiesl
Manager, Product Presentation
The Vapor Corporation

O. B. Capps
Manager, Railroad Service
Edgewater Steel Co.

Bernard J. Eck
Director Product Development
Griffin Wheel Co.

John Zupez
Retired
Abex Corporation

Harry M. Jones
Asst. Chief Engineer, Railroad Division
The Timken Co.

Charles F. Roselius
Vice President, Engineering
Ellcon National Co.

Robert B. Love
Asst. Vice President
American Steel Foundries

Carl E. Tack
President
Trans Dyne Inc.

D. J. Albanese
Vice President, Technical and Overseas Operations, National Castings Division
Midland Ross Corporation

William D. Wallace
Vice President, Technical Sales
Cardwell Westinghouse Co.

V. Terrey Hawthorne
Vice President, Rail Dynamics, Inc.

Grant Arrasmith
Director of Railroad Test Laboratories
Abex Corporation

J. P. Van Overveen
Mechanical Engineer
Bay Area Rapid Transit District

FIGURE CREDITS

We wish to thank the following organizations and persons for allowing us to use their illustrations in this book.

ARTICLE 1

Figures	*Credit*
16, 17	D. G. Blaine
15	C.T.A.
21	E. T. Harley
12, 13	F. A. King, DMIR Ry.
1-3, 7, 8	Nordbok (*Lore of the Train,* pp. 40, 61, 94, 80, 93)
4-6, 9-11	Simmons-Boardman Publishing Corp. (*1878* and *1888 Car Builders' Dictionary*)
14	(*Railway Age*)
19	UPRR
18	WABCO
20	WABCO-AMTRAK

ARTICLE 2

14	G. M. Best
24, 25	DMIR
42, 43	Hawker-Siddeley
27	R. M. Co.
3, 4, 6	Simmons-Boardman Publishing Co. (*1879 Car Builders' Dictionary*)
11, 12	(*1888 Car Builders' Dictionary*)
2, 5, 7, 8, 10, 13, 15-23, 26, 28-35, 36-40	WABCO
44	WABCO-AMTRAK
9	WABCO (*Blackall's Air Brake Catechism–1894*)
1	US & S Division of WABCO (*Air Brake Test*)

ARTICLE 3

1-3, 5, 7-9, 11, 12, 14-17, 19, 21, 22, 24, 26-29, 31, 32	E. T. Harley
33A-33C	PRR (*Modern Locomotives and Cars – 1939*)
20A, 20B, 23, 25, 30A-30C	Simmons-Boardman Publishing Co.
4, 6, 10, 13, 18	C. M. Smith

ARTICLE 4

1-44	Electro-Motive Division, G.M.C.

ARTICLE 5

1-1 – 4-11	General Electric, Erie Plant

ARTICLE 6

Introductory figure, 1-7	Canadian Pacific Railway

ARTICLE 7

Figures	*Credits*
10	Alcan
3-6, 11, 12, 14, 16, 17	Canadian National Railways
1, 2, 8, 13, 15, 18, 19	Canadian Pacific Railways
9	Hawker Siddeley Canada
7	Railfare

ARTICLE 8

3B	D. G. Blaine
3C	F. A. King, DMIR Ry.
1-3A, 4-72	Pullman Standard

ARTICLE 9A

1-13	Pullman Standard

ARTICLE 9B

1	Boston and Main Railroad
7	Hall of Records
3	Illustrated London News, April 10, 1852, p. 285
12	Norfolk and Western
2	Peale Museum
10	Pullman Car & Manufacturing Corp.
4-6, 8, 9, 11, 14	John White, Jr., Smithsonian Institute
13	Dang Worman

ARTICLE 10

25, 26, 28, 29, 33	General Steel Industries
22, 23	St. Louis Car. Co.
1-21, 23, 24, 27, 30, 31	U. S. Patent Office

ARTICLE 11

1-16	Vapor Corp.

ARTICLE 12

1-8	Edgewater Steel Co.

ARTICLE 13

1-4	Griffin Wheel Co.

ARTICLE 14

1-9	Abex Corp.

	ARTICLE 15
1-12	The Timken Roller Bearing Co.
	ARTICLE 16
Figures	*Credits*
6	Ajax Consolidated
5, 9	D. G. Blaine
14	D. G. Blaine (Ajax Consolidated)
1-4, 7, 8, 10-12	C. F. Roselius
13	Universal Railway
	ARTICLE 17
2-10, 12, 14, 16, 18-20, 22, 23	American Steel Foundries
11, 13, 15, 17, 21A-21D, 24	Simmons Boardman Publishing Co. (*Car Builders' Dictionary*)
	ARTICLE 18
1-15	Midland Ross Corp., National Castings Division
	ARTICLE 19
3-8, 10, 11, 13, 14, 16-21	Cardwell Westinghouse
1, 2	Simmons-Boardman Publishing Co. (*1879 Car Builders' Dictionary*)
9, 12, 15	WABCO
	ARTICLE 20
1-7	Rail Dynamics, Inc.
	ARTICLE 21
1-17	Abex Corp.
18-21	D. G. Blaine
	ARTICLE 22
25, 26	BARDT
31, 32	Boeing-Vertol
27-29, 33	Harold E. Cox (*Electric Railway Car Trucks*)
7-13, 15, 16, 21-23	Harre W. Demoro
17-19	G. Krambles, C.T.A.
C9	London Transport Board
C6-C8	London Transport Executive
C10	London Transport Scientific Adviser
C1-C5	L.P.T.B. Railways
24	NYCTA
20	Pullman Standard
14	Herman Rinke
35	Rockwell National, Atchison Division
34	Rockwell – Pullman Standard
Introductory figure, 1-6, 36	J. P. Van Overveen
30	Westinghouse, Buffalo Division

Contents

1 Introduction and Overview 1
D. G. Blaine

2 Train Braking and Speed Control Systems 21
D. G. Blaine

3 The History of the Steam Locomotive in the United States 57
E. T. Harley

4 The History of the Diesel Electric Locomotive in the United States 103
A. N. Addie

5 One Hundred Years of Progress in Railway Mechanical Engineering–Electric Locomotive Development 137
P. H. Baker, A. C. Bieber, J. C. Caldwell, R. E. Rinehart, and M. A. Schleigh

6 The Development of Remote Multiple Unit Locomotive Control 161
Charles W. Parker

7 Canadian Rolling Stock Developments 171
Kelly J. Arrey and W. G. Blevins

8 The Evolution of Rail Freight Vehicles in the United States 185
William P. Manos

9A Railroad Passenger Cars. A Century of Mechanical Engineering 217
William Van der Sluys

9B A Short History of the Railroad Passenger Car 231
John H. White, Jr.

10 The History and Development of Railway Passenger Car Trucks in the United States 245
F. T. Burgess and K. L. Jackson

11 Temperature Comfort in Rail Transportation Vehicles 275
O. M. Chiesl

12 Wrought Steel Passenger Car and Locomotive Wheel Development 293
O. B. Capps

13 Cast Steel Freight Car Wheel Development 301
Bernard J. Eck

14 Technological History of Plain Journal Bearing Development 307
John Zupez

15 Railroad Roller Bearing Development 313
Harry M. Jones

16 100 Years of Progress in Railway Car Hand Brakes 323
Charles F. Roselius

17 History of the Railway Freight Car Truck in the United States 337
Carl E. Tack and Robert B. Love

18 The History and Development of Standard Railroad Car Couplers 357
D. J. Albanese

19 An Abbreviated History of Draft Gears for American Freight Cars 371
William D. Wallace

20 100 Years of Progress in Railway Mechanical Engineering–Special Freight Car Cushioning 393
V. Terrey Hawthorne

21 Railway Friction Materials

Development of Brake Shoes: Part I – Cast Metal 399
Grant Arrasmith

Development of Brake Shoes: Part II–High Friction Composition 409
D. G. Blaine

22 Rapid Transit Cars, Trucks, Propulsion–100 Years Development 415
J. P. Van Overveen

I Introduction and Overview

D. G. BLAINE

In 1880, when the American Society of Mechanical Engineers was founded, Railway Mechanical Engineering had just emerged from what would be called an art to a true professional engineering discipline. The beginnings of the profession have their roots entwined in the development and application of the steam power plant to the land transportation field in England and Wales during the early 1800's. Importation of the railway to the U.S. occurred in the 1820's. It was then recognized by many that the great continental overland distances and forbidding mountain ranges were not amenable to conquest by a river and canal system.

Early railway engineers were rather broad gauged in their thoughts and actions. Railway engineers of the period had to be concerned with the layout and construction of the railway route and line as well as the provision for the necessary stations, shop maintenance and fueling facilities. In addition, they had to be concerned with the details of locomotive and car construction.

THE LOCOMOTIVE ENGINE

Major thought and consideration went into the design, construction and maintenance of the steam locomotive engine. Reasonably reliable motive power was essential for a practical transportation system to serve the public needs.

Following the lead of Richard Trevethick and George Stephenson, North American locomotive engineers quickly adapted the horizontal fire tube boiler to the rather rigid confines of the railroad clearance diagram. Some experimentation with vertical boilers on the Baltimore and Ohio showed that the heating surface and steaming rate couldn't be met well with this type. The width of the locomotive was limited to about 9 ft and the height to about 12 ft. The width was limited by the practical considerations of real estate required. Track center spacings were from 12 to 13 ft. Clearance on curves for passing trains was a prime consideration. The British standard gauge of 4 ft 8½ in. (1435 mm) soon became the general standard, although the original Erie R.R. was built with a 6 ft track gauge (1829 mm). Several Southern Lines used a 5 ft gauge (1523 mm).

The cramped quarters for the fire box and boiler were not conducive to optimum combustion and heat transfer efficiency. Thus it became common to force fire a locomotive with the steam from the cylinder exhaust used as a blast aspirator up the stack. This created a considerable forced draft through the fire bed. Wood was the most common fuel in North America since it seemed there was no end to the hard and softwood forest stands.

This solution to producing enough steam to fill the cylinders at modest speed plus the cramped boiler and fire box dimensions obviated use of the condensing principle. Incomplete combustion was the rule of the day. Flaming red hot sparks spewing forth from the stack created early public liability damage claims by farmers and others within about 500 ft of the tracks. This led to a whole panoply of inventions for spark arresting gear which didn't spoil the draft through the firebox. The familiar antique locomotive print and picture usually has the front of the engine adorned by an enormous ballon, diamond or V shaped stack complete with iron netting and a spark ash collector. These became the rule of the day by the 1840's. Wood fired locomotives persisted until after 1900 in heavy lumbering and timber country. In the British Isles scarcity of wood led early to the use of coal as the predominant locomotive fuel.

The steam locomotive engine had other characteristics which worked against optimum efficiency of the power plant. Cold water was introduced into the boiler from the tank on the tender by feed water pumps usually driven from the piston crossheads. Another factor involved the wide variation in speed required for the locomotive to perform its normal work. High tractive effort was required at low speeds with very little horsepower and high boiler horsepower was required at moderate and high speeds.

Boiler and fire box design occupied the time of many engineers. The water leg fire box became universal. The 50-60 psig (4.6-5 bars) boiler pressures gave way to 80-100 psig (6.5-7 bars) by the 1860's and jumped to 120-140 psig (9.5-10.9 bars) range by the late 1870's and early 1880's.

Much inventive thought went into the development of valve gears. The two cylinder double acting engine with the cylinders outside the frame and wheels became the North American standard. By the mid 1870's the predominant valve motion was the relatively compact Stephenson valve gear mounted inside the wheels. This was driven by main axle eccentrics and generally implied a curved slotted link for easy reversing and variable cut off control. The Walschaerts valve gear mounted outside the driving wheels was developed with lap and lead control.

The basic size of the locomotive was primarily controlled by the weight (mass) of the trains which had to be handled to meet the throughput (service) demands of the public. Second, the size of rail and general quality level of the track structure itself limited the weight on driving axles. Up until the 1870's a locomotive producing 10,000 lb starting tractive force was adequate for most railroads. The generally poor horizontal and vertical rail alignment coupled with soft dirt ballast and sharp curves required a very flexible locomotive. The 4-4-0 wheel arrangement having a four wheel equalized pony or guiding truck and two driving axles came to be called the "American" type. This was the rule of the day. The boiler could produce enough steam for about 200-400 horsepower at the rail. This was adequate to handle 200-600 tons of cars over most terrain at 15-25 mph. Trains of from 10-30 eight-wheeled double truck cars of that day might have reached 1000 ft in length.

For heavy dense traffic areas and for long significant mountain gradients, locomotives with three or four driving axles were used. However, axle loads had to be kept to about 20,000 lb (9.5 MT). Weight of the "T" headed rail ranged from about 45-60 lb/yard (23-29 kg/m). Steel rail was far too expensive until the broad introduction of the Bessemer process in the late 70's and 1880's.

FREIGHT AND PASSENGER CARS

Early development of freight and passenger car rolling stock in North America departed from prevailing practice in Britain and Continental Europe. On the eastern side of the Atlantic, two axle and three axle rigid frame car designs sufficed for the bulk of passenger and freight stock. Quality and maintenance of their track structure was far superior to that prevailing in most of North America. Train lengths were short and rarely exceeded 40-60 car axles.

Another important factor, not well understood, is the fact that the British Isles and Western Europe were densely populated and the emphasis was upon heavy, frequent passenger service. Freight transport in this area was and still is fundamentally by waterway. Two-thirds of the freight tons travel most of their way by ship and barge. There is hardly any major population center more than 100 miles (160 km) from a navigable waterway.

Responding to the North American conditions, railway engineers quickly found a car with two swivelling trucks (4 axles-8 wheels) rode far better than a train of rigid frame two- or three-axle cars on North American tracks. Generally, prior to the American Civil War, there was little need for any but the simplest suspension systems. Many freight cars had no springs at all, and others had a coil or semi-elliptic spring above each journal bearing box. However, passenger cars generally had equalized trucks which had a primary coil spring system. To improve lateral ride, the swing hanger bolster had been developed which included a second set of springs. This gave acceptable ride quality at 40-60 mph speeds.

By the 1870's, the freight car truck in North America generally had a rigid frame sometimes of wood but generally wrought iron. Springing was introduced between the truck bolster and spring plank at the outer ends of the bolster. The truck bolster bore the weight of half the car on a swivelling center plate. Some freight car trucks made use of the swing motion principle, which was by 1880 the common standard for passenger cars. Cast iron wheels were commonly used with about as many designs as there were foundries.

The coupling system of the 1870's was still the simple link and pin type in the center of the car. This was not only dangerous to operate, but introduced considerable free slack. By the 1870's many passenger cars used early forms of the semiautomatic type. There was no energy absorbing draft gear. European practice had gone to the hook and screw link type coupling with spring loaded side buffers to eliminate free slack and give a more stable lateral-dynamic ride condition.

INTERCHANGE

North American railways found it was quite advantageous to interchange freight and passenger cars as the experience of the National Military Railroads had proven during the 1861-1865 war between the states. Problems surrounded the dozens of individual railway "standards" which then existed. Repair, maintenance and replacement of cars worn or damaged on a "foreign" road created a confused situation and endless acrimonious debate. This prompted formation of the first industry wide standards group in 1867 – the Master Car Builders Association. This voluntary association of U.S. and Canadian railway car engineers and managers set about to bring order out of chaos. A few years later, the American Railway Master Mechanics Association was constituted to bring about similar standardization for locomotive design and components. The Master Car Builders Association ultimately became the Mechanical & Research Division of the Association of American Railroads.

Both associations found their primary duties involved the pragmatic day-to-day problems of performance, maintenance and repair of rolling stock from well over 100 private companies. However, they managed to accomplish several far reaching development and standardization projects. These engineering accomplishments included wheel, axle, journal bearing design, the automatic car coupler, and the quick action automatic air brake.

At the time it was necessary to rely on the engineering and development capabilities of a few major railroads and many of the suppliers including car and locomotive builders. Among the railroads who maintained sizeable design and test laboratory capabilities were the Pennsylvania, New York Central and Southern Pacific. The dynamometer car was developed in this period to autographically record locomotive and train performance.

With the introduction of steel rail in the 1870's it was possible to make major strides forward in the size and weight of locomotives and cars and trains to better serve the burgeoning transportation needs of the continent. The 15-20 ton capacity freight car of the 1870's gave way to the 25 ton, 30 ton and 35 ton sizes. Wood was being pushed to the limit as train tonnages pushed upward from about 500 tons to as much as 1500 tons. Rail size increased to 80-90 lb/yard (40-47 kg/m) allowing 20,000 to 25,000 lb (10-11 tonnes) axle load.

Unfortunately, individual company or personal efforts at improvements were not producing results or equipment which was broadly applicable. Colleges and universities recognized the need for formal engineering curricula in the mechanical area plus laboratories for deliniation of principles, development of new concepts and overall performance evaluation.

Notable but not exclusive in pioneering railway mechanical engineering efforts are Purdue University, University of Illinois, Penn State and Iowa State.

ASME AND RAILROAD DEVELOPMENT

Formation of the ASME in 1880 provided a means and forum to assemble collective engineering thought and progress. This had a major impact and benefit in the steam power and fluid power industry by establishing standards for pressure vessels, piping, fastening systems and performance codes for steam engines, power plants, locomotives, compressors, fans, etc.

During the first decade of our history there were about ten railway engineering papers and discussions. In the '90's the number almost doubled. By the first decade of the new century the number of papers had risen to about 25. Perhaps coincidental, but with Mr. Westinghouse's ASME

Presidency in 1910, the number of papers in transactions more than doubled to over 50 during the "teens" of the twentieth century. Mr. Westinghouse had strong interests in railroad electrification, signalling and automatic train control and the overall subject of improving railway train braking systems.

Electric street railways blossomed out in almost every city over 20,000 population. Interurban electric railway networks blanketed the midwest and northeast. Major cities established and built subway and elevated railways using the electric multiple unit control system developed by Sprague and Westinghouse.

It was during the period of the first two decades of this century that North American railroads cast the die for long and heavy freight train operation to meet the public needs with essentially single track routes over very long distances. This was the only way that privately owned railroads could possibly provide the train movement capacity when several passenger trains a day had to operate intermixed with freight.

The 3000 tractive horsepower steam locomotive became the fundamental building block for motive power consists. Rapid introduction of steel in car construction enabled trains nearly a mile in length and of 4000-10000 tons to be handled. Peak traffic density in World War I required an overall average of 14 freight and passenger trains per day for each of the 290000 route miles in the U.S.A. and Canada.

In 1920 the Railroad Division of ASME was formed giving recognition of the many ASME members active in this field.

The "Roaring 20's" saw consolidation of trends in previous decades since 1900. Locomotive horsepower grew into the 4000-5000 range and freight trains lengthened out to 150 cars with up to 12,000 trailing tons. In very dense traffic areas, separate tracks were maintained for freight and passenger service.

Rail of 130-150 lb/yard (68-79 kg/m) was developed to give reasonable life under 30-50 million gross tons of traffic per year. The private automobile was beginning to eat into railroad passenger service and spelled the end of the electric interurban.

Then followed the Great Depression of the 1930's. In this period railway mechanical engineering activity was focused on the very successful high speed streamlined passenger train, running at 100+ mph (160+ kmh). Most streamliners made use of the then new diesel electric power plant—previously thought of only as an economical way to perform switching services in environmentally critical areas. The diesel electric was a resounding success in high speed passenger service giving most of the advantages of true electrification.

For freight service the steam locomotive reached its finest hour with articulated speedsters capable of generating 5000-6000 rail horsepower at 50 mph. However, the oil or coal fired horizontal fire tube boiler, even at 300 psi and high superheat, couldn't compete at 5 to 6 percent average thermal efficiency with the diesel electric, which had four to five times the overall thermal efficiency.

The super overtime requirements of World War II, with traffic almost twice the peak of the 1920's, simply wore out the steam locomotive fleet by 1946. Then began one of the most astonishing metamorphases seen in modern technology. The diesel electric freight locomotive completely captured the field in the short space of 10 years. About 20,000 diesel electric units replaced almost twice their number in steam locomotives. Virtually all North American railroads were almost completely dieselized by 1960.

Considerable experimentation was made with gas turbine-electric locomotives in the late 50's and early 60's. However this type of prime mover could not compete with the controlability and economics of the diesel electric locomotive power plant.

Interestingly, the success of the diesel electric was a major stumbling block in the road to major expansion of mainline electrification. Electrification had looked very promising in the 1915-1935 period.

In fact, several hundred miles of electrification have been taken out of service since 1960. The ease of multiple uniting with diesel electrics allows almost any practical tractive horsepower to be assembled in a train and with remote radio control placed under the control of one engineman. This means 20,000-25,000 horsepower can be used and trains run as close together as signal spacing will allow. By using helpers, trains with 30,000-35,000 horsepower can be successfully and safely operated. In this case a second engineman is most practical, or radio remote unit control can be used with remote units spaced 55-100 cars back in the train.

This remote multiple unit train horsepower capability is difficult to meet with electric locomotives due primarily to practical limitations of electrical section kilowatt capacity. While within present engineering capabilities, the random demand for slugs of 50,000-100,000 kW of power in remote locations far from commercial networks presents serious economic problems. However, the fuel oil economics could change this picture drastically in the next decade.

The private automobile and airplane finally spelled the demise of once fine intercity rail passenger service in North America.

Personal rapid transit for everyone was national policy based on a seemingly infinite supply of low cost fuel oil and gasoline. Glad to be rid of money loosing empty passenger trains, railroads turned their attention to improving freight operations.

During the last 15 years the Rail Transportation Division has maintained an active interface with the Land Transportation Group of IEEE. We jointly sponsor a two-day Rail Conference each year which covers mass transit rail and other electrical engineering developments of interest to railroads and the commuter transportation business. The division changed its official name from "Railroad" to "Rail Transportation Division" to reflect this broadened coverage. Generally our division has eight technical sessions a year where 30 to 40 papers are presented on technical matters.

Many changes have taken place on the face of railroading in the past decade or so. The composition of the freight car fleet has changed drastically with the infusion of almost a million new cars whose loaded weights are 110 to 157 tons. The 100 ton capacity freight car is the rule of the day. About 80,000 piggyback and container service cars are in place today where there was a mere handful in the late 1950's. Box cars have grown to 90 ft in length with 10,000 cu ft (283 m^3) interiors. ASME railway engineers are in the forefront of these developments.

Improved braking and power control systems allow the operation of 7000-10,000 ton trains at 50 to 70 mph safely and satisfactorily. Very long distance unit coal trains are being operated with equipment utilization five to ten times the common averages of a decade and a half ago.

ASME members are active in working toward solutions for operating and equipment problems which relate to train track dynamics with modern and future services. Many developments are being carried out in cooperative engineering efforts of the Association of American Railroads, Railroad Progress Institute and the U.S. DOT Ground Transportation Research Test Facility at Pueblo, Col.

With assists from the federal governments in the U.S., Canada, and Mexico, major improvements in passenger services are underway. The realities of the finite petroleum supply and rising fuel costs make a revitalization of intercity rail passenger service a necessity. This area is providing opportunities for major engineering contributions from the Rail Transportation Division. The day of the 100-125 mph (160-200 kmh) passenger train will return soon. Operating speeds of 150 mph (250 kmh) can be practically accomplished.

The following sections review the landmarks, developments and major advancements in railway mechanical engineering for cars and locomotives from the inception of railroad engineering in North America 150 years ago and highlight developments since the founding of ASME in 1880.

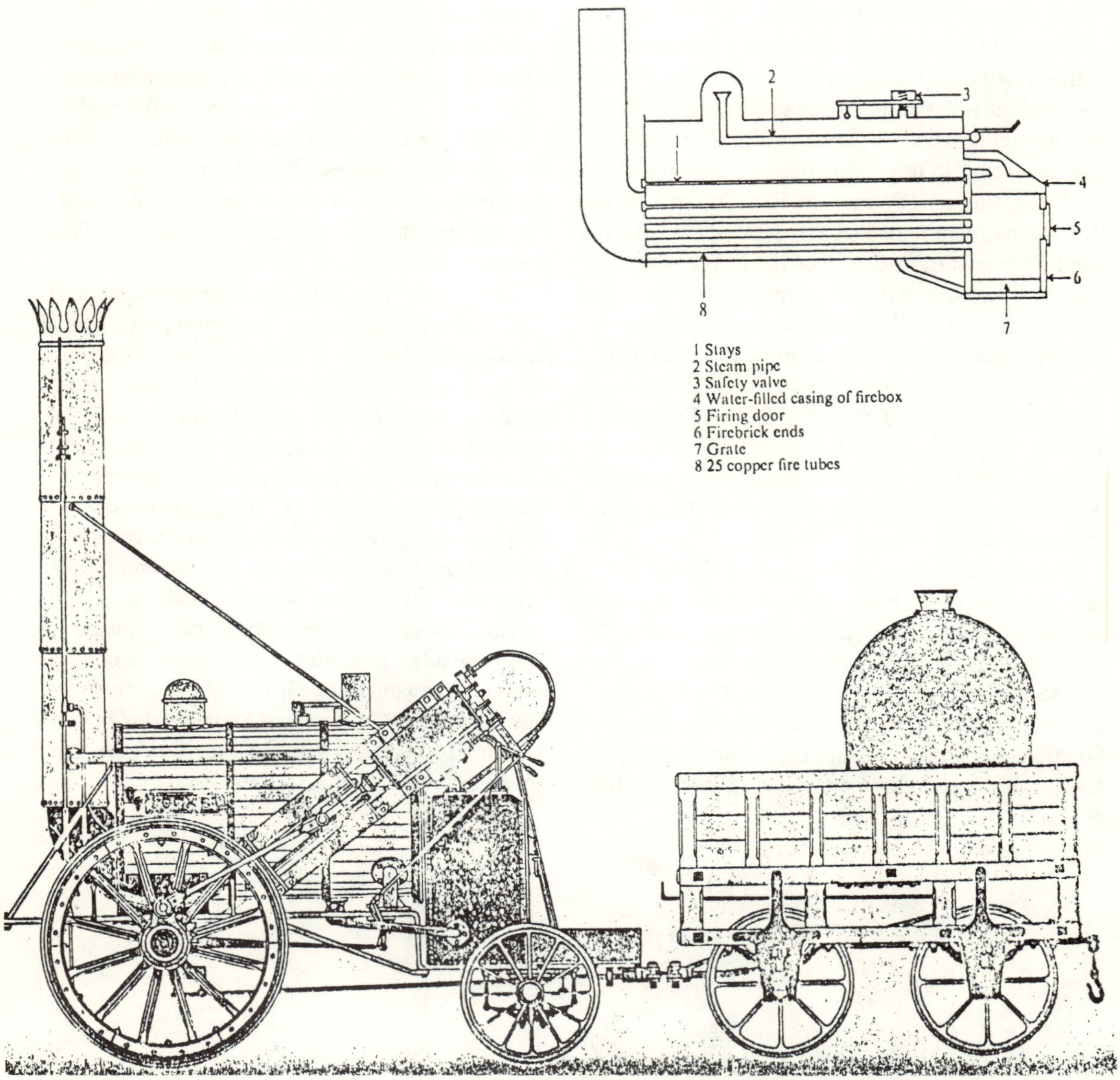

FIG. 1 STEPHENSON'S "ROCKET"

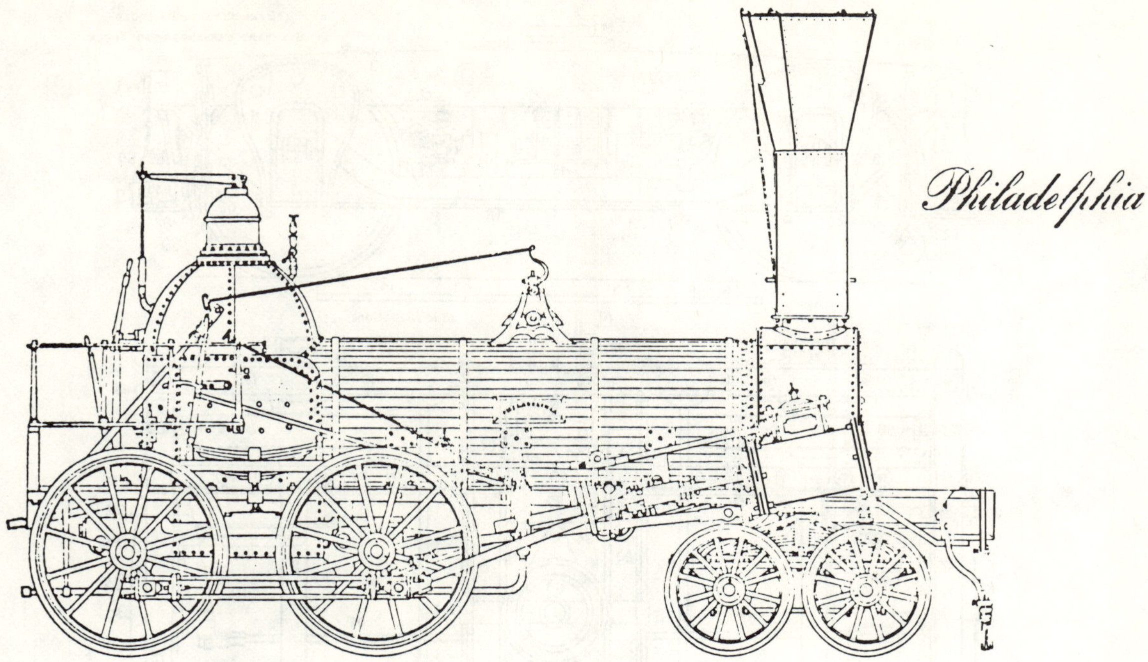

THE [illegible] PHILADELPHIA EXEMPLIFIES WILLIAM NORRIS'S "IMPROVED EIGHT-WHEEL LOCOMOTIVE" WHICH HAD THE BURY BOILER AND BAR FRAMES FROM ENGLAND, PLUS THE AMERICAN LEADING TRUCK AND TRAILING COUPLED AXLE.

FIG. 2 THE "PHILADELPHIA" BY NORRIS – 4-4-0 OR AMERICAN TYPE

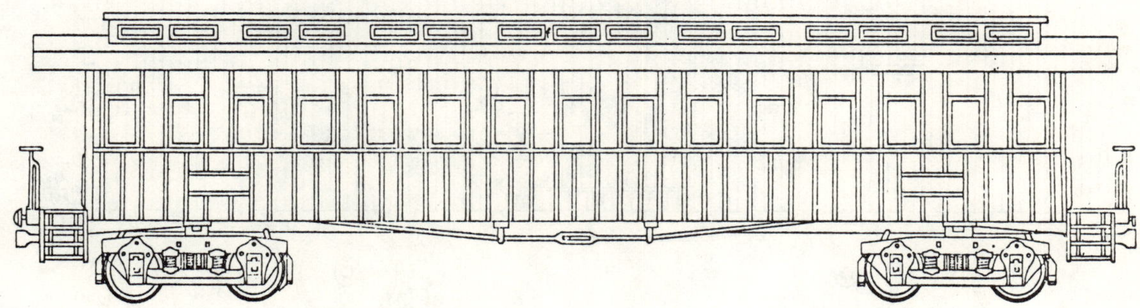

This 1865 American car was in general use on most railroads. It was 50 feet in length and weighed just under 16 tons.

FIG. 3 TYPICAL AMERICAN PASSENGER CAR

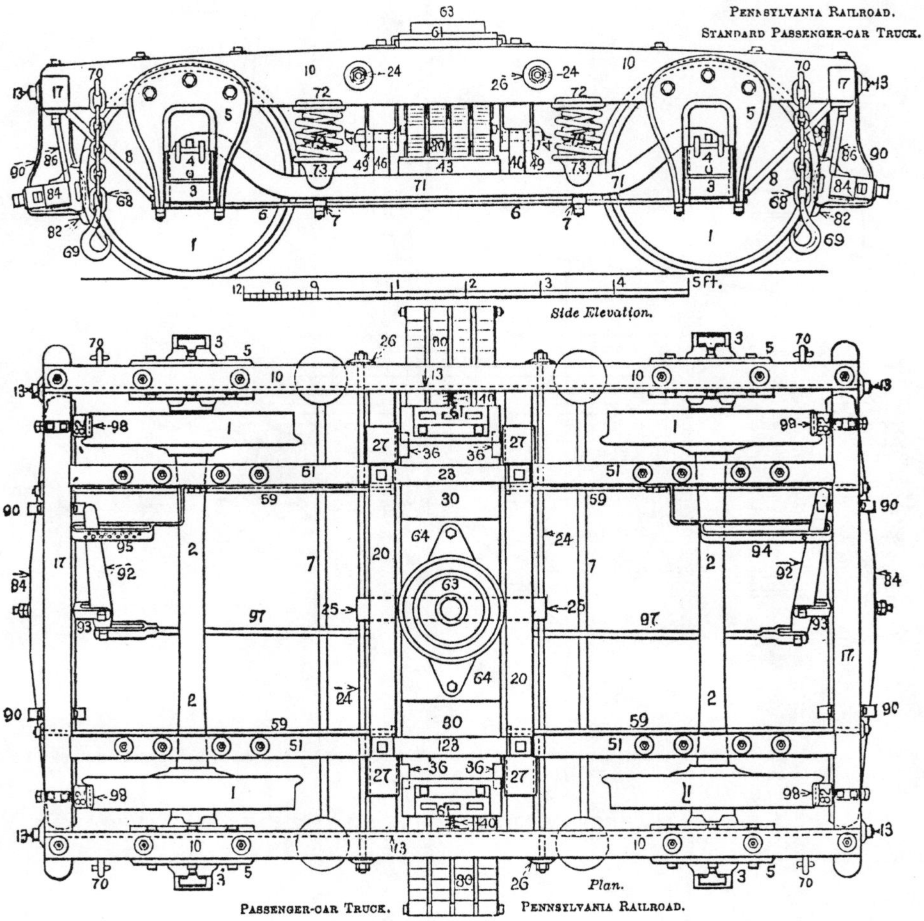

FIG. 4 EQUALIZED, SWING HANGER PASSENGER CAR TRUCK

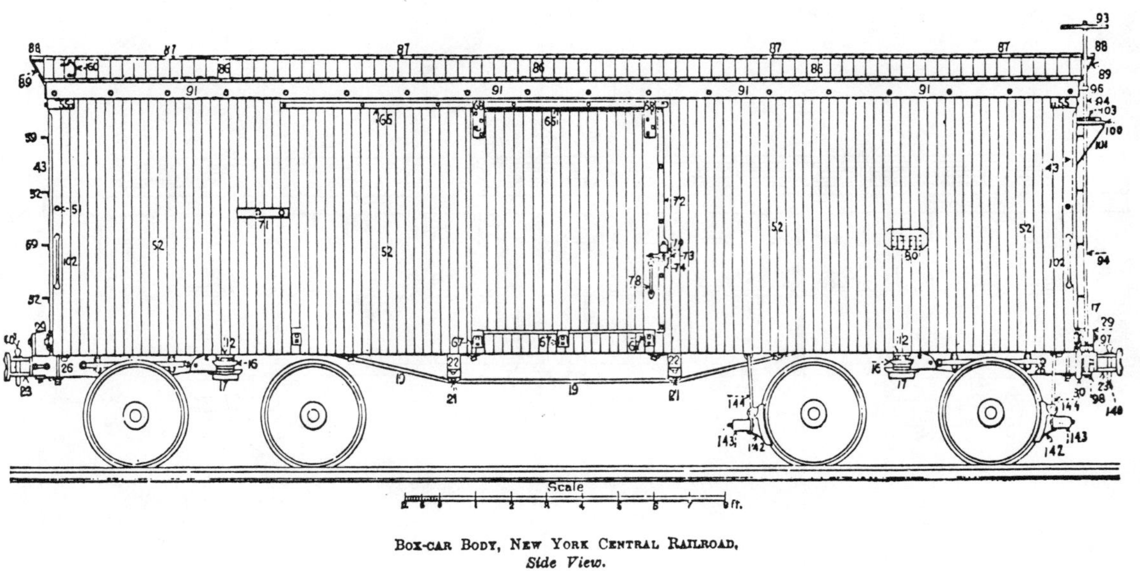

FIG. 5 TYPICAL NORTH AMERICAN BOX CAR 1870's

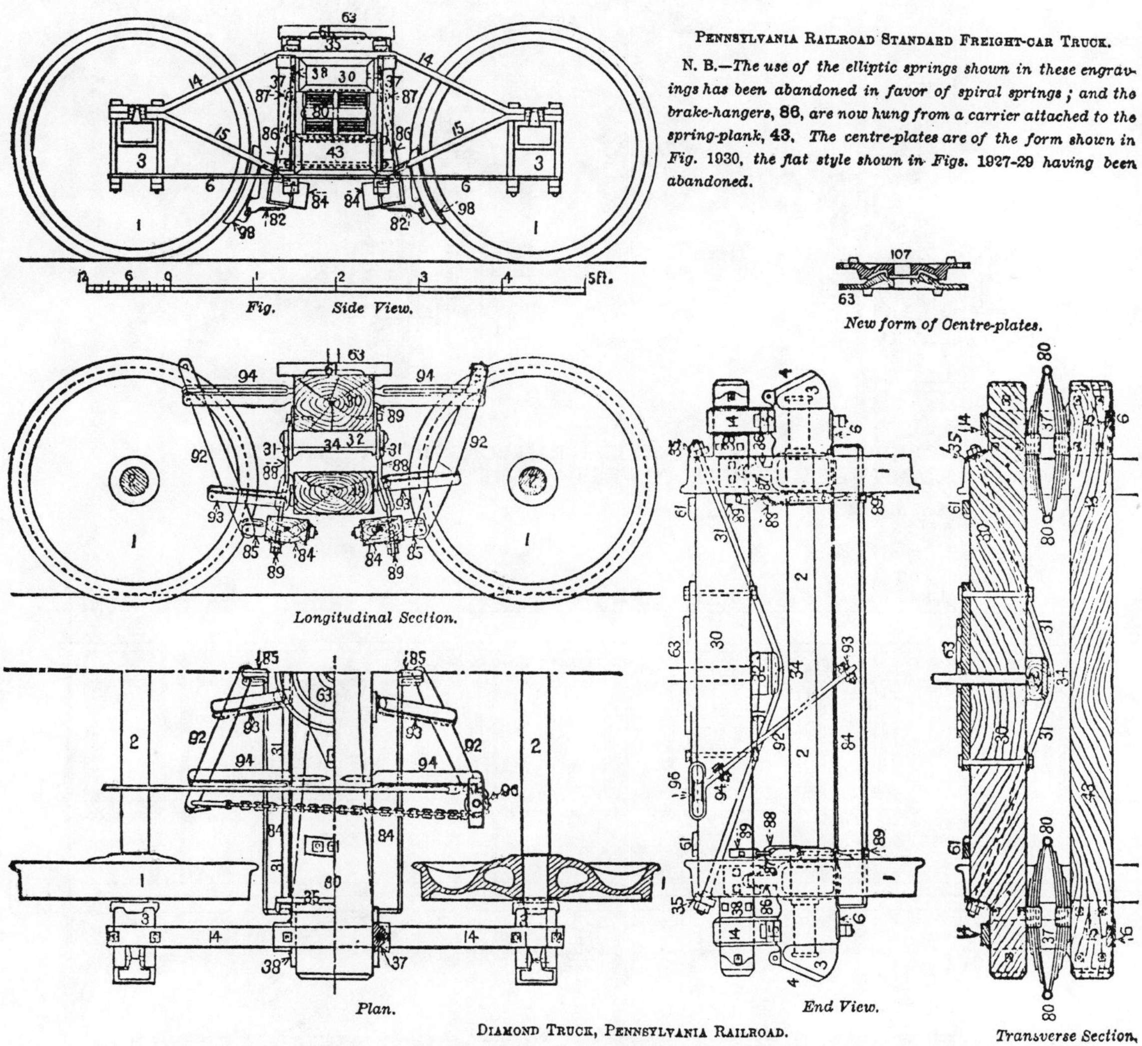

FIG. 6 DIAMOND ARCH BAR TRUCK. COMMON STANDARD FREIGHT TRUCK TYPE 1870's UNTIL THE CAST STEEL TRUCK FRAMES OF 1915 AND LATER PERIODS

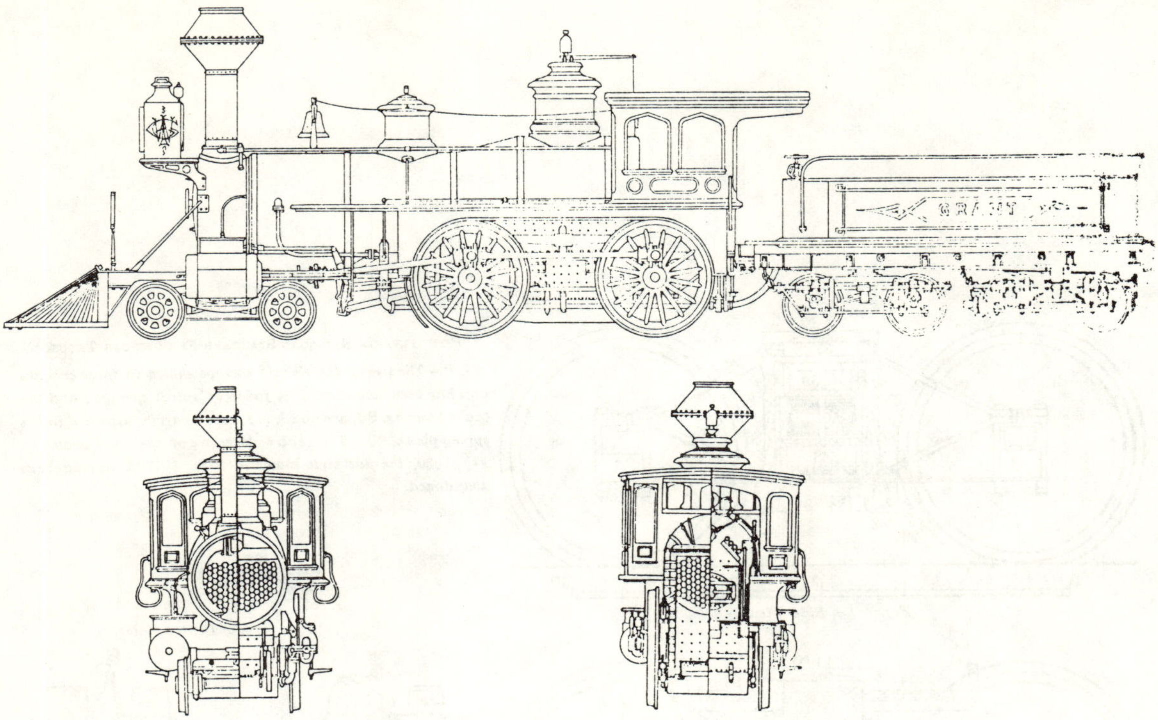

FIG. 7 WORKHORSE OF NORTH AMERICAN RAILROADS IN THE 19TH CENTURY – AN EXAMPLE OF A GRANT LOCO WKS 4-4-0 IN 1873

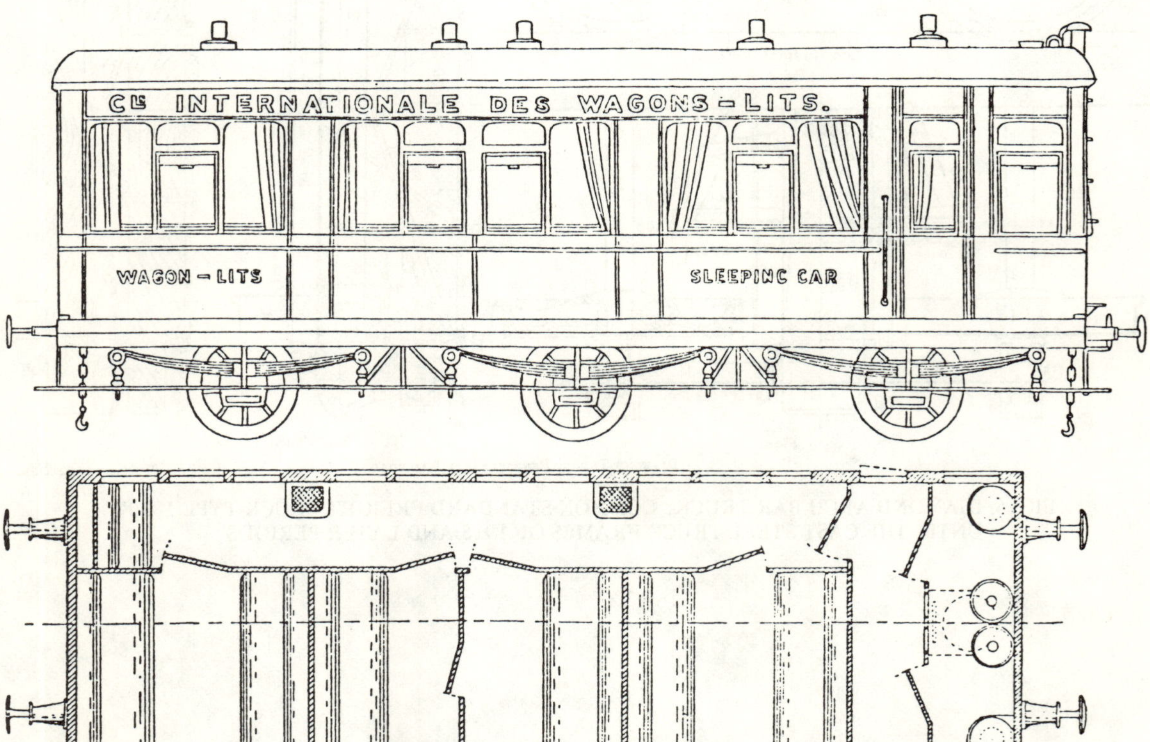

FIG. 8 3-AXLE EUROPEAN SLEEPING CAR

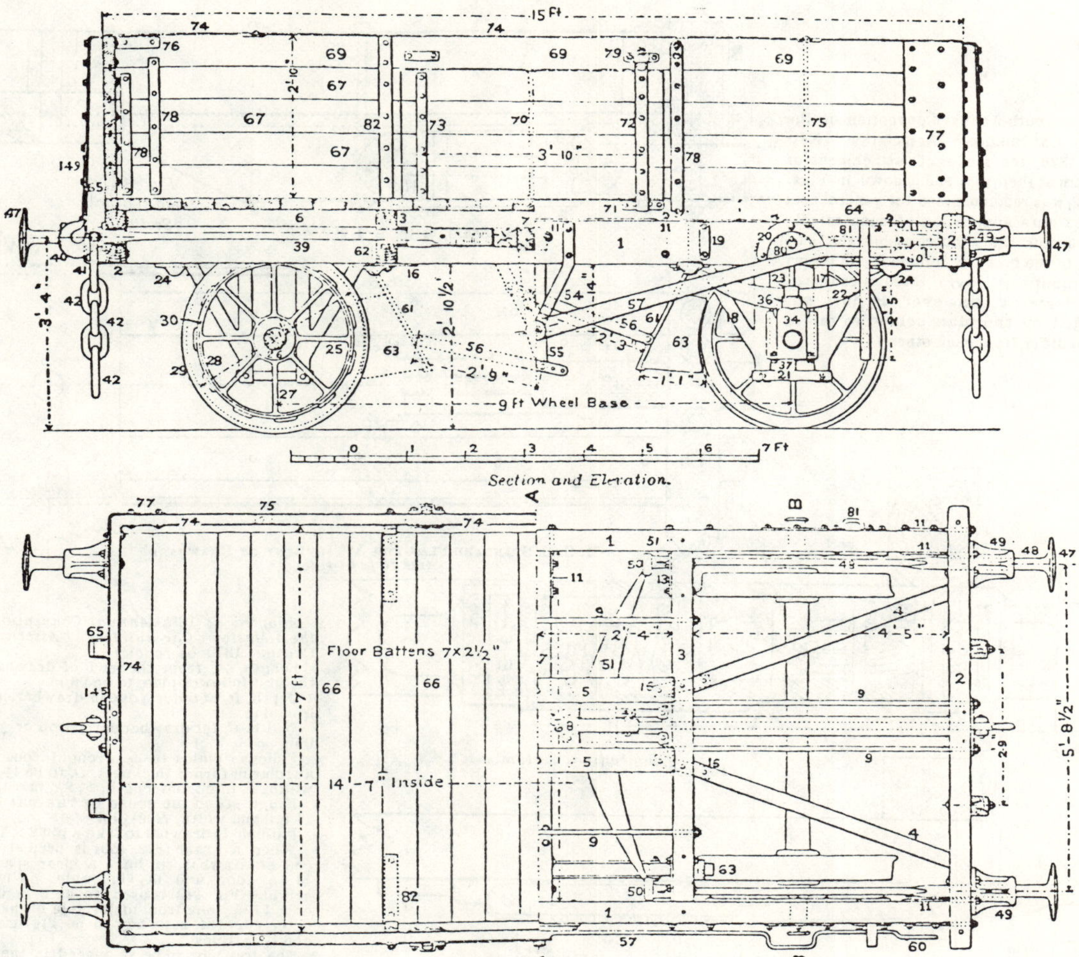

Section and Elevation.

Plan, showing Framing

English Open "Goods Wagon."

Average weight 10,700 lbs. Maximum capacity. 18,000 lbs

General Classification of List of Parts

1-21. *Underframe.* 22-38. *Running-gear.* 39-52, and 145, 146. *Buffing and Draw-gear.* 53-63. *Brake-gear, or Brake-work.* 64-78. *Body.*

FIG. 9 4-WHEEL BRITISH FREIGHT WAGON 9 TON CAPACITY

(At the Tenth Annual Convention of the Master Car-Builders' Association, New York, 1876, the mode of attachment of drawbars at their rear end, shown in Figs. 401-402, was recommended for general use; and all future application of such attachments below the timber under the car-bodies, in the manner "alluded to in report of Committee," were discountenanced. This drawing, however, is now superseded by the plan below, so far as the two differ from each other.)

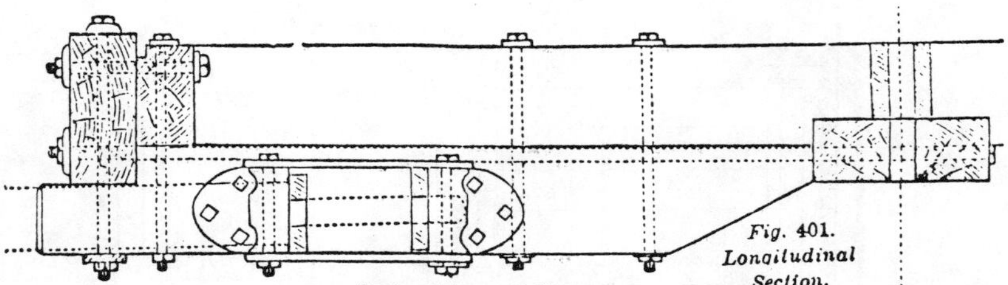

Scale, ¾ in. = 1 ft.

Fig. 401. Longitudinal Section.

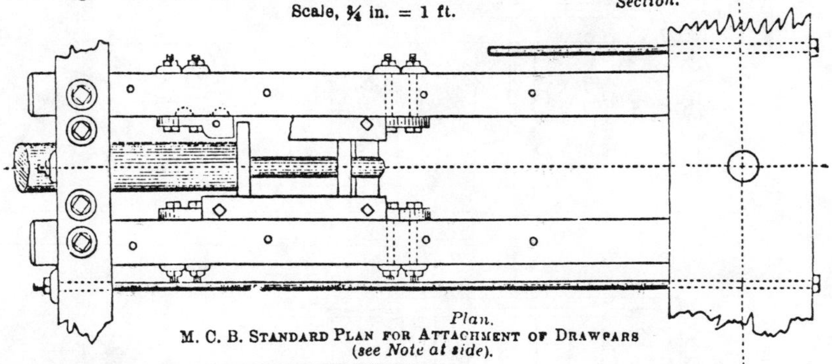

Plan.
M. C. B. STANDARD PLAN FOR ATTACHMENT OF DRAWPARS (*see Note at side*).

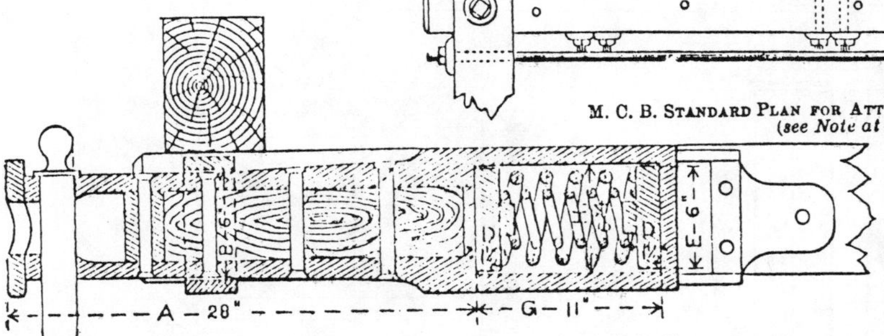

Longitudinal Section.

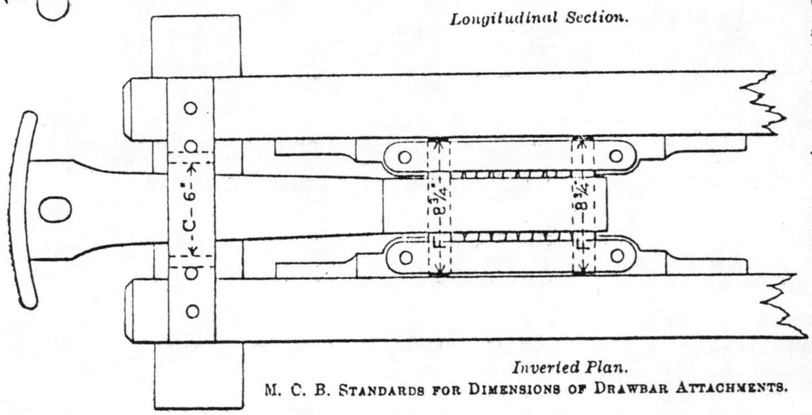

Inverted Plan.
M. C. B. STANDARDS FOR DIMENSIONS OF DRAWBAR ATTACHMENTS.

Adopted at 13th Annual Convention of the Master Car-Builders' Association, Chicago, 1879, as follows:

Length, *A*, from the end of drawbar to the first follower-plate to be 28 in.

Depth, *B*, at outer yoke in drawbar carry-iron, 6 in.

Width, *C*, for drawbar carry-iron or yoke, 6 in.

Follower-plates to be wrought-iron, rectangular in form; thickness, *D*, to be 1½ in.; width, *E*, 6 in., and length, *F*, 8¼ in.

Length, *G*, of the HOUSE OF THE DRAWBAR, 11 in., and width, *H*, 6¼ in.

Pin-hole in drawbar to take a pin 2 × 1½ in.

When a STRAP (Fig. 450) is used at back end of drawbar, to have a clear space of 11 in. long and 6¼ in. wide. When a SPINDLE (Fig. 449) is used, to be 3 in. diameter, 17 in. long from under side of head to outer edge of key; key to be 2⅛ in. wide and ½ in. thick.

The drawbars to be so placed in the cars as to locate the coupling-pins not less than five inches from the centre to the face of the dead-woods.

(*As to the distance, K, the intention of the action taken by the Convention is doubtful.*)

At the same meeting it was recommended that the CAPACITY OF DRAW- AND BUFFER-SPRINGS should not be less than 18,000 lbs. Also, that when the CONTINUOUS DRAWBAR (Figs. 424-7) is used, the length and size of drawbar shall be the same as above.

FIG. 10 LINK AND PIN COUPLING OF 1880's – M.C.B. STANDARD INCLUDING DRAFT SPRINGS

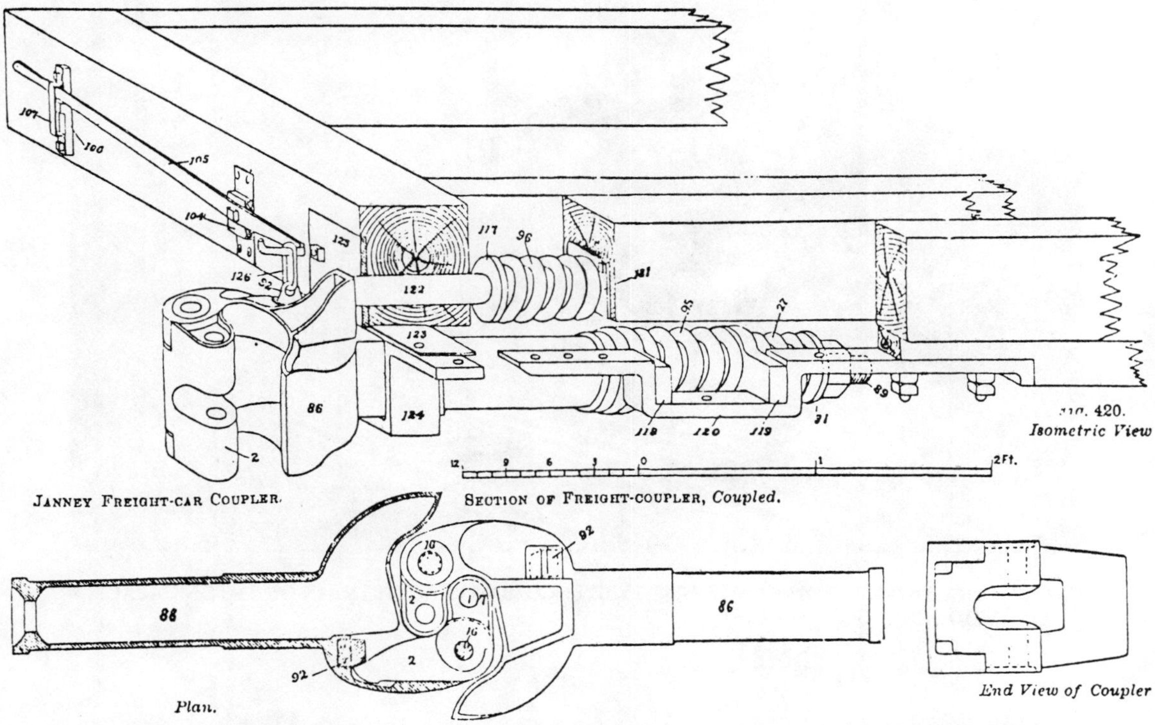

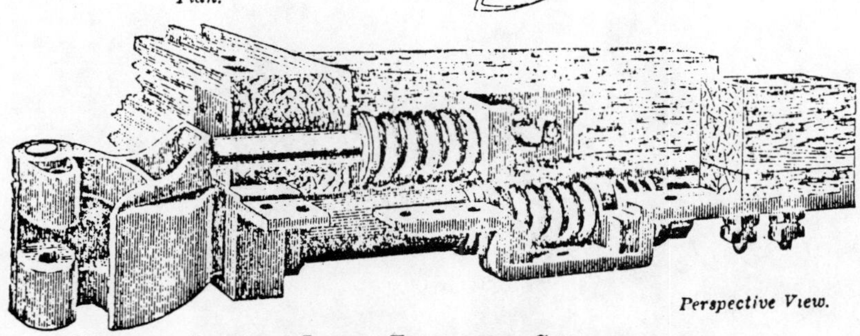

89. *Freight Draft-bolt.**
92. *Freight Coupler-pin.**
95. *Freight Main Draft-spring.**
96. *Freight Buffer-spring.*
104. *Freight Lever Fulcrum-plate*
105. *Freight Lever.*
106. *Freight Lever-catch.*
107. *Freight Lever-plate.*
117. *Freight Buffer-washer.*
119. *Freight Follower.*
120. *Freight Draft-iron.*
121. *Freight Buffer-spring Plate.*
122. *Freight Buffer.*
123. *Freight Chafing-plate.*
124. *Freight Stirrup.*
125. *Freight Face-plate.*
126. *Drop-pin Link.*

NAMES OF PARTS OF JANNEY FREIGHT-CAR COUPLER;

2. *Janney Knuckle.*
16. *Janney Knuckle-pin.*
17. *Janney Coupling-pin.*
27. *Janney Centre Buffer-spring.*
31. *Janney Draft-washer.*
86. *Freight Coupler.**

FIG. 11 JANNEY TYPE AUTOMATIC COUPLER 1880's–GENERIC STYLE BECAME M.C.B. STANDARD

FIG. 12 DULUTH AND IRON RANGE 4-8-0 TYPE LOCOMOTIVE WITH TRAIN OF 25-TON CAPACITY WOODEN ORE CARS – CIRCA 1900.

FIG. 13 DM&N CLASS C-1 CONSOLIDATION NO. 305 BLASTS UP PROCTOR HILL IN 1904 WITH A TRAIN OF 28 ORE EMPTIES. THE DULUTH DOCKS CAN BE SEEN IN THE DISTANCE. THE ENGINE WAS ALMOST NEW, HAVING ARRIVED FROM THE DICKSON WORKS OF THE AMERICAN LOCOMOTIVE COMPANY ONLY TWO YEARS BEFORE. THE RAPID CHANGE TO "BIG" 50-TON CAPACITY STEEL CARS IS UNDERWAY. MOST OF THE TRAIN IS 50 TON STEEL CARS WITH SOME VINTAGE 1895 30-35 TON CAPACITY ALL WOODEN CARS.

FIG. 14 HIGH SPEED PASSENGER TRAIN CIRCA 1900, PENNSYLVANIA RAILROAD, CLASS E-2

FIG. 15 FIRST ELECTRIC MULTIPLE UNIT CAR TRAIN (1898) EQUIPPED RAILROAD – SOUTH SIDE RAPID TRANSIT CHICAGO, ILL.

FIG. 16 HEAVY ELECTRIC INTERURBAN TRACTION – PORTLAND, EUGENE & EASTERN CIRCA 1920

FIG. 17 LIGHT WEIGHT ONE MAN SAFETY CAR CONTROL – STREET CAR TYPICAL OF 1920's – FRESNO CALIFORNIA STREET RAILWAY

FIG. 18 HARBINGER OF THINGS TO COME – LIGHT WEIGHT MULTISECTION CAR TRAINS ON NEW YORK SUBWAY LATE '30s AND EARLY '40s

FIG. 19 BIG STEAM POWERED FREIGHT TRAIN OF LATE 1920's TO 1940's

FIG. 20 MODERN HIGH SPEED DIESEL ELECTRIC FREIGHT TRAIN WITH TWO 6600 HP UNITS–UPRR NEAR SINCLAIR WYOMING

FIG. 21 MODERN HIGH SPEED PASSENGER TRAIN (AMTRAK)

2 Train Braking and Speed Control Systems

DAVID G. BLAINE

INTRODUCTION

In early descriptions and illustrations of rail guided way transport dating from the late 1700's, there are a few mine and stone tramway cars with simple forms of handbrakes. These appear to be adaptations and modifications of the types of friction brakes then in common use on horse drawn passenger coaches and freight or "goods" wagons. Generally, all used some form of linkage so the driver or his assistant could apply brake blocks of wood to the wrought iron wheel tires (tyres) shrunk onto the spoked wooden wheels of the vehicle. There was another method used with the spoked cast iron wheels of early coal tramway cars. This was called spragging and consisted of a large stick or pole thrust between the spokes to cause the wheel to lock and slide, providing some retarding force. The latter method was only usable if speed was held to a brisk walk!

By the middle 1820's, rapid progression of the steam locomotive engine power plant allowed early railway trains in England to attain speeds of 10-25 mph (16-40 kmh). In order to stop a train, the only method available was to shut off steam and let speed drift down to 5-7 mph (7-10 kmh) and then signal the brakemen to apply the handbrakes on the few cars equipped. To assist in stopping at stations, a short sharp upgrade was usually graded into the track profile approaching the station. This also helped departing trains to accelerate.

Interestingly, the first dean of British railway engineers, George Stephenson, found it necessary to invent the first known power brake. This was a valve and a small steam cylinder which pushed brake blocks against the cast iron driving wheel treads, creating retarding force to slow the train and hold it stopped when on grades. Although not mentioned specifically in texts of the time, engravings of the 1830-1833 period show this obvious steam brake cylinder on a few locomotives.

The railway transportation idea spread like wildfire in Britain and in western Europe during the 1840's and 1850's. Generally high quality road bed and track structure was the rule rather than the exception. Very soon railway passenger trains were operating at 30, 40, and even 50 mph (50-80 kmh). Obviously, at these speeds some positive braking arrangement was needed to insure that a following train could stop short of a collision in case the train ahead made an unscheduled stop. Most passenger cars were equipped with a high leverage handbrake which could be applied by a trainman upon receipt of a proper signal from the driver.

On high traffic density routes British railroads soon found it necessary to divide the track into specific sections or "blocks" with the entrance to each block controlled by a signal operated by a signalman in a "block house" or signal "tower" spaced every few miles.

However, in Britain and Europe, freight or "goods" trains remained rather short, being made up of a few simple four-wheeled 4-6 tonnes capacity cars. Speeds were kept low to moderate and few if any freight cars had any form of brakes. The need for long distance very high volume freight shipment per train was and still is not required because of the easy availability and access to water transportation.

As passenger train speeds inched up into the 60 mph range (100 kmh), in Britain and Europe it became apparent that some form of train power brake was desirable and in some cases essential to insure safe operation. It was realized that the steam locomotive was capable of exerting negative or braking tractive effort. The Le Chatelier or counter pressure brake was developed in which the valve gear was reversed and hot water at boiler pressure introduced into the steam chest. This flashed into wet steam which was drawn into the cylinders and compressed with the valve gear in reverse. This system was widely applied in Europe where it was a very useful adjunct to the hand braked cars. However, the retarding effect varied with the weight (mass) of the train. This first true locomotive dynamic brake was applied in mountain regions of North America where it was generally called the "water brake".

In England and France a form of continuous power brake was applied to some passenger trains in the 1860s. This was a vacuum brake with all cars connected by a vacuum brake pipe to the locomotive. The driver (engineman) could apply and release brakes on all wheels of the train by varying the vacuum. Vacuum was created by a steam ejector or aspirator on the locomotive. Maximum practical differential across any vacuum brake cylinder piston was only about 10 psi (2/3 atm or approximately 0.7 bar). Therefore, vacuum brake cylinders were very large and cumbersome.

Railway development in the U.S. and Canada proceeded differently than in Britain and Europe due to major differences in population density, distribution and geography. In North America the need was to transport very large tonnages of freight over long overland distances, often crossing one or more high mountain ranges. The track quality was relatively poor and curves or grades very sharp. This led to early adoption of the double truck freight or passenger car having two swivelling usually four-wheel trucks. Freight car capacities in the 1850s and 1860s were in the 10-15 ton range. Many freight cars had handbrakes on one or both trucks. Passenger cars almost always had the handbrake arranged to operate all eight wheels from either end of the car. It was expected that brakemen would proceed over the tops of the cars of a freight train in motion to apply the handbrakes when called for by the engineman's "down brakes" whistle signal–regardless of weather.

A few applications of vacuum brakes were made to passenger rolling stock on the eastern seaboard. However, it was soon learned that the vacuum brake lost effectiveness as altitude increased. This was a severe limitation for North America.

The great Civil War Between The States of the early 1860's clinched the basic railroad operating principle of interchange of rolling stock in general and freight cars in particular. This quickly brought about the formation of the Master Car Builders Association in the U.S.A. and Canada. This is the forerunner of the Mechanical Division of the Association of American Railroads and started to operate in 1867. The purpose was to standardize rolling stock so that maintenance, repair and construction of cars would permit easy interchange of safe operating equipment.

THE AIR BRAKE

In 1869 George Westinghouse (later president of ASME) invented and applied the first continuous compressed air brake system in the world. This was demonstrated and applied on the Pennsylvania Railroad. The system was called a straight air brake in which the engineman introduced compressed air into a train lined brake pipe to apply the brake cylinders and brake shoes on all cars and locomotives in the train. The brake pipe was connected by hoses between the ends of adjoining vehicles in the train. Exhausting brake pipe pressure allowed brakes to release.

Three years later the automatic air brake made its appearance to provide much better operational safety in case the train accidentally parted. Undesired train separations were a rather common occurrence with the link and pin coupling systems of the day. A control valve, called a triple valve, and an auxiliary reservoir on each vehicle were charged with air pressure to release the brakes. If the train brake pipe pressure were reduced by the engineman's brake valve–or by an undesired train parting–the brakes on all portions of the train applied without any action required by the train crew. The fail-safe function of the automatic air brake became established engineering practice. By the late 1870's, most passenger rolling stock was equipped with some form of air brakes.

In the early 1880's, sizeable numbers of freight cars were being equipped with automatic air brakes, following invention of the brake cylinder pressure retaining valve. The retainer allowed safe operation on long and heavy grades, since the brake system could be recharged from time to time without losing air brake retarding force. The air brake system made possible the safe operation of much longer trains at considerably higher speeds.

Another development of note which also contributed to safer higher capacity train operation was the electric signal system. Signal towers controlling the entrance to the next track block sections were usually connected by telegraph. Use of the telegraph to dispatch trains had been started in 1851. Complex manually operated mechanical interlocking plants with switches and signals interlocked had been developed in England prior to 1860. However, it was not until 1871 that the practical electric track circuit was invented by Dr. Andrews in New Jersey. This indicated presence of a train in a particular block and automatically set signals at each end of the block to "stop". Trains were thus required to be able to stop from authorized running speed before passing the next signal, which might be "stop".

By the early 1880's Westinghouse had integrated compressed air with electric circuit controls to operate switches and signals in complex switching interlocking arrangements. Switches and signals several miles away from the control tower could be reliably operated in proper sequence to establish a "safe route" for the train.

FREIGHT TRAIN BRAKE SYSTEMS

The Bessemer steel-making process revolutionized manufacture of railway rails and permitted the enormous expansion of railroad lines across the continent. The stronger, tougher less expensive rail led to the use of bigger cars and more powerful locomotives. By the middle 1880's, North American railroads recognized the need for an acceptable standard continuous power brake system on all interchange rolling stock. The Master Car Builders organized and held a two-year test program to evaluate various systems proposed.

Several versions of electro-pneumatic brakes were tried along with air brakes and mechanical or inertia systems. Tests were made on the CB&Q near Burlington, Iowa and were notable for the instrumentation developed to record speed, acceleration, deceleration, impact buff and draft forces within the trains of 50-25 ton capacity freight cars. In 1889 the MCB and the Association of Railway Master Mechanics adopted the quick action automatic air brake system as the interchange standard.

Of interest are the test train experiences at the Burlington Trials which demonstrated the need for energy absorbing friction draft gear and the interlocking automatic coupler. Development and adoption followed shortly after. (See chapters on draft gear and coupler development.)

Introduction of steel into freight car construction in the 1890's made possible the truly large, heavy tonnage freight trains to serve the needs of the North American market. Five thousand to 6000 ton trains became possible in the early 1900's. Train lengths stretched out to the 4000-5000 ft range (1200-1500 m). Back to the laboratory went braking engineers and the MCB to better understand the gas (air) fluid dynamics. By 1908 the brake system was reengineered to suit the 80-100 car trains now possible with steel underframes, automatic couplers and draft gear. Since long freight trains with up to a mile of brake pipe required unique control valves, the passenger and freight types parted company for good in 1908-1910. Passenger brake control needed graduated release for normal service application and high pressure in emergency. However, graduated release would not work reliably with over 2500-3000 ft of 1¼ in. brake pipe. Passenger control also needed a positive high brake cylinder pressure in emergency applications. Of interest, European railroads adopted North American pipe sizes for their air brake work and still use 1¼ in. for their train brake pipe.

The necessity to operate safely with trains of 120 to 150 or more cars brought about a broad engineering investigation which lasted from 1925-1933. This resulted in the "AB" automatic air brake system capable of safely handling trains of 10,000 to 15,000 tons with lengths over 1½ miles (around 2.5 km). The AB brake system on trains of this size had to safely and adequately control some 25,000 to 30,000 friction brake horsepower spread out over 150 or more cars.

In the past 20 years the response to shippers' needs has seen another revolution in freight car design with the development of the 100 and 125 ton capacity cars weighing 131.5 and 157.5 tons loaded (119-143 tonnes). Axle loadings are about 33-40 tons (30-36 tonnes), far beyond the 22-ton axle loads now allowed in Europe. At present, over half the North American freight car fleet consists of cars which weight from 110-157 tons loaded.

The freight car air brake control valves have been improved in recent years to include much faster transmission of service release and application functions embodied in the ABD and ABDW types. Faster operation and somewhat higher braking levels are needed to preserve train stopping ability as greater numbers of heavier cars are operated in trains.

A 14,000-18,000 ton train of modern cars requires finely tuned control of from 30,000-50,000 friction braking horsepower. Improved control of the in-train forces generated with the bigger capacity cars is being progressed with a cooperative research program involving close interaction of the AAR, Railroads, Suppliers and the Federal Railway Administration.

The physical effects of gravity on the generally undulating profiles of track over the majority of route miles and typical mixes of empty and loaded cars can have the in-train forces minimized by use of the recently designed empty-load brake equipments. These are uniquely designed to suit North American type freight cars. With empty-load brake equipment the modern high capacity light weight cars can be braked high enough to provide improved stopping and heavy grade operation without overbraking the car when empty.

In the past 10-15 years, advanced engineering studies by brake system and locomotive manufacturers have brought about a practical use of radio remote control of locomotives placed back in the train. The engineman on the lead locomotive can control the power, dynamic and air brake on units located ½–¾ mile away to the rear. This makes possible the use of 25,000-35,000 tractive horsepower per train. Major improvements in railway route throughput and capacity can be obtained with this system of radio remote control. The air brake system benefits because response times are only half to one third of a train with brake control only from the lead locomotive brake valve.

The evolution to very long trains of bigger cars was also dependent upon the multiple unit locomotive control principle. This was made possible by the development of the diesel electric locomotive, which inherently can have up to about ten large units operated from one control stand. Of course, this could be accomplished by electric locomotives if the electric power distribution system could deliver 25,000-30,000 kW to each train in the electrical section.

PASSENGER TRAIN BRAKING SYSTEMS

As mentioned above, by 1910, passenger and freight control valves had parted company because of the fundamentally different functions and degree of retardation required for the long freight train which could easily be 4000 to 5000 ft from locomotive to caboose. Passenger brake control had to contend with train speeds two to three times freight maximums. Retarding forces had to be three or four times that on a freight car. Graduated release was essential to obtain smooth and accurate stops of passenger trains. These factors, requiring stops from 80 or so mph within 5000-7000 ft long signal blocks, needed fast uniform application.

Road test investigations by railroads and suppliers working with the MCB developed the Universal Passenger Car Control System. This could be operated electropneumatically or pneumatically following the principles of the automatic air brake. Thus, even the latest feature controls could always operate in harmony with existing equipment. This is a cardinal rule of the need to interchange rolling stock.

The electropneumatic Universal Control valve was widely applied to cars on electrified routes. Development of a motor driven air compressor small enough to fit under the car floor permitted transit versions of the Universal Equipment to be broadly used on subway and interurban electric trains. In the latter services, load sensitive variable load equipment was developed to insure uniform performance regardless of the passenger load.

The advent and rapid adoption of the high speed diesel powered streamliners of the late 1930's, 1940's and early 1950's brought about 100-120 mph (160-200 kmh) train speeds on many railroads. "HSC" or high speed control brake equipment was engineered to get the maximum retardation abilities within practical rail adhesion limits. Speed governor control automatically varied the brake cylinder pressure to obtain the most from the cast metal friction material of the brake shoes. Antiwheel slip or decelostat control were developed to prevent wheel sliding, loss of retarding effort, and damage to wheels and track.

During this period the North American design of the railway disc brake made its appearance and was applied on many cars and trains. The disc brake used composition friction materials and was much simpler than the speed governor control required with cast metal brake shoes.

In the late 1950's the private auto and airplane were luring away most of the remaining railway intercity passengers. At this time a practical high friction composition tread brake shoe was developed. This allowed considerably smaller and more simplified braking force equipment. A combination of this and disc brakes has proved to be the most beneficial in shoe and wheel life and gives optimum wheel-rail adhesion.

Of interest, the major application of high friction composition brake shoes has been to freight cars and locomotives. Well over half the 2 million cars in the North American fleet are now equipped. equipped.

In the mass transit field the years from the mid-60's onward have seen a succession of new developments in braking equipment. Electric brake pipe analog equipment was applied successfully to work better with semiautomated train control systems. These systems blend friction and dynamic or other braking means automatically in response to brake intensity demand. Electronic logic, or electro-hydraulic logic and force systems are operating well on many properties. Generally the muscle for friction braking force is supplied by pneumatic means, although several hydraulic or hydro-pneumatic types are in service. Automatic blending of friction and dynamic is the rule in this type service.

Late developments in the 1970's have seen AMTRAK take over most of the intercity passenger service and make significant strides to regain rail passenger business. Their new rolling stock uses a modern electropneumatic brake system, disc brakes, and all axles are equipped with an electronic anti-wheel slip equipment. Locomotives, both diesel electric and electric, are equipped with an automatic blending of the dynamic and air brakes.

FIG. 1 STEPHENSON LOCOMOTIVE WITH STEAM DRIVER BRAKES

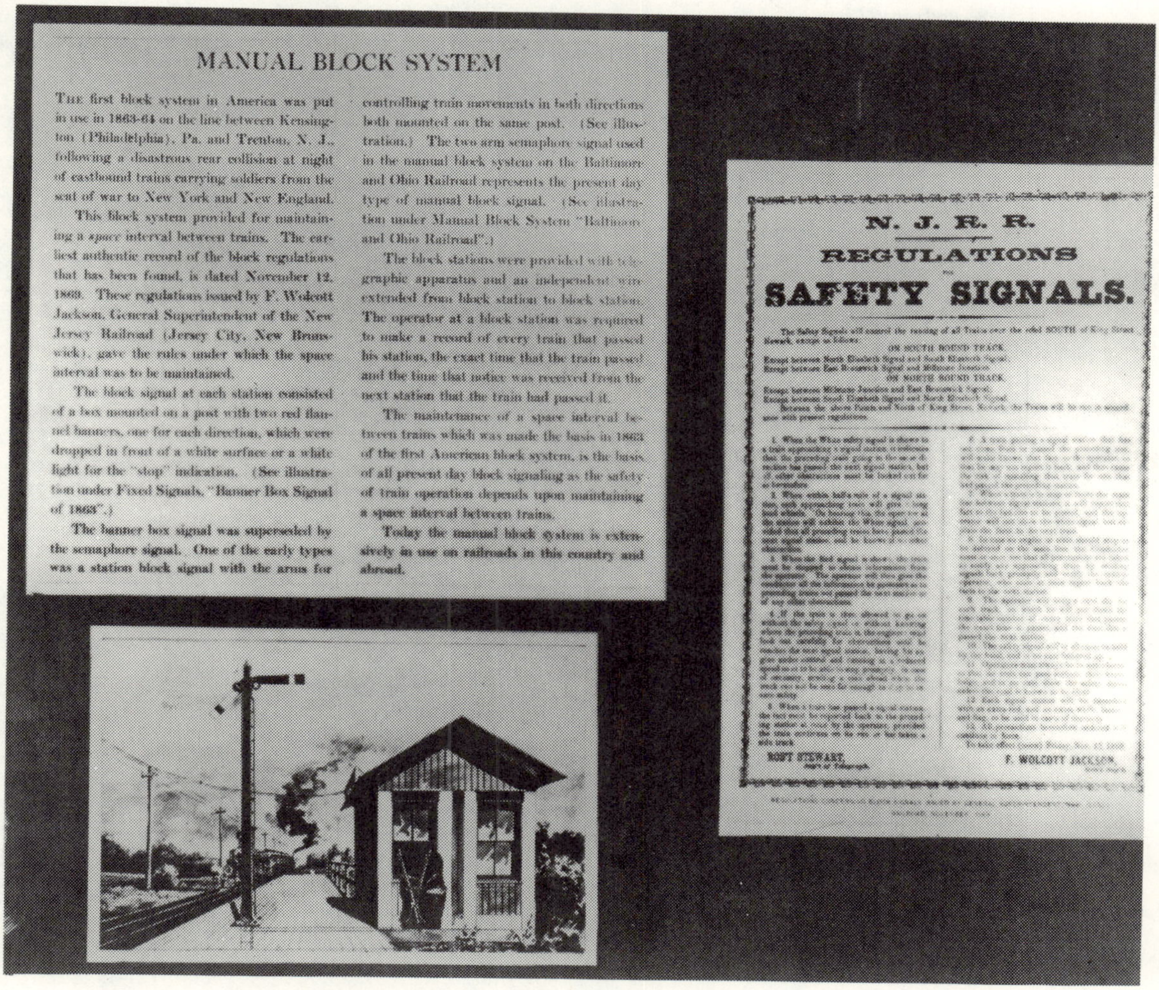

FIG. 2 MANUAL BLOCK SYSTEM

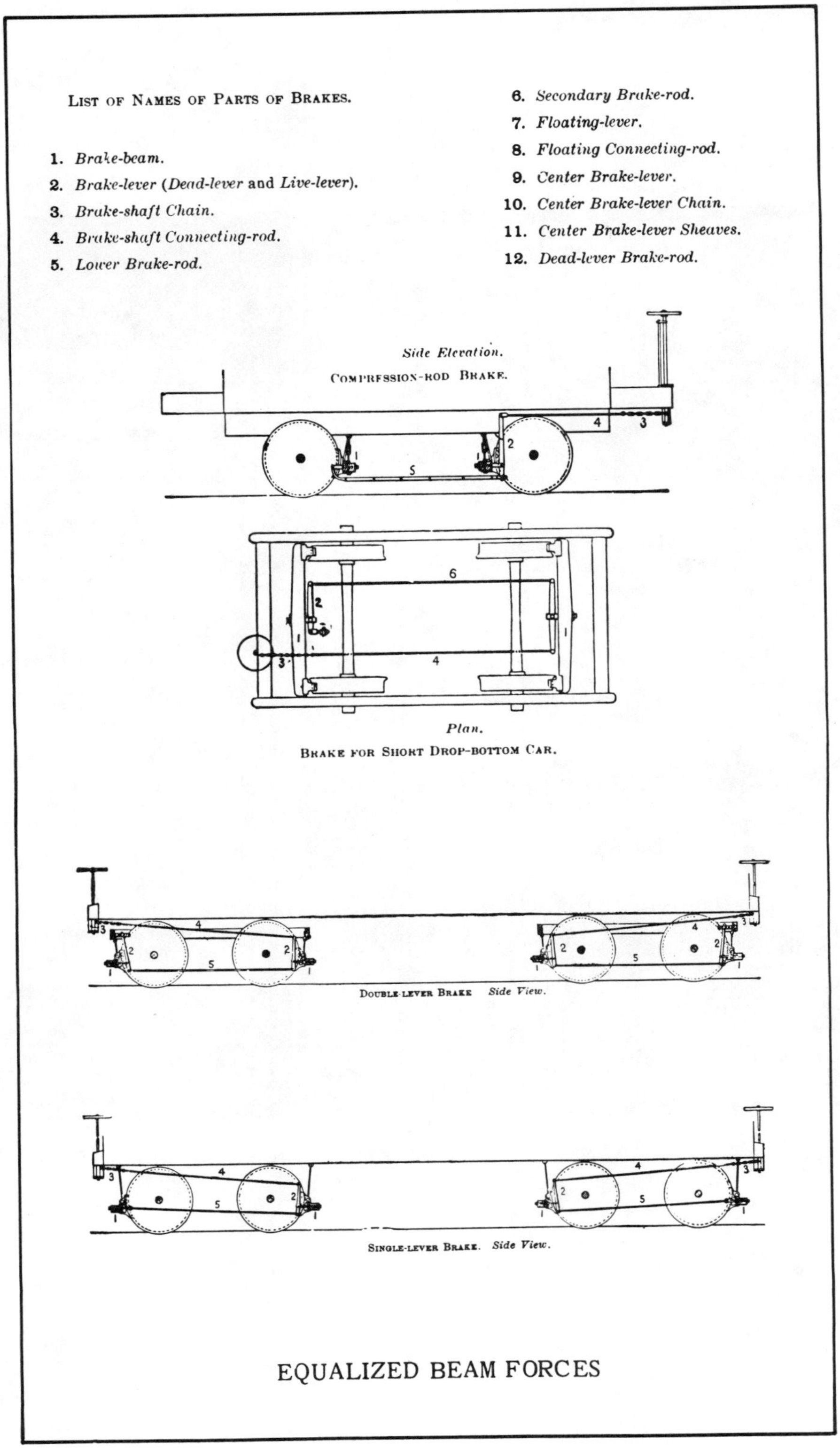

FIG. 3 FREIGHT CAR HAND BRAKE RIGGING, 1870's-1880's

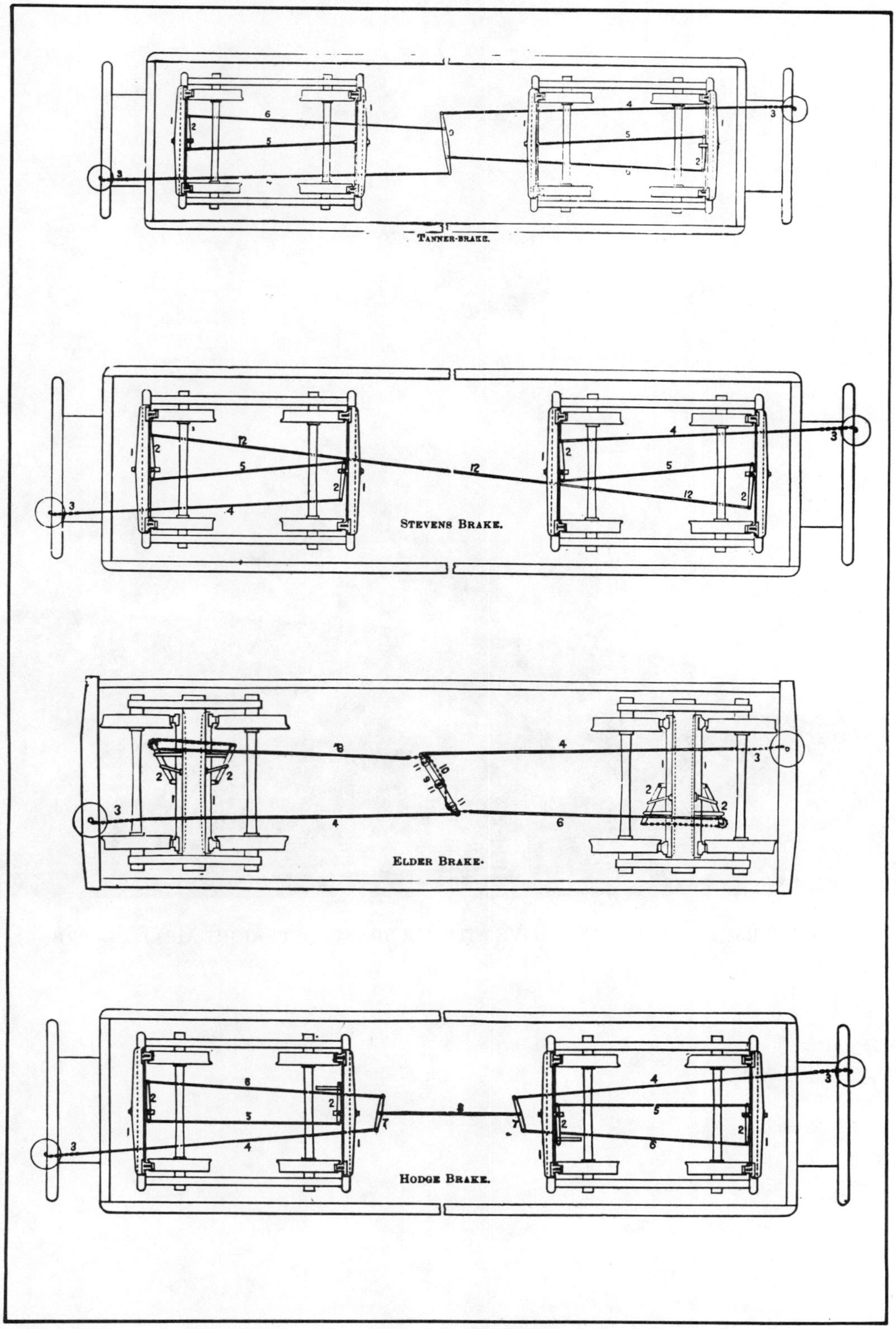

FIG. 4 PASSENGER CAR HAND BRAKE RIGGING, 1870's-1880's

FIG. 5 FIRST CAR AND LOCOMOTIVE WITH AIR BRAKES (STRAIGHT AIR) 1869 - PRR

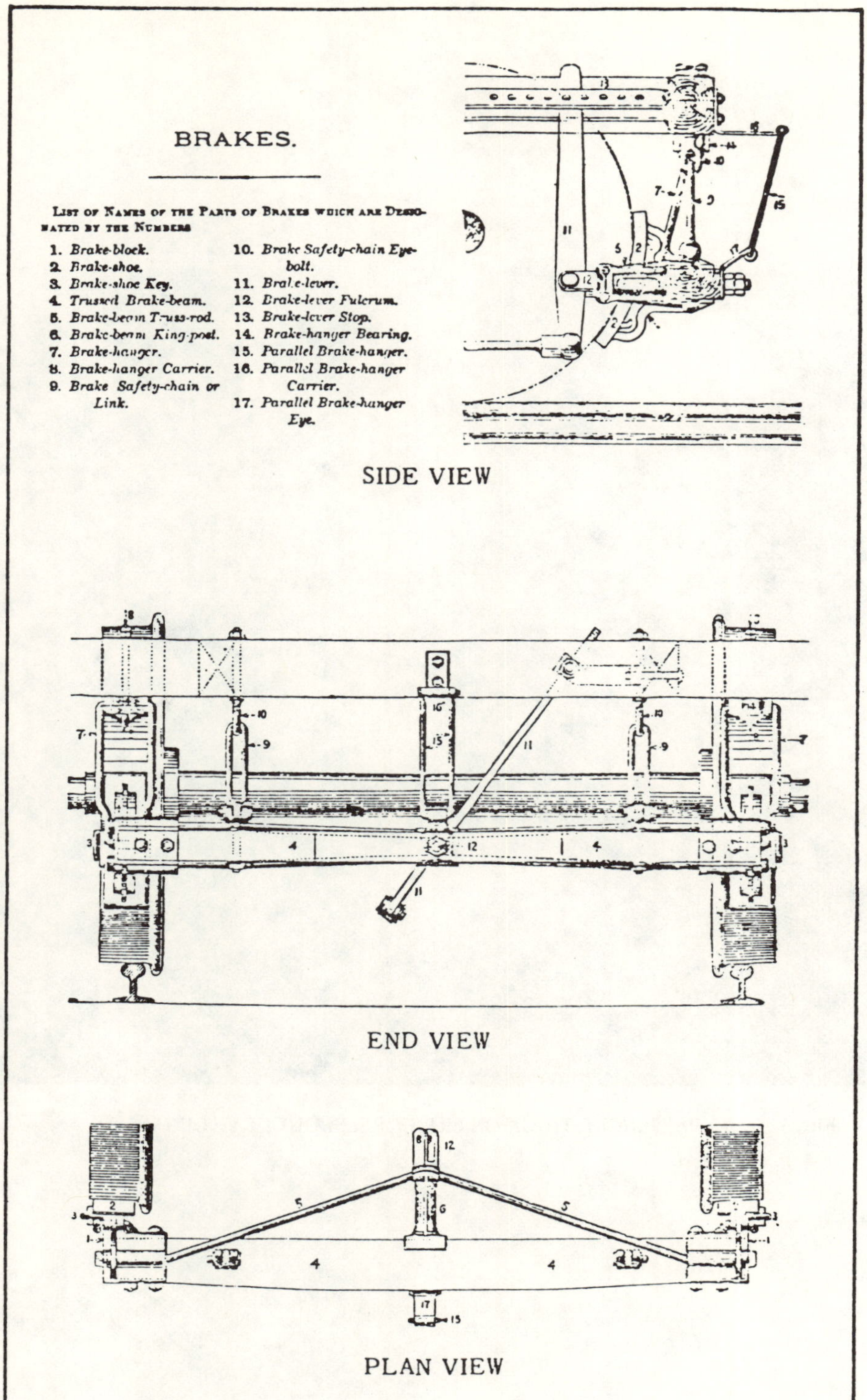

(Composite type beams after Westinghouse 1873 Patent)

FIG. 6 TRUSSED BRAKE BEAMS, 1870's

FIG. 7 5-CAR PASSENGER TRAIN ON PRR HORSESHOE CURVE CIRCA 1869

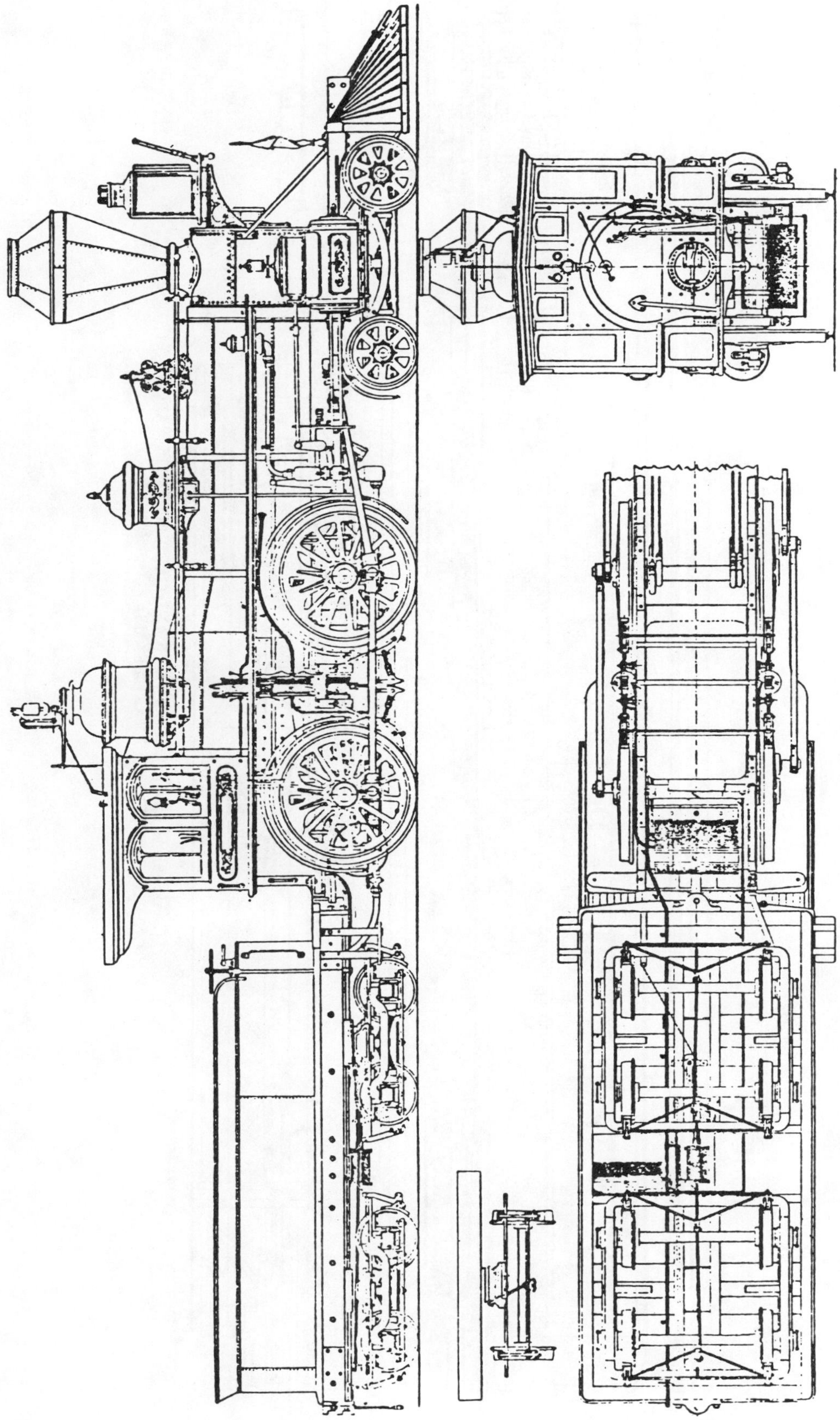

FIG. 8(a) AUTOMATIC AIR BRAKES ON LOCOMOTIVE AND CAR FROM 1876 WABCO CATALOG (NOTE IRON TRUSSED TENDER BRAKE BEAMS)

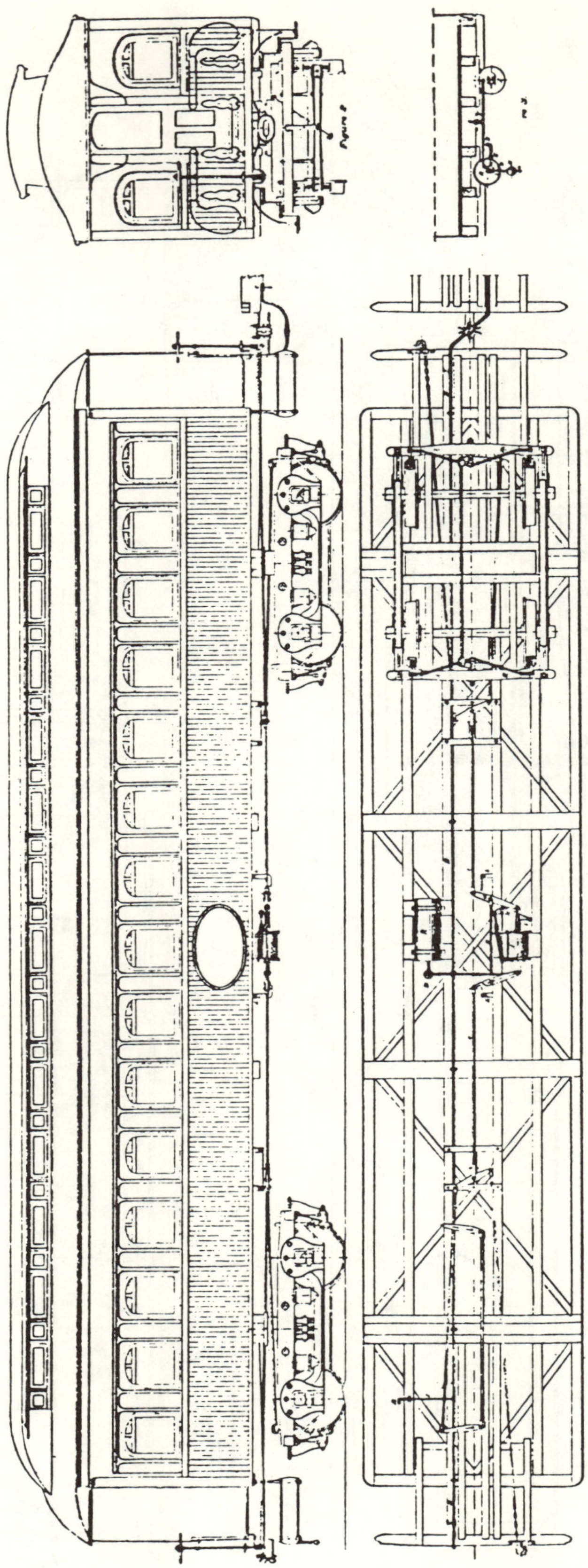

FIG. 8(b) AUTOMATIC AIR BRAKES ON LOCOMOTIVE AND CAR FROM 1876 WABCO CATALOG
(NOTE IRON TRUSSED TENDER BRAKE BEAMS)

AIR-BRAKE CATECHISM.

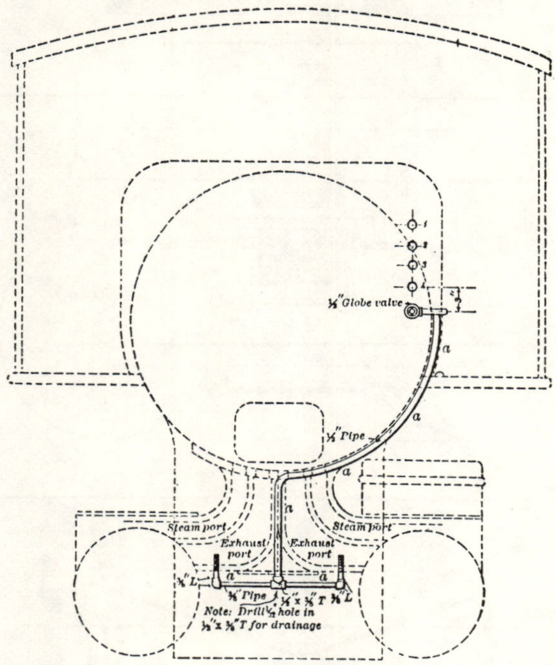

FIG. 9 LE CHATELIER OR WATER BRAKE (STEAM COMPRESSION OR DYNAMIC BRAKE)

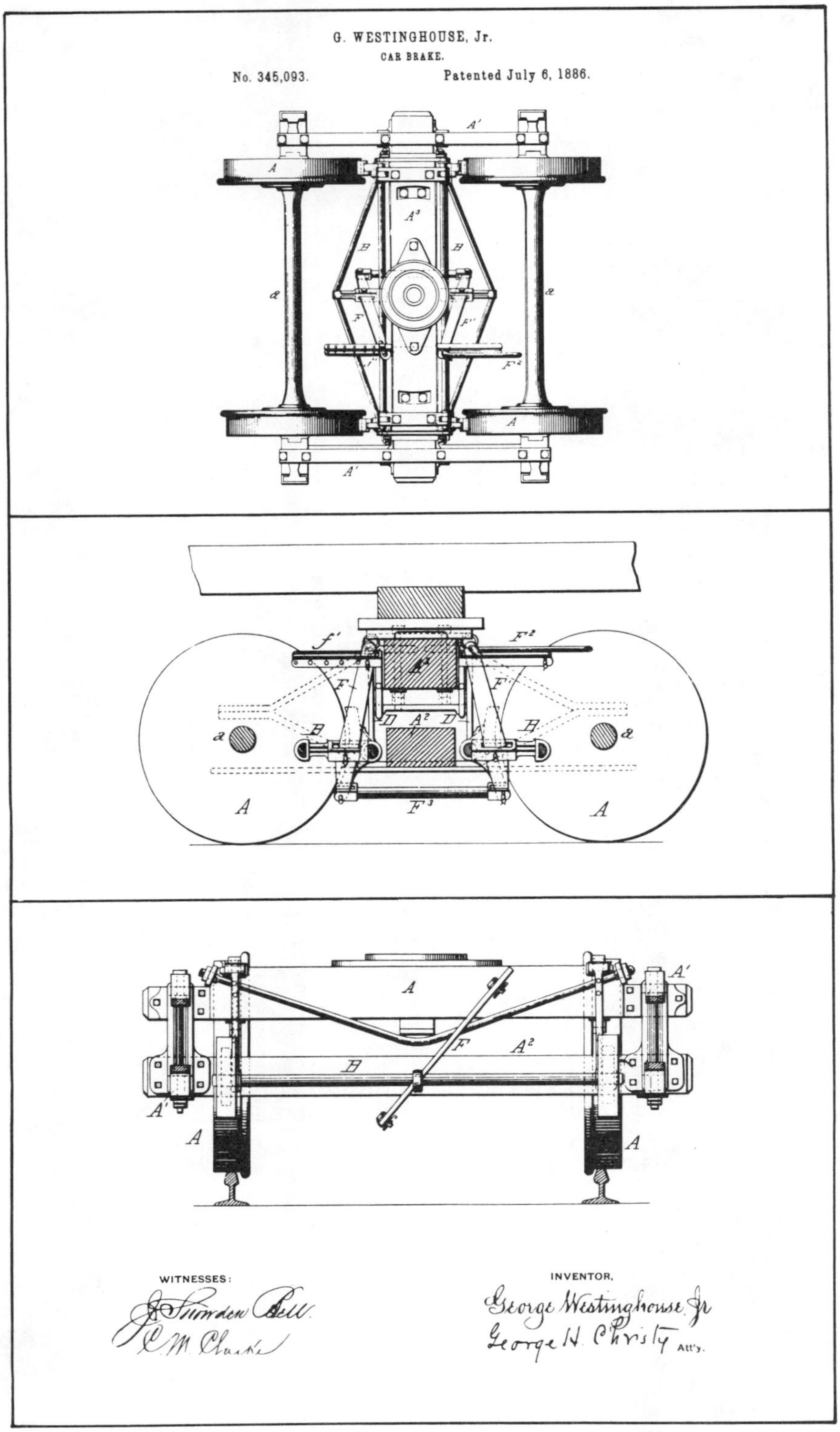

FIG. 10 WELDED TRUSSED BRAKE BEAM

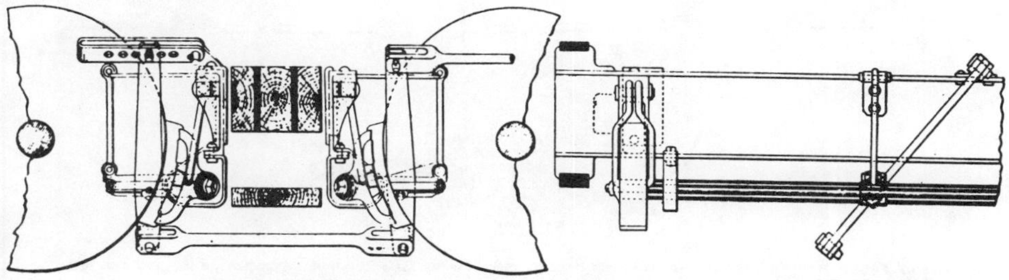

Sectional Side Elevation. *Sectional End Elevation.*

BRAKE-GEAR, FREIGHT, FOR DIAMOND TRUCK.
Inside-hung from Channel-beam Transomes.

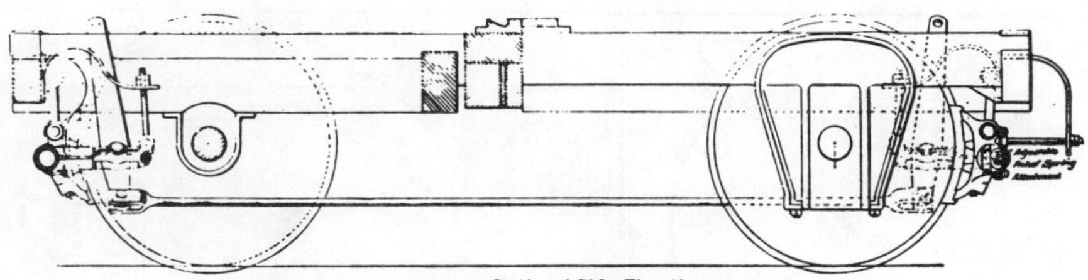

Sectional Side Elevation.
BRAKE-GEAR, PASSENGER, AS APPLIED TO FOUR-WHEELED TRUCK.
Wheels Braked upon One Side Only.

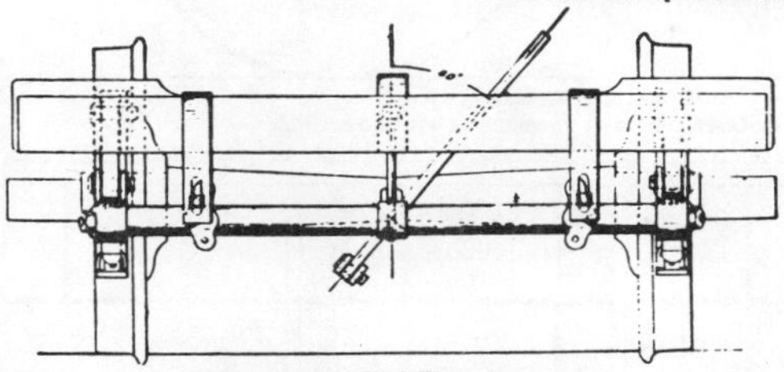

End Elevation.
BRAKE-GEAR, PASSENGER, AS APPLIED TO FOUR-WHEELED PASSENGER TRUCK.
Wheels Braked upon One Side Only.

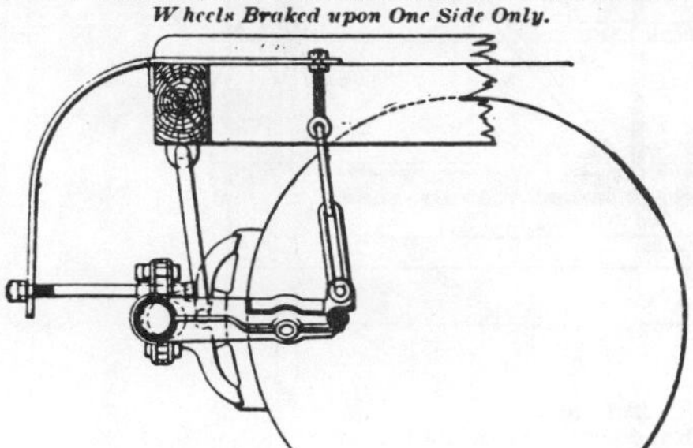

Part Side Elevation.
BRAKE-GEAR, PASSENGER, SHOWING ANOTHER MANNER OF HANGING-BRAKES.
Brakes Hung upon One Side of Wheel.

NAMES OF PARTS.

1. *Upper Brake-lever Connecting-rod.*
2. *Upper Brake-lever Connecting-rod Clevis*
3. *Live Truck-lever.*
4. *Brake-lever Connecting-rod.*
5. *Truck Brake-lever.*
6. *Upper Lever-connecting-rod.*
7. *Dead Truck-lever.*
8. *Brake-beam.*
9. *Brake-beam Strut.*
10. *Brake-head.*
11. *Brake-shoe.*
12. *Brake Release-spring.*
13. *Brake Safety-hanger.*
14. *Brake-beam Adjusting-hanger Carrier.*
15. *Brake-beam Adjusting-hanger.*
16'. *Brake-lever Guide and Stop.*
16. *Brake-hanger.*
17. *Brake-hanger Bracket.*

FIG. 11 PASSENGER CAR TRUCK BRAKE RIGGING DETAILS – 1890's

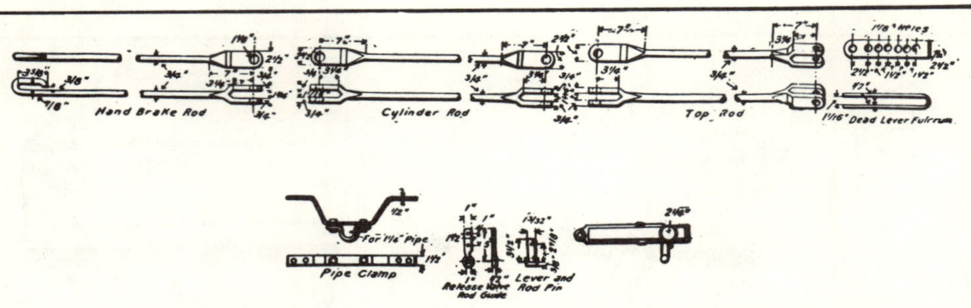

DETAILS OF FOUNDATION FREIGHT BRAKE GEAR ADOPTED BY THE MASTER CAR BUILDERS' ASSOCIATION 1889, 1890 AND 1891.

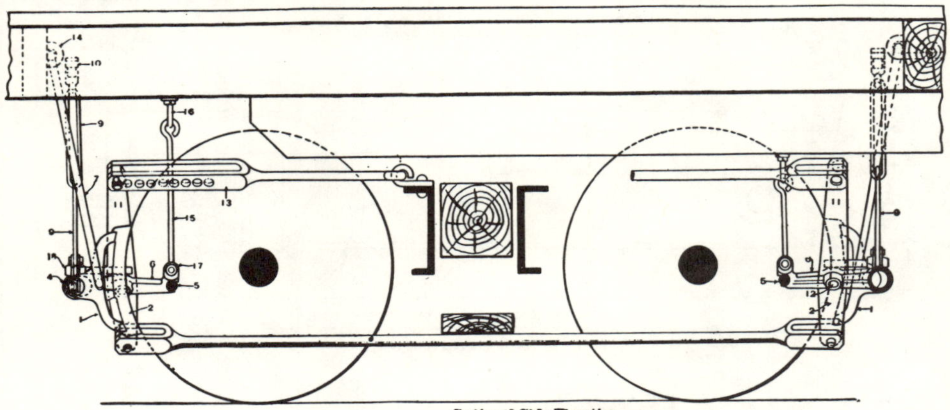

Sectional Side Elevation.
DIAMOND TRUCK, BRAKES OUTSIDE-HUNG FROM THE CAR-BODY.

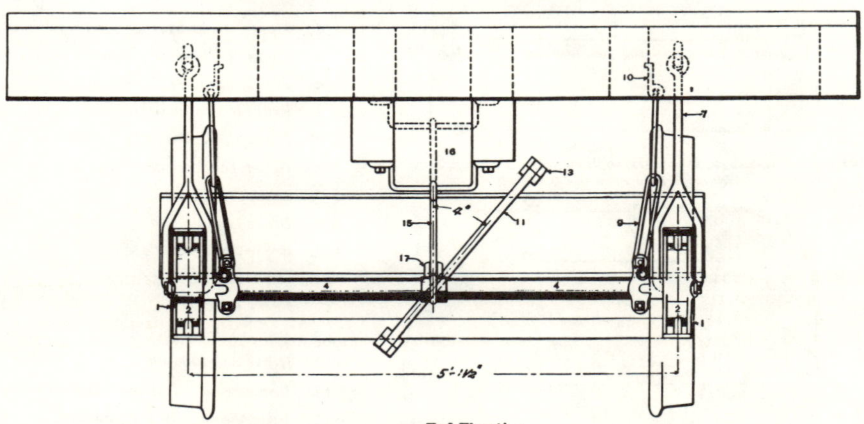

End Elevation.
DIAMOND TRUCK, BRAKES OUTSIDE-HUNG FROM THE CAR-BODY.

NAMES OF PARTS OF FREIGHT BRAKE GEAR.

1. *Brake-block.*
2. *Brake-shoe.*
3. *Brake shoe Key.*
4. *Trussed Brake-beam.*
5. *Brake-beam Truss rod.*
6. *Brake-beam King-post.*
7. *Brake hanger.*
8. *Brake-hanger Carrier.*
9. *Brake Safety-chain* or *Link.*
10. *Brake Safety-chain Eye-bolt.*
11. *Brake lever.*
12. *Brake lever Fulcrum.*
13. *Brake-lever Stop.*
14. *Brake-hanger Bearing.*
15. *Brake-beam Adjusting-hanger.*
16. *Brake beam Adjusting-carrier.*
17. *Brake-beam Adjusting-hanger Eye.*

FIG. 12 FREIGHT CAR TRUCK BRAKE RIGGING DETAILS – 1890's

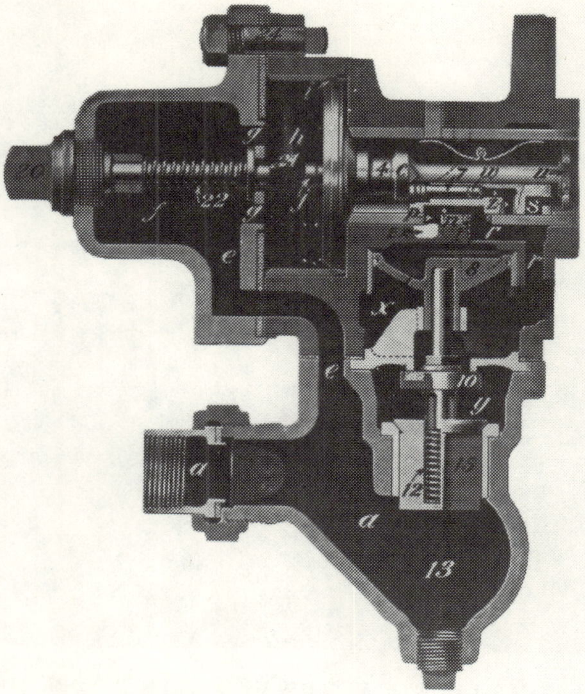

FIG. 13 QUICK ACTION OR "H" TRIPLE VALVE – MCB STANDARD 1889

FIG. 14 SOUTHERN PACIFIC FIRST 50 CAR TRAIN OF NEW CARS WITH QUICK ACTION OR H TRIPLE VALVES - 1891 NEAR SACRAMENTO, CALIFORNIA

FIG. 15 PENNSYLVANIA 50 CAR TRAIN OF MIXED WABCO AND NEW YORK AIR BRAKE QUICK ACTION TRIPLES NEAR BOLIVAR, PA. CIRCA 1895

FIG. 16 K-2 TRIPLE VALVE – 1908 QUICK SERVICE AND QUICK ACTION FOR 80-100 CAR TRAINS

FIG. 17 SHIFT CHANGE WABCO PLANT – 1909 WILMERDING, PA. LAID OUT AND BUILT BY GEORGE WESTINGHOUSE STARTING IN 1883

FIG. 18 FIGHTING SNOW IN THE ROCKIES CIRCA 1920 – 6-2-8-0 LOCOMOTIVES PUSHING THE ROTARY PLOW, DRGW

FIG. 19 100 CAR TRAIN ASCENDING THE ALLEGHENIES WITH AUTOMATIC BLOCK – EARLY CTC LATE 1920's

CAPACITY
(Throughput)

$$C_{MAX} = \frac{\text{TIME SPAN}}{\text{TRANSIT TIME per TRAIN}} \times \text{TONS/TRAIN} = \text{THROUGHPUT}$$

$$C_{OPR} = \frac{\text{HOURS} \times 3600 \text{ (Seconds/Hour)}}{\dfrac{\text{D Block (ft.)} \times \text{N (No. blocks apart)}}{\text{SPEED (fps)}}}$$

$$= \frac{H \times 3600 \times V \times 1.467}{D_b \times N}$$

$$= \frac{1 \times 3600 \times 60 \times 1.467}{9000 \times 3} = 11.7 \text{ Trains per hr}$$

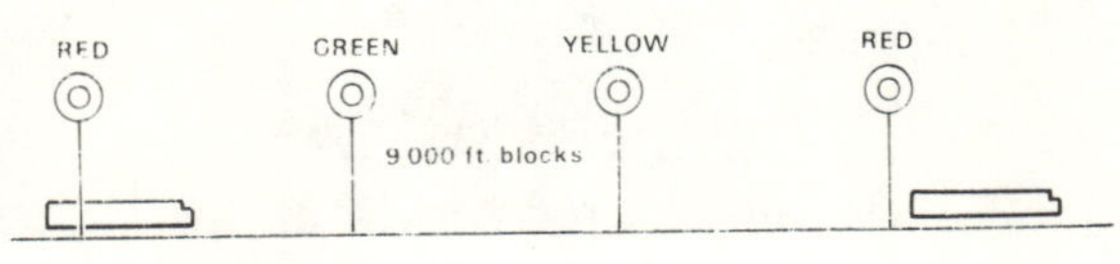

FIG. 20 RAILROAD CAPACITY (THROUGHPUT)
SINGLE DIRECTION RUNNING

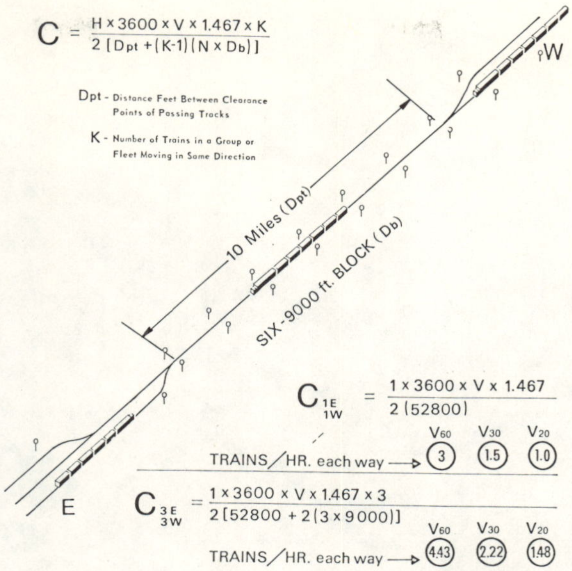

SINGLE TRACK CAPACITY

FIG. 21 RAILROAD CAPACITY SINGLE TRACK
TWO DIRECTION RUNNING

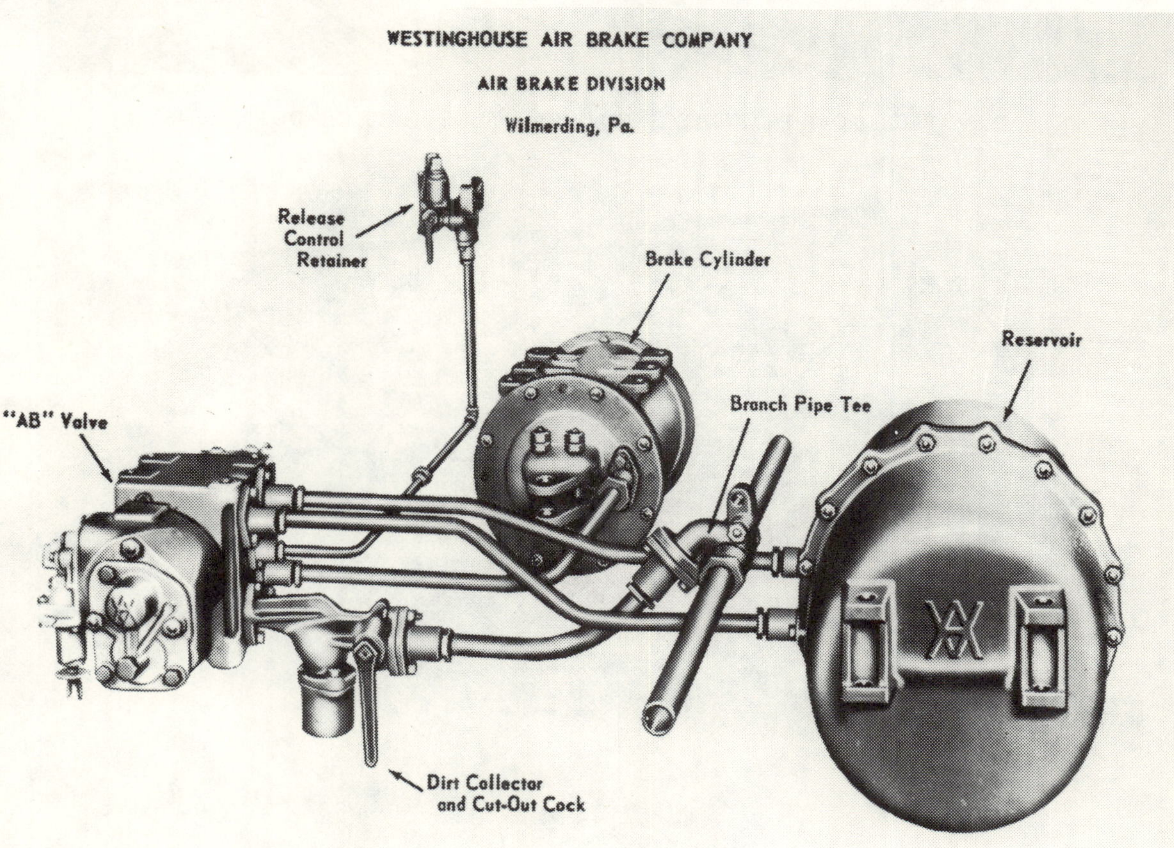

FIG. 22 AB BRAKE FOR RAILROAD FREIGHT CARS (1930's)

FIG. 23 ABLC 150 CAR TRAIN DESCENDING HORSESHOE CURVE GRADE PRR, 1948

FIG. 24 70 TON ORE CAR AB AUTOMATIC EMPTY LOAD BRAKE 1937

FIG. 25 190 CAR LOADED ORE TRAIN (20,000 TONS) WITH DM&IR 4-8-8-4 LOCOMOTIVE. "TWILIGHT OF BIG STEAM POWER" - 1955

FIG. 26 145-100 TON CARS OF PELLETS DOWN ERIE MINING CO'S 2 PERCENT GRADE TO TACONITE HARBOR. LOCOMOTIVE AND ALL CARS WITH EMPTY-LOAD EQUIPMENT AND HIGH FRICTION COMPOSITION SHOES

FIG. 27 150 CAR RESERVE MINING CO. ORE TRAIN NEAR BABBIT, MINNESOTA – EMPTY LOAD WITH AB, AC AND ABD VALVE CARS IN MIXED CONSIST

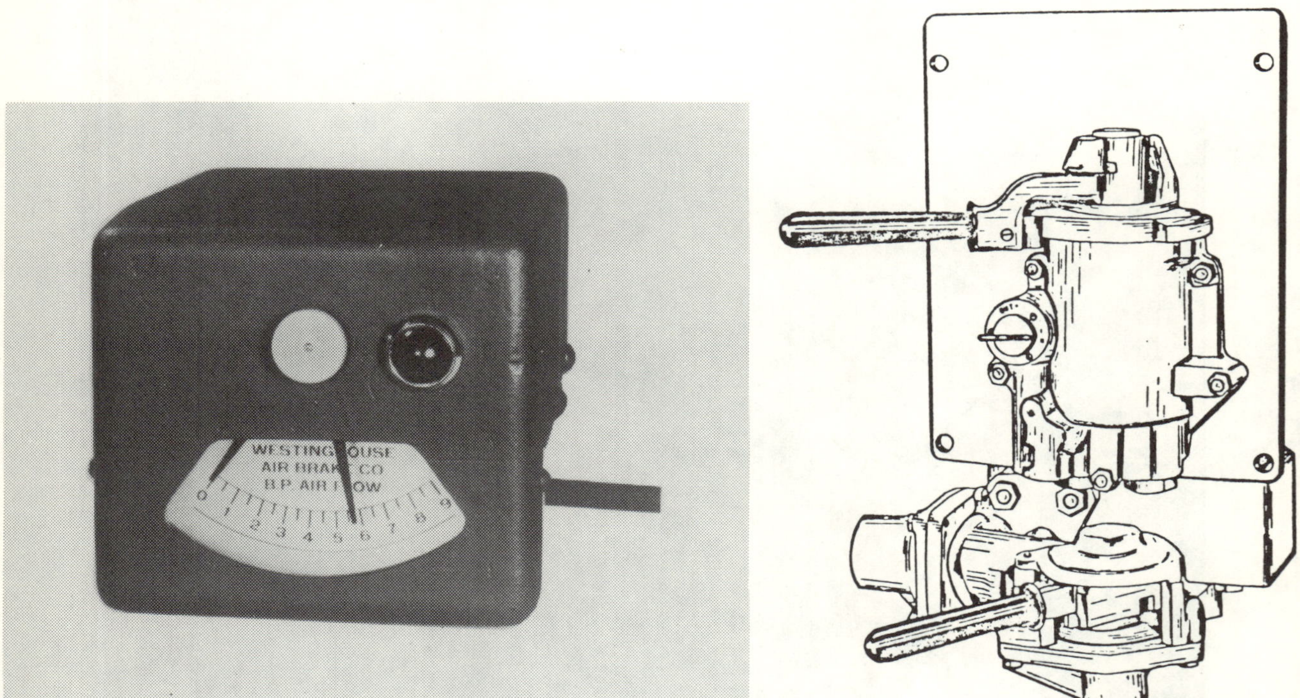

FIG. 28 MODERN PRESSURE MAINTAINING LOCOMOTIVE – 26-C BRAKE VALVE AND BRAKE PIPE FLOWMETER

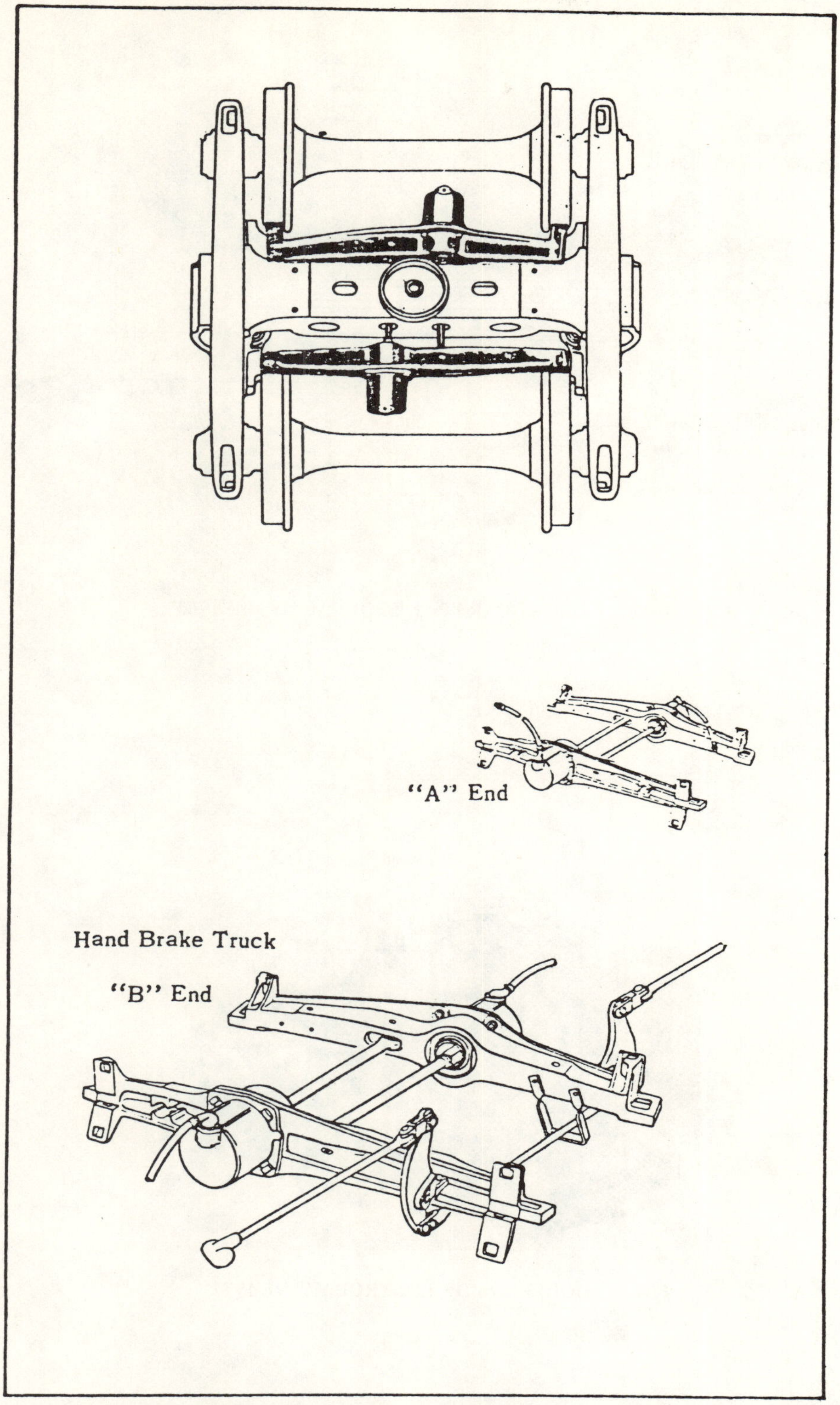

FIG. 29 TRUCK MOUNTED DIRECT ACTING WABCOPAC OR NYCOPAC BRAKE UNITS FOR FREIGHT CARS

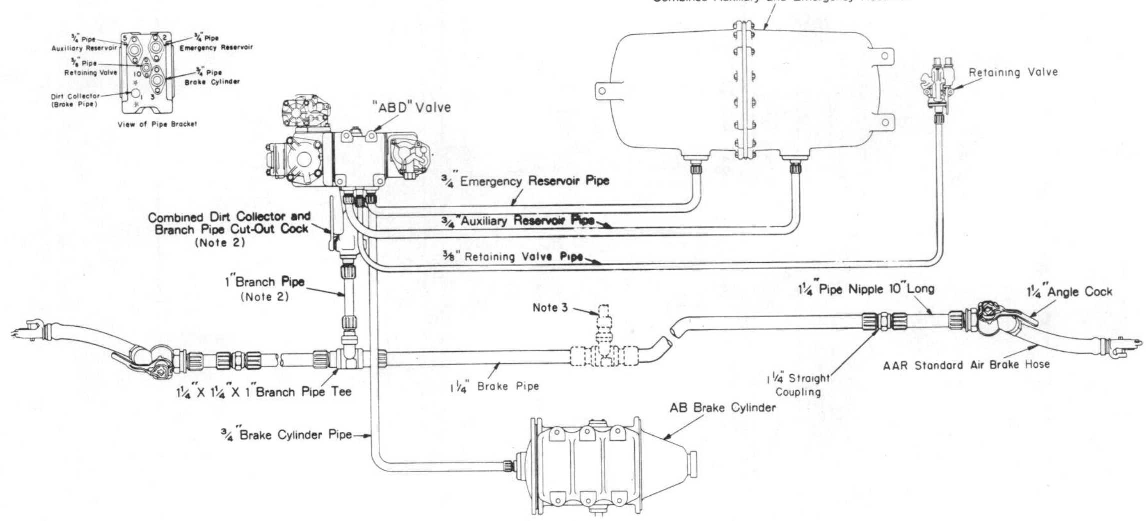

FIG. 30 ABD FREIGHT CAR BRAKE EQUIPMENT MID 1960's

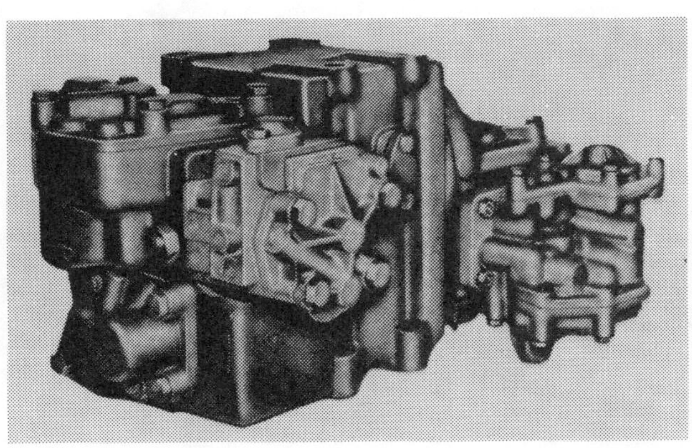

FIG. 31 MODERN ABDW CONTROL VALVE 1977

FIG. 32 SC-1 EMPTY LOAD ON 100 TON CAR

FIG. 33 MODERN 100 TON CAPACITY FREIGHT CAR

FIG. 34 FIRST GRADUATED RELEASE PASSENGER BRAKE TYPE LN (CIRCA 1908)

FIG. 35 UNIVERSAL PASSENGER CAR CONTROL VALVE (1914) WITH ELECTROPNEUMATIC AUTOMATIC BRAKE OPTION

FIG. 36 BURLINGTON ROUTE – PIONEER ZEPHYR AT PLANO ILL. – MODERN HIGH SPEED GOVERNOR CONTROL EQUIPMENT DEVELOPED ON THE ZEPHYRS

FIG. 37 D-22-BR CONTROL VALVE APPLIED TO MOST HIGH SPEED STREAMLINED CARS IN 1940'S AND 1950'S

FIG. 38 CNW-UPRR-SPRR CITY OF SAN FRANCISCO TRANSCONTINENTAL STREAMLINER CROSSING GREAT SALT LAKE LATE 1940'S. HSC BRAKE EQUIPMENT WITH ELECTROPNEUMATIC, SPEED GOVERNOR CONTROL AND DECELOSTAT ANTIWHEEL SLIP CONTROL

FIG. 39 STREAMLINED PASSENGER CAR TRUCK WITH TRUCK MOUNTED BRAKE CYLINDERS AND CLASP BRAKE RIGGING WITH IRON SHOES ARRANGED FOR SPEED GOVERNOR AND DECELOSTAT ANTIWHEEL SLIP CONTROL

FIG. 40 PRR CONGRESSIONAL LIMITED WITH FAMED GG-1 ELECTRIC LOCOMOTIVE – WASHINGTON TO NEW YORK SERVICE

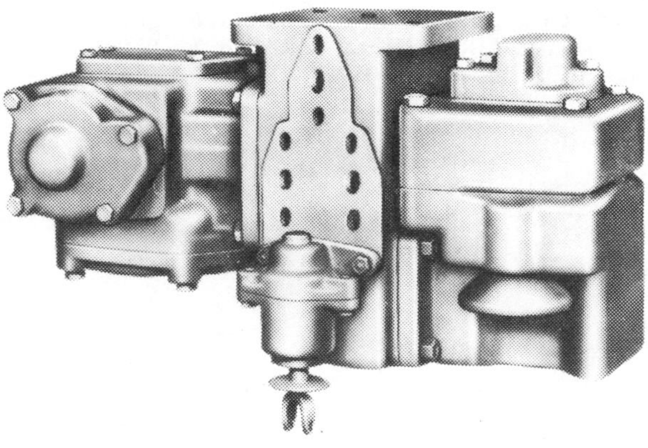

FIG. 41 MODERN 26-C PASSENGER CAR CONTROL VALVE USED ON MOST NEW MAIN LINE AMTRACK AND RAILROAD SUBURBAN COMMUTER SERVICES IN USA, CANADA, MEXICO

FIG. 42 MODERN INSIDE FRAME SUBURBAN CAR TRUCK WITH AIR SPRINGS AND COMBINATION WHEEL TREAD AND DISC BRAKES (TATOA-HAWKER SIDDELEY)

FIG. 43 BI-LEVEL SUBBURBAN CARS IN TORONTO CANADIAN AREA (TATOA)

FIG. 44 MODERN AMFLEET CAR WITH CS-1 E.P. BRAKES, DISC BRAKES, E-5 ELECTRONIC WHEEL SLIP CONTROL

3 The History of the Steam Locomotive in the United States

E. T. HARLEY

INTRODUCTION

It is difficult to contemplate any vehicle of commerce or for that matter any machine devised by man that is more beloved than the steam locomotive. Songs are sung about its lonesome whistle in the night. The entertainment arts use it to set the scene for almost anything involving commercial travel from the early nineteenth century to World War II. Historical societies and many publications still study its every detail. Children (and adults) yearn for models of it which range all the way from toy electric trains under the Christmas tree to operating live-steam scale reproductions sometimes weighing a ton or more. A few brave souls even maintain and operate restored full-sized steam locomotives at great expense and personal sacrifice. Yet this compassion is well founded because the railroads were the most important single factor in the development of modern America and the steam locomotive was the railroad's prime mover for 120 years.

Amazingly, the steam locomotive was displaced by the diesel-electric in the short time span of little more than 10 years following World War II. The transition would probably have occurred sooner, but proven designs of all types of diesel-electric locomotives were not available in quantity until the war was over. While the reasons at the time were largely economic and operational, it is difficult to think of an engine of transport that would fit more poorly in today's scene, if by some chance the diesel-electric had not been developed. The smoke, fumes, sparks, soot, cinders, ashes and noise inherent in its operation would have made it an environmental disaster–akin to a volcanic eruption in Central Park.

In addition, its thermal efficiency of only about 6 percent would have branded it as a waster of energy, with appropriate economic and other penalties assessed for such a wanton disregard of a limited asset.

Yet, the steam locomotive existed in a different time – a time when smoke both from locomotives and factories represented employment and lack of it represented hunger and poverty. It existed in a time when we had not reached the limits of our energy resources. Most of the latter-day U.S. steam locomotives burned coal, a plentiful, low cost and readily available fuel in most locations.

Still, we in the industry get occasional inquiries from individuals wondering why the railroads aren't returning to the steam locomotive to get away from our dependence on imported petroleum fuels. The answer is that no U.S. railroad could afford not to dieselize in the post-World War II era that was at least not hostile to the steam locomotive. Since that time labor costs, energy costs and environmental regulations have only reinforced the correctness of that decision. If indeed railroads are forced by public policy, economics or supply constraints to go to nonpetroleum energy sources, it will very possibly be to electrification, which can utilize any commercially feasible energy supply now available, or which may be developed in the future.

Strangely, electrification was available as an alternative to steam 30 years before the development of the diesel-electric, but financial constraints have largely blocked its adoption in the privately owned U.S. railroad structure. The diesel-electric possessed many of its advantages without the awe-inspiring head-end capital costs of a major electrification.

To this day, no engineering developments, even in this accelerated age of progress, have shown any evidence that some breakthrough will revive the steam locomotive in any form. In the last days of steam there were brief bursts of hope that the steam turbine (both a direct drive version and in several designs utilizing an electric transmission) would keep steam in operation, but none of them even threatened the diesel-electric and most of them were not even the equal of contemporary reciprocating steam designs.

The gas-turbine electric threatened the diesel-electric on the Union Pacific Railroad at the end of the steam era. These locomotives burned a cheap residual petroleum fuel similar to the fuel utilized by many steam locomotives operating in the same region. The local refiners, however, learned new techniques such as catalytic cracking and hydro-forming so that they no longer had to produce these cheap fuels. The gas turbines could not economically survive on higher priced diesel fuel and the highly variable railroad duty cycle, which is anathema to the gas turbine, so they were scrapped and replaced by diesel-electrics. The Union Pacific also built an experimental coal-burning gas turbine, but rapid blade erosion caused by the fly ash rendered it totally uneconomic.

THE EARLY DAYS OF STEAM

It all began in England with advances in the basic approach to technology made possible by the Industrial Revolution and concurrent developments by such pioneers as Newcomen and Watt in the eighteenth century. The steam locomotive, however, was a creation of the nineteenth century. The Tramways were already in place at many locations in both England and the U.S. They used largely wooden and iron track structures with animal, gravity or even stationary steam engine cable haulage. The Tramway offered a smooth, hard-surfaced roadbed with less rolling resistance than a conventional road surface; thus, less power was required for movement. In addition, guidance to multiple vehicles in tandem was created by either flanged rail (plateway) or by flanged wheels (railways). These advantages remain to this day in the low energy requirements of railroads compared to highway haulage.

Richard Trevithick, utilizing a very crude steam locomotive, proved in 1804 in Cornwall (England) that a smooth-wheeled locomotive on a plateway could haul a 10 ton load (along with 70 passengers) in 5 wagons. He also damaged the plateway with the weight of his locomotive and it was retired to stationary steam hammer operation. This was the first of many problems arising from lack of systems coordination between the early civil engineers who built the track and structures and the mechanics and mechanical engineers who built the locomotives. Criticism of Trevithick for breaking up the track on this pioneering effort would, however, be akin to criticizing the Wright Brothers for allowing their plane to be damaged by the wind after the historic Kitty Hawk flights. Such events have occurred only a few times in the history of engineering.

For the next 25 years about 50 locomotives were built by such English pioneers as Trevithick, Murray, Blenkinsop and Hedley. It was not until the Rainhill trials of 1829, however, that steam traction really proved itself. George and Robert Stephenson's "Rocket" met all of the requirements of this difficult test and incorporated some of the features of a steam locomotive that were to last for the next 120 years – a horizontal fire tube boiler utilizing induced draft from the cylinder exhaust to regulate output and two outside cylinders rod connected to cranked driving wheels which were quartered so that the locomotive would start even if one of the cylinders was stopped at "dead-center". This locomotive was later modified with cylinders more nearly horizontal to produce better ride characteristics – also a feature of later locomotives.

Meanwhile, in the United States, John Stevens fought for a railroad charter as early as 1811 and even built a demonstration model railroad at his estate at Hoboken, N.J. in 1826, but this was not a commercial application. The first locomotive to operate in the U.S. was the "Stourbridge Lion," imported from England in 1829 by Horatio Allen of the Delaware and Hudson Canal Co. As in the case of Trevithick's 1804 experiments, its 7 ton weight proved too heavy for the track and it too was retired to stationary service. Peter Cooper built a tiny vertical boilered demonstration locomotive of less than one ton in 1829 called the

"Tom Thumb" to interest the directors of the Baltimore and Ohio and it actually became the first American built locomotive to pull a carriage of passengers. Regular U.S. railroad service began in 1831 on the South Carolina Railroad and later in the same year on the Mohawk and Hudson (N.Y.) and the Camden and Amboy (N.J.) railroads. The year 1831 also marked the beginning of the locomotive-building career of Matthias Baldwin who was to found the largest steam locomotive building company in the U.S. which bore this name to the end of "steam".

THE AMERICAN STANDARD ERA (Figs. 2-4)

Another pioneer of this era was Henry Campbell of the Philadelphia, Germantown and Norristown Railroad, who in 1836 patented the first "American Standard" 4-4-0 locomotive (4 leading wheels in an engine truck under the cylinders, 4 driving wheels and no trailing wheels). This design, when improved by Harrison in 1838 with equalization between the driving wheels to make it more flexible, became the U.S. standard locomotive replacing the lighter capacity 4-2-0 (Jervis type) which accounted for two-thirds of all U.S. locomotives in 1840. One by one the great builders of the time – Norris, Rogers and finally Baldwin in 1845 accepted the 4-4-0 and over 25,000 were built for domestic service as late as the early 20th century. The great advantage of this locomotive (and its predecessor the 4-2-0) was 3-point suspension with the engine truck and the equalized driving wheels on each side providing the three points. This enables the American Standard to negotiate poor track – the hallmark of early U.S. railroading. This is the locomotive that fought in the Civil War on both sides, although the North had the great advantage of having the manufacturing facilities. This is also the locomotive that "won the west" – the two locomotives touching pilot to pilot at Promentory Point, Utah, on May 10, 1869 were "American Standards". This is the locomotive that burned readily available wood in its early days with a giant "balloon stack" spark arrestor. After the wood ran out following the Civil War, it burned coal and even oil. Its weight (and capacity) rose as track improved from 12 tons in Campbell's 1837 locomotive to 87 tons on the last one built for a major domestic railroad (Reading) in 1914. This type, however, reached its zenith about 1870 when it constituted nearly 85 percent of all U.S. locomotives, and from that point on it was slowly pushed out by larger and more capable wheel arrangements. Strangely, in Europe the 4-4-0 was not universally accepted because of the excellence of civil railway construction; therefore, more latitude was available to their designers.

LARGER LOCOMOTIVES
(Figs. 5-16, 33a, b, c)

As engineering improved and more adequate materials and manufacturing techniques became available, steam locomotives grew larger and more capable. The environmental changes surrounding the steam locomotive cannot be understated – steel rail (made economically possible by the Bessemer process in 1856), ballasted track, bridges that no longer failed under load, the automatic knuckle-type coupler, which was installed beginning in 1882, and the U.S.'s greatest single contribution to railroading, the automatic air brake introduced by George Westinghouse in 1872, all contributed to allowing an increase in equipment and train size.

When you consider that the first steam locomotives were made out of wrought iron, copper, brass, cast iron and wood and the last ones were nearly all steel, just the changes in material usage were revolutionary – similar to the introduction of aluminum to aeronautics.

The American Standard was replaced in turn by the Ten Wheeler (4-6-0) which was similar to the 4-4-0 with additional driving wheels. Types with two wheel engine trucks were made possible by the use of a lateral motion bolster – the Mogul (2-6-0) and Consolidation (2-8-0).

By the twentieth century these locomotives were in turn displaced by the Mikado (2-8-2), Santa Fe (2-10-2), Atlantic (4-4-2), Pacific (4-6-2), Mountain (4-8-2) and others which utilized a two wheel trailing truck to support the firebox which had grown steadily larger to produce more power output. By the 1920's the demand for more power produced even larger fireboxes which required 4-wheel trailing trucks for support. The locomotives utilizing this truck such as the Hudson (4-6-4), Berkshire (2-8-4), Northern (4-8-4), and Texas (2-10-4), along with the Pennsylvania Railroad's Duplex types, while constituting the most powerful nonarticulated types in service at the end of the steam era, totalled only approximately 2600 steam locomotives in a U.S. fleet size which reached its peak of nearly 70,000 in 1924. This fleet had shrunk to 29,000 by 1949 when dieselization was in full swing and commercial steam locomotive production was halted (with the exception of the Norfolk and Western Railway which continued to produce locomotives at their Roanoke, Virginia facility until 1953). Of these 29,000 locomotives, only 8,000 were built after 1925. Thus it can be seen that the diesel revolution largely displaced economically worn-out motive power—not modern high capacity steam locomotives.

SWITCHING LOCOMOTIVES (Figs. 17, 33c)

Switching locomotives developed with no engine or trailing trucks to maximize their tractive effort. These locomotives could dispense with the engine truck because they operated at low speed and the trailing truck because they did not require large fireboxes for high power output. The 0-6-0 and 0-8-0 wheel arrangements were most common. Approximately 20,000 switcher locomotives were constructed for domestic service between 1831 and 1950. The geared locomotives (Shay, Climax, Heisler, etc.) did not receive general acceptance for railroad switching purposes, although they were used extensively by other industries which utilized railroad equipment, particularly the logging business.

THE ARTICULATED LOCOMOTIVE (Figs. 1, 18, 19, 22, 31)

While the U.S. articulated locomotives included the largest steam locomotives ever built, their numbers – approximately 2,800 built for domestic service between 1904 and the end of steam – indicates their very limited usage. The first applications were Mallet compound 4-cylinder types of the 0-6-6-0 and 0-8-8-0 wheel arrangement which were largely restricted to yard work because of their inability to operate at any reasonable road speeds and the 2-6-6-0, 2-8-8-0, 2-6-6-2, and 2-8-8-2 wheel arrangements which could operate on the road at modest speeds up to about 25 mph with heavy trains. The design was expanded experimentally up to the 2-10-10-2 type and in six cylinders to the 2-8-8-8-2T and 2-8-8-8-4T wheel arrangements, but never repeated in significant quantities due to many limitations. The 4-cylinder simple articulated which was built in production quantity starting in 1924 was designed to overcome the slow speed characteristics of the Mallet, but less than 500 were built. The more modern designs with 67-70 in. driving wheels were capable of 60 mph operation with an indicated power output in the area of 7,000 hp being not uncommon. The ultimate size and power output was reached with the Union Pacific Alco-built 4-8-8-4's "Big Boys" and the Chesapeake and Ohio and Virginian's Lima-built 2-6-6-6's with much heavier axle loads and 6-wheel trailing trucks. These locomotives were built during the World War II era and they weighed in excess of 1,100,000 lb including tenders. Baldwin also built some extremely large 2-8-8-4's with 63-64 in. drivers for slower speed service.

LOCOMOTIVE BOILERS (Figs. 20a, b)

The basic locomotive boiler was developed in England by 1830 – a horizontal fire-tube type with self-regulated induced draft from the steam exhaust system. U.S. practice soon diverged from British practice in using thinner wrought iron boiler plates simply because U.S. track would not tolerate the weight of a British boiler. Much later – around 1860 – the U.S. abandoned the British practice of using copper for firebox and tube construction and finally adopted steel as the standard material for boiler construction, although steel did not completely displace wrought iron until after 1900. Early locomotive boilers soon reached a practical limit of 48 in. in diameter using ¼ to 5/16 in. wrought iron construction and an operating pressure of 100 psi. This was dictated by desire to have the center of gravity of the locomotive low for safe operation on marginal track. Fireboxes were located as low as possible between the frame members and on the American Standard (4-4-0) between the driving axles which limited the grate area to no more than about 16 ft^2. Such boilers were easily hand fired with wood and later with coal even at firing rates greatly exceeding stationary practice. Efficiency was not a primary consideration, so it was not unusual to see boiler efficiency in the 40-50 percent range at maximum output. As boilers got larger the diameter grew to over 100 in., pressures rose to 300 psi, and grate areas regularly exceeded 100 ft^2. Total heating surfaces rose as much as 10 fold from the 400-700 ft^2 range of early locomotives. Firebox design utilizing stayed water legs remained remarkably uniform over the years in various shapes and sizes including the Bury, Wootten, Crown Bar, Belpaire and the most commonly used Radial-Stayed designs. The water tube firebox was tried in several configurations but never proved totally successful. As the firebox grew larger it could no longer be contained between the frame members and driving wheels so it eventually (around 1900) was relocated behind the driving wheels with a trailer truck to support its weight. The brick arch was adapted (starting in 1845) to lengthen flame travel (and reduce smoke) and the final adoption was as a suitable combustion chamber to increase the firebox volume, which was made more successful by the general introduction of electric welding to firebox construction in the 1920's. Fully welded boilers were developed in the 1930's, but were never used universally at least in part because of the necessity of annealing the entire boiler shell in one operation.

MECHANICAL STOKERS (Figs. 21, 22)

As locomotives became larger, the capability of hand firing a coal burning locomotive of over 1500-2000 hp with a shovel became a problem in human endurance. While designs of mechanical stokers were developed as early as 1905 on the Pennsylvania Railroad, satisfactory models were not in general use until the 1920's, which was just in time for the high-capacity steam locomotives of that era. The Interstate Commerce Commission ordered stokers applied on passenger locomotives of 160,000 lb or more on driving wheels and freight locomotives of 175,000 lb or more on driving wheels in 1938. Stokers were developed with capacities as high as 45,000 lb/hr for use on the Northern Pacific's 2-8-8-4 articulated locomotives with the world's largest fireboxes (182 ft^2 grate area), but such a rate could not be considered a continuous duty.

SUPERHEATING (Fig. 23)

While superheating was tried to a limited extent in the early days, the general adoption of the Schmidt return bend element design in 1910 probably represented the most important single advance made during the 20th century. It immediately obsoleted compounding except in Mallet designs and produced an increased power output of 25-30 percent as compared to a saturated locomotive. Superheat eventually reached the level of 300-350 F when suitable designs and lubricants were developed. It obsoleted the slide valve which scored at high temperature and hastened the development of the piston valve which could operate at these temperatures.

BOILER FEED EQUIPMENT (Figs. 24, 25)

The earliest locomotives used direct driven pumps which operated from the running gear with output in direct proportion to speed. Thus, if a locomotive encountered a particularly difficult operation at low speed, it was necessary to uncouple it from the train and run back and forth to raise the boiler water level and avoid a boiler explosion due to low water exposing the top of the firebox (crown sheet) to overheating. The steam injector which was developed by Giffard in France during the 1850's came into general use in the 1860's at least in combination with direct driven pumps and by the 1890's had displaced them entirely. It is interesting to note that the direct-action steam driven water pump, which received general acceptance in marine and stationary service following its development in 1849 by Worthington, did not receive railroad acceptance until its usage was coupled with workable open and closed type feedwater heaters starting about 1920.

STEAM DISTRIBUTION (Fig. 26)

While the use of lapped valves to reduce steam consumption by cutting off the flow of steam at a fixed point well before the end of the piston stroke was well known even before the first locomotives were built, it remained until 1842 before a completely successful variable cut-off valve gear was developed in England by the firm of Robert Stephenson and Co. This constituted another of the most important milestones in steam locomotive development because it permitted greater efficiency when maximum tractive effort was not required. As steam locomotives became larger, the room available for the inside application of Stephenson valve gear diminished and it was necessary to go to the more accessible outside-mounted Walschaerts valve gear starting about 1900. An alternate to the Walschaerts arrangement was the Baker valve gear developed about 1912. While many other arrangements were tried over the years of steam operation, the three designs – Stephenson, Walschaerts and Baker – were used on the vast majority of steam locomotives built in the U.S.

CYLINDER ARRANGEMENTS (Figs. 27, 28)

The majority of U.S. steam locomotives utilized two simple outside mounted cylinders, generally with slide valves in the 19th century and piston valves in the 20th century. Attempts were made to increase efficiency in the presuperheat era by utilizing compound arrangements such as Cross, Vulclain, Tandem and Balanced Compound systems, but they increased the complexity and maintenance to the point that they were never generally accepted. In the 1920's there was a brief interest in 3-cylinder simple drives utilizing the Gresley conjugated motion or an additional set of valve gear on one side, but again mechanical complexity forced its discontinuance. Finally, in the 1930's there was a brief interest in extremely high pressure boilers utilizing multiple expansion cylinder arrangements, but only experimental quantities were built. Cylinder sizes and capacity increased steadily over the years, reaching a maximum diameter of 4 ft on low pressure compound cylinders during the Mallet era. A maximum piston thrust of 219,000 lb was developed in 2-10-4 locomotives built for the Santa Fe starting in 1938. These cylinders utilized a 30 in. diameter, 34 in. stroke and a maximum boiler pressure of 310 psi. Any greater capacity required 4-cylinder designs.

FRAMES (Fig. 29)

The earliest frames were wood and strap iron (occasionally with cast iron pedestals) rigidly attached to the boiler. By 1850 the outside cylinder arrangement had become standard and built-up wrought iron frames were used. The frame was rigidly attached to the cylinders which in turn were rigidly attached to the boiler at the cylinder saddle-smoke box connection. All frame-to-boiler connections elsewhere utilized flexible or sliding support to allow for boiler expansion. As locomotives became larger, wrought iron was displaced by steel and finally, in 1925, the one-piece cast steel bed frame was developed by the General Steel Castings Corporation. These mammoth single castings eventually included integral cylinders and integral supports for most elements that were attached to the frame. They were sufficiently rigid to allow the introduction of roller main journal bearings (and later even roller bearing equipped driving rods) because they maintained the excellent alignment necessary for these elements and did not shake loose in service.

RUNNING GEAR COMPONENTS

Driving wheels started out like wagon wheels with composite designs utilizing cast iron (hubs), wood (spokes) and wrought iron (tires). Cast iron spoked drivers next evolved with wrought iron or chilled cast iron tires. The final designs were cast steel in spoked or disc configurations with wrought steel tires. Driving wheel diameter depended on the service desired. It was not uncommon for passenger locomotives to have driving wheels up to about 7 ft in diameter, but locomotives used in freight service used smaller wheels to achieve higher tractive effort at the expense of speed. The driving wheels were driven by a piston, piston rod, crosshead and a system of driving rods which had to be properly counterbalanced to avoid damaging the track. It is sufficient to say that design in this area was critical because weight had to be reduced to a minimum consistent with good practices and the characteristics of materials available at the time. Axles, crankpins and driving boxes were also subjected to high forces and had to be properly designed.

ENGINE AND TRAILING TRUCKS

Engine trucks were necessary to guide steam locomotives at speed. The early designs in the U.S. were generally the 4-wheel type because the 2-wheel type did not receive general acceptance until it was given a lateral motion bolster about 1860. The 4-wheel type was also greatly improved by giving it controlled lateral motion and it was continued in use until the end of steam, generally being preferred for high speed service. The Trailing Truck was developed to carry the larger fireboxes and boilers required for higher power output. It finally developed in 2, 4 and even 6 wheel outside bearing designs with controlled radial motion. In modern high speed steam locomotives it was not unusual to have as much as half of the locomotive weight (excluding the tender) carried by these trucks.

TENDERS (Figs. 8, 10, 19, 28, 30c, 33a)

The tender must be considered in any discussion of steam locomotives because of its importance to successful operation. Tenders were kept as small as possible because they represented a nonrevenue load. The original designs utilized ratios of about 1 lb of fuel for every 6 or 7 lb of water to equalize usage and simple pedestal mounted wheel sets. As tenders got larger, they were mounted on two sets of four-wheel trucks but still utilized the "U" shaped water cistern surrounding the fuel space. The final evolution of tenders was to enormous sizes (sometimes in excess of 400,000 lb – fully loaded), which occasionally outweighed the hauling engine. One-piece cast steel water-bottom tender beds formed the underside to which were mounted 6-wheel or 8-wheel trucks, or five sets of pedestal mounted wheels with a 4-wheel "engine truck" for guidance. Fuel to water ratios were modified to normally provide two water stops for every fuel stop, except where water could be scooped from a track pan as was common on the busiest lines of the New York Central and Pennsylvania Railroads. In this case fuel space could be further increased at the expense of water space.

GENERAL ARRANGEMENT (Figs. 30a, b, 31)

Even though steam locomotives increased greatly in size over the years, their general arrangement changed very little. While the first locomotives had no cabs to protect the crews, and a few of the earlier designs placed them on top of or straddling the boiler, it was found that the optimum location was to place the fuel supply as close as possible to the cab and firebox. The Southern Pacific reversed the engine vs the tender on their cab-in-front articulateds, which greatly improved visability in snow sheds and tunnels, but such a design was only practical using oil as a fuel, because it could be piped the entire length of the engine. The location of the stack and domes did not vary greatly, but they got lower, longer and wider as the top of the boiler approached the clearance limits. The headlight and pilot (cowcatcher) naturally stayed in front, and the cylinders remained under the smoke box and stack where the simplest drafting arrangements were made possible.

COMMERCIAL ASPECTS (Figs. 10, 11, 32)

Unlike England, which had its great locomotive "Shops" operated by the railroads, the locomotive industry in the U.S. was largely centered around the commercial builders. The only possible exception to this was the Pennsylvania Railroad, which produced about 6500 locomotives during an 80 year period starting in 1866. While this was not in the same class with Baldwin (over 75,000 locomotives), it represented a sizable output. The total production of all locomotive builders including the railroad shops was about 175,000 locomotives from 1831 to 1950 of which 37,000 locomotives were exported to other countries. Steam locomotives were imported in much smaller quantities and in later years very rarely. The locomotive builders depended, in turn, particularly in later years, on specialty suppliers for such items as steel castings, including the very large cast engine and tender beds, wheels, roller bearings, stokers, injectors, air brake equipment and hundreds of other items. Some of these items were single source and involved patents such as superheaters, feed water heaters, engines to tender radial buffers, boosters (a small auxiliary steam engine geared to an axle to increase the starting tractive effort), etc. Some of the locomotive builders also produced components themselves which could also be bought elsewhere, such as axles, springs, bolts, air reverse gear (an early servomechanism to actuate the valve gear for forward and backward motion at various levels of cut-off), forgings, etc. Locomotive prices held at about $8000 per unit for most of the 19th century, but with inflation and an enormous increase in size and complexity, eventually ended up costing $200,000 or more in the final years. In the early days there were as many as 40 locomotive builders, but at the end, exclusive of railroad shops, there were only three commercial builders of large steam locomotives (Alco, Baldwin and Lima). Standardization of design was almost nonexistent, except during the brief period of the World War I era when the U.S. Government controlled the railroads under the United States Railroad Administration. Most of the railroads thought their own designs were necessary for their own requirements. At times a railroad would simply furnish the builder with a bundle of blueprints and say "built it" – something that would be unheard of in today's world of standardized diesel-electric locomotive designs from each of the two major domestic builders. Many components were developed by railroad personnel and sometimes sold under royalty agreements. The test of success in such endeavors, however, was the sale to railroads other than those that were associated with the inventor. It has been said, with some overstatement, that the steam locomotive standard was "N.T.A." (no two alike).

SUMMARY

While it can be noted that the bulk of major developments in steam locomotive design came from elsewhere (primarily Europe), the U.S. required something different in order to colonize its vast expanses. We needed a simple wood-fired locomotive that would stay on the rails even on the poor track that was being pushed to the West. The answer was the American Standard 4-4-0. Many years later, when we improved our facilities, we built the largest and most powerful steam locomotives ever built. Yet compared to the steam locomotive engineering excellence of France (with their latter-day design genius Andre Champelon who built some of the most efficient steam locomotives ever constructed), our locomotives were still relatively crude and simple. Strangely, the last steam locomotives to operate in France were U.S. built (World War II rehabilitation) 1-4-1 R's, which are about as far from a Champelon design as it is possible to get. It has also been noted that steam locomotives that have been built in recent years in inaccessible places in the world, such as the People's Republic of China, are essentially U.S. designs of the 1940's with appropriate political slogans painted on them. It would thus appear that the relatively simple and more easily maintained designs developed in the U.S. were on the "right track", even though it took over 120 years to prove it.

REFERENCES

[1] Bruce, Alfred W., "The Steam Locomotive in America," W. W. Norton, 1952.

[2] White, John H. Jr., "American Locomotives – An Engineering History 1830-1880," Johns Hopkins Press, 1968.

[3] Wiener, Lionel, "Articulated Locomotives" (originally published 1930), reprinted by Kalmbach, 1970.

[4] Johnson, Ralph P., "The Steam Locomotive," Simmons-Boardman, 1942.

[5] "The Locomotive Dictionary," 1906, 1909, 1912, 1916, 1919 and "The Locomotive Cyclopedia," 1922, 1925, 1927, 1930, 1938, 1941, 1944, 1947, 1950, 1956, Simmons-Boardman Publishing Corp. (dates as indicated).

[6] Nock, O. S., "Locomotion," Charles Scribner's, 1975.

[7] Reed, Brian, "Locomotives in Profile," Doubleday, Vol. 1 (1971) and Vol. 2 (1972).

FIG. 1 CHEASAPEAKE AND OHIO 2-6-6-6 SIMPLE ARTICULATED NEAR WHITE SULPHUR SPRINGS, WEST VIRGINIA – SEPTEMBER 1947. THIS REPRESENTS A MAXIMUM-SIZE STEAM LOCOMOTIVE AT FULL POWER OUTPUT.

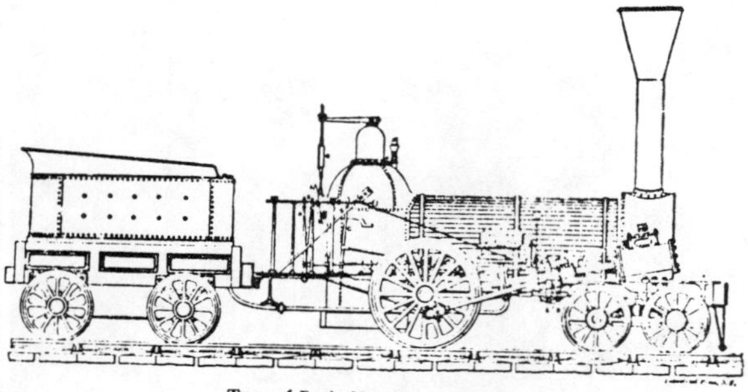

Type of Early Norris Locomotive

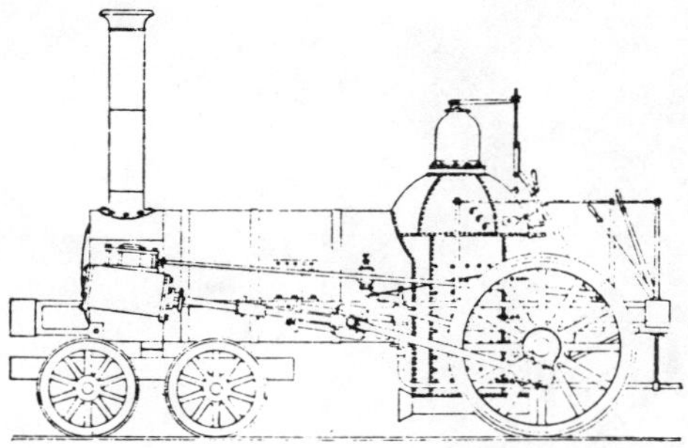

From a Drawing by C. H. Caruthers

The "Black Hawk," built for the Philadelphia & Trenton R. R. by The Baldwin Locomotive Works, 1835. The first Baldwin Engine with outside Cylinders

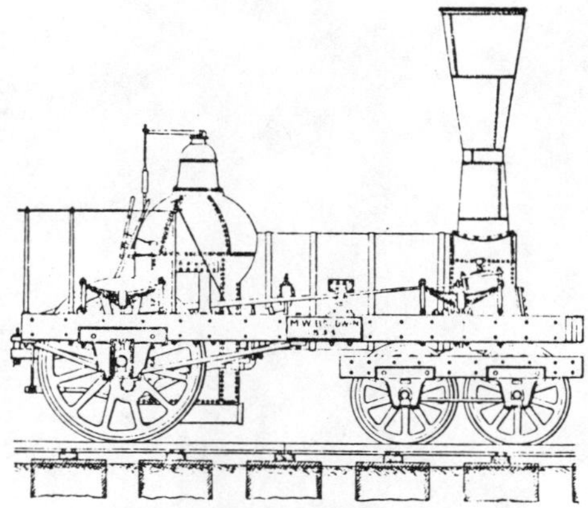

Type of Baldwin Locomotive built for the Philadelphia & Columbia R. R., 1834

Cylinders	17" x 22"
Drivers, diam.	56"
Weight on drivers	40,850 lb.
Weight, total engine	64,550 lb.

Equipped with hook motion valve gear and independent variable cut-off. Used in smoke burning tests of 1859, and worn out in 1876.

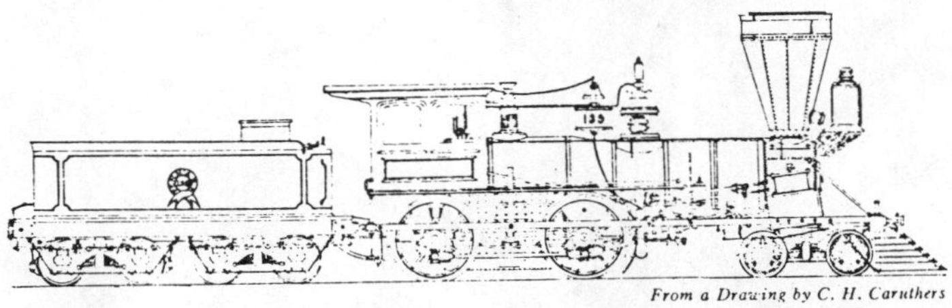

From a Drawing by C. H. Caruthers

American Type Freight Locomotive, built by The Baldwin Locomotive Works, 1857

Cylinders	16" x 24"
Drivers, diam.	66"
Weight on drivers	40,675 lb.
Weight, total engine	63,800 lb.

Smith's patent boiler with short tubes and 54" combustion chamber.

Used in smoke burning tests, 1859.

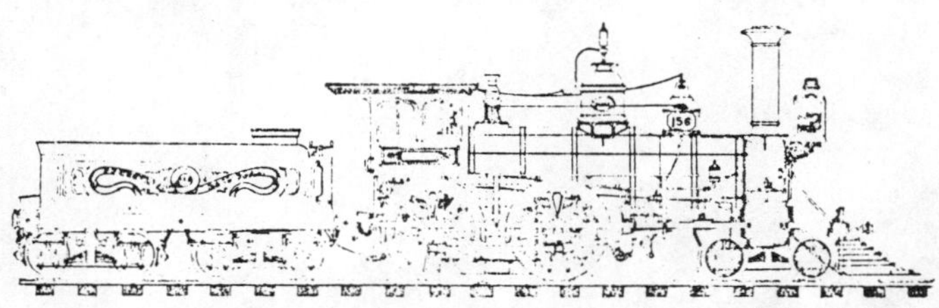

From a Drawing by C. H. Caruthers

American Type Passenger Locomotive, built by The Baldwin Locomotive Works, 1859

Cylinders	16" x 24"
Drivers, diam.	60"

Equipped with Stephenson link motion outside driving wheels.

Used in express passenger service between Jersey City and New Brunswick. Cut up by Pennsylvania R. R. in 1882.

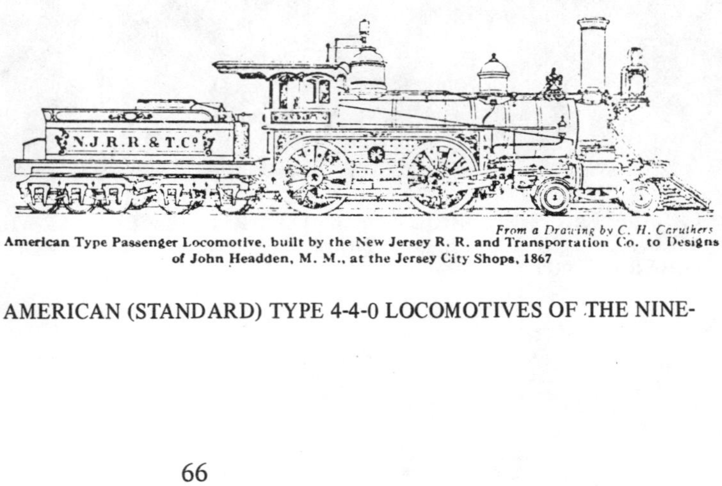

From a Drawing by C. H. Caruthers

American Type Passenger Locomotive, built by the New Jersey R. R. and Transportation Co. to Designs of John Headden, M. M., at the Jersey City Shops, 1867

FIG. 2 JERVIS TYPE 4-2-0 AND AMERICAN (STANDARD) TYPE 4-4-0 LOCOMOTIVES OF THE NINETEENTH CENTURY

FIG. 3 AMERICAN (STANDARD) 4-4-0 LOCOMOTIVE BEING "HOISTED" AT ALTOONA SHOP. THE USE OF AIR BRAKE EQUIPMENT AND THE EXTENDED SMOKE BOX ON THE LOCOMOTIVE WOULD PLACE THE DATE AT ABOUT 1880.

AMERICAN LOCOMOTIVE COMPANY,

NEW YORK.

Class 440–152 Road Number, 329

BUILT FOR THE N. Y. C. LINES (P. & L. E. R. R.)

GAUGE OF TRACK	CYLINDERS		DRIVING WHEEL DIAMETER	BOILER		FIRE BOX		TUBES		
	Diam.	Stroke		Diameter	Pressure	Length	Width	Number	Diameter	Length
4′-8½″	19″	26″	72″	64″	200#	108″	40″	334	2″	12′–2″

WHEEL BASE			WEIGHT IN WORKING ORDER—POUNDS.			
Driving	Engine	Engine & Tender	Leading	Driving	Engine	Tender
8′–6″	23′–7″	54′–2¾″	51000#	100900#	151900#	143150#

FUEL	HEATING SURFACES, SQUARE FT.				GRATE AREA SQ. FT.	MAXIMUM TRACTIVE POWER	FACTOR OF ADHESION
Kind	Tubes	Fire Box	Water Tubes	Total			
Soft Coal	2113.15	172.4	21.2	2306.75	30	22200#	4.54

Tender, Type 8-Wheeled. Capacity, Water 8400 Gals. Fuel, 10 Tons

NEGATIVE NO. 496

FIG. 4 NEW YORK CENTRAL LINES (P&LE) AMERICAN "STANDARD" 4-4-0. WHILE THIS LOCOMOTIVE REPRESENTS A LATER AND HEAVIER VERSION OF THE MUCH COPIED TYPE, IT MAINTAINS THE EARLIER DESIGN FEATURES – NARROW FIREBOX LOCATED BETWEEN THE DRIVING WHEELS. STEPHENSON VALVE MOTION AND SLIDE VALVES.

NEGATIVE NO. 2464

BALDWIN LOCOMOTIVE WORKS

Class 10 36 D 427 — Road No. 330

Built for Colorado & Southern Ry.

Gauge,	4′ 8½″
Cylinder,	21″ x 28″
Valve,	Balanced Piston.
Boiler.	
Type,	Wagon Top
Material,	Steel.
Diameter,	67⅜″
Thickness of sheets,	$\frac{11}{16}$″ & ¾″
Working Pressure,	200 lbs.
Fuel,	Soft Coal.
Staying,	Radial.
Fire Box.	
Material,	Steel.
Length,	120⅛″
Width,	39¼″
Depth, front,	78$\frac{11}{16}$″
" back,	66$\frac{11}{16}$″
Thickness of sheets, sides,	⅜″
" " " back,	⅜″
" " " crown,	⅜″
" " " tube,	½″
Water Space.	
Front,	5″
Sides,	4½″
Back,	4½″
Tubes.	
Material,	Iron.
Wire Gauge,	No. 12.
Number,	294.
Diameter,	2″
Length,	14′ 10″

Heating Surface.	
Fire box,	194.5 sq. ft.
Tubes,	2270.5 sq. ft.
Total,	2465.0 sq. ft.
Grate Area	32.6 sq. ft.
Driving Wheels.	
Diameter, outside,	69″
" center,	62″
Journals, main,	10″ x 12″
" others,	9″ x 12″
Engine Truck Wheels.	
Diameter,	30½″
Journals,	6″ x 10″
Wheel Base	
Driving,	15′ 0″
Rigid,	15′ 0″
Total Engine,	26′ 2″
" " & Tender	54′ 11″
Weight.	
On Driv. Wheels,	145,250 lbs.
On Truck,	42,200 lbs.
Total Engine,	187,450 lbs.
" " & Tender	about, 340,000 lbs.
Tender.	
Wheels, Number,	8.
" Diameter,	33½″
Journals,	5½″ x 10″
Tank Capacity,	8000 gals.
Fuel Capacity	10 tons.
Service,	Passenger.

Drawing ***No.*** 28.

FIG. 5 THE TEN WHEELER 4-6-0, THE SUCCESSOR TO THE AMERICAN STANDARD 4-4-0. THIS LOCOMOTIVE RETAINS A NARROW FIREBOX OF A DESIGN SIMILAR TO THE 4-4-0 ALTHOUGH SOMEWHAT LONGER, BUT HAS THE MORE MODERN PISTON VALVES ACTUATED BY STEPHENSON INSIDE VALVE MOTION.

AMERICAN LOCOMOTIVE COMPANY,
NEW YORK.

Class 280-244 | Road Number, 5564

BUILT FOR THE L. S. & M. S. R. R.

GAUGE OF TRACK	CYLINDERS. Diam.	CYLINDERS. Stroke	DRIVING WHEEL DIAMETER	BOILER Diameter	BOILER Pressure	FIRE BOX Length	FIRE BOX Width	TUBES Number	TUBES Diameter	TUBES Length
4′-8½″	23″	30″	57″	80″	200 lbs.	108⅛″	73¼″	518	2″	15′-6″

WHEEL BASE Driving	WHEEL BASE Engine	WHEEL BASE Engine & Tender	WEIGHT IN WORKING ORDER—POUNDS. Leading	Driving	Engine	Tender
17′-3″	26′-5″	60′-11″	29700	214300	244000	153150

FUEL Kind	HEATING SURFACES, SQUARE FT. Tubes	Fire Box	Water Tubes	Total	GRATE AREA SQ. FT.	MAXIMUM TRACTIVE POWER	FACTOR OF ADHESION
Soft Coal	4181	202	29	4412	55.8	47300 lbs.	4.55

Tender, Type 8-Wheeled. Capacity, Water 7500 Gals. Fuel, 12 Tons.

NEGATIVE NO. B-644

FIG. 6 NEW YORK CENTRAL LINES (LS&MS) 2-8-0 CONSOLIDATION TYPE. THIS WAS THE ALL TIME "CHAMPION" STEAM LOCOMOTIVE WHEEL ARRANGEMENT IN THE U.S., AT LEAST IN PART BECAUSE LOCOMOTIVE PRODUCTION PEAKED WHEN THIS DESIGN WAS POPULAR.

AMERICAN LOCOMOTIVE COMPANY
NEW YORK

Class, 282 S 285 | Road Number, 406

BUILT FOR THE GREEN BAY & WESTERN.

GAUGE OF TRACK	CYLINDERS Diam.	CYLINDERS Stroke	DRIVING WHEEL DIAMETER	BOILER Inside Dia.	BOILER Pressure	FIRE BOX Length	FIRE BOX Width	TUBES Number	TUBES Diameter	TUBES Length
4′-8½″	22″	30″	64″	72½″	245 lbs.	114″	76¼″	53 132	2″ 3½″	18′-0″

WHEEL BASE Driving	WHEEL BASE Engine	WHEEL BASE Engine & Tender	WEIGHT IN WORKING ORDER POUNDS Leading	Driving	Trailing	Engine	Tender
16′-9″	36′-11″	72′-2⅞″	29500	204000	51500	285000	176000

FUEL Kind	EVAPORATING SURFACES. SQ. FT. Tubes	Flues	Fire Box	Arch Tubes	Total	SUPERHEATING SURFACE SQUARE FT.	GRATE AREA SQ. FT.	MAXIMUM TRACTIVE POWER	FACTOR OF ADHESION
Bit. Coal	496	2164	251	26	2937	1184	60.4	47250 lbs.	4.32

Tender Type, 8 Wheeled. Capacity, Water, 9000 Gals. Fuel, 14 Tons

ORDER No. S-1815
October, 1939

FIG. 7 GREEN BAY AND WESTERN 2-8-2 MIKADO TYPE LOCOMOTIVE. THIS IS A LIGHT AXLE-LOAD (51,000 LB) DESIGN WHICH RESEMBLES LOCOMOTIVES BUILT FOR FRANCE (WW II REHABILITATION) AND IS REMARKABLY SIMILAR TO LOCOMOTIVES RECENTLY BUILT IN THE PEOPLE'S REPUBLIC OF CHINA.

NEGATIVE NO. 8990

THE BALDWIN LOCOMOTIVE WORKS

2-10-2 TYPE LOCOMOTIVE

Class 14-54-¼-F, 297 — Road No. 6100

Built for **Baltimore & Ohio R. R.**

Railroad Co's Class S-1

Gauge	4′ 8½″
Cylinders	29¾″ x 32″
Valves	Piston, 14″ diam.

BOILER

Type	Straight top
Diameter	90″
Working pressure	220 lb.
Fuel	Soft coal

Firebox

Material	Steel
Staying	Radial
Length	132⅛″
Width	96″
Depth, front	91¾″
Depth, back	77⅞″

Tubes

Diameter	5½″	2¼″
Number	53	232
Length	23′0″	23′ 0″

Heating Surface

Firebox	262 sq. ft.
Combustion chamber	85 sq. ft.
Tubes	4881 sq. ft.
Firebrick tubes	42 sq. ft.
Total	5270 sq. ft.
Superheater	1512 sq. ft.
Grate area	88 sq. ft.

Construction No. 57,214

DRIVING WHEELS

Diameter, outside	64″
Diameter, center	56″
Journals, main	13½″ x 15″
Journals, others	11″ x 13″

ENGINE TRUCK WHEELS

Diameter, front	33″
Journals	6″ x 10″
Diameter, back	46″
Journals	9″ x 14″

WHEEL BASE

Driving	22′ 4″
Rigid	22′ 4″
Total engine	42′ 11″
Total engine & tender	89′ 10⅞″

WEIGHT
in working order

On driving wheels	347,230 lb.
On truck, front	31,570 lb.
On truck, back	57,710 lb.
Total engine	436,510 lb.
Total engine & tender	735,000 lb.

TENDER

Wheels, number	12
Wheels, diameter	33″
Journals	6″ x 11″
Tank capacity	15,800 U. S. gal.
Fuel capacity	23 tons
Tractive force	84,260 lb.
Service	Heavy freight

Drawing No. 17

Equipped with superheater, stoker, and air brake on all driving and tender wheels, with two 8½″cross-compound pumps.

FIG. 8 2-10-2 (SANTA FE) TYPE USED FOR HEAVY HAULING AT RELATIVELY LOW SPEEDS. THIS LOCOMOTIVE IS EQUIPPED WITH A VANDERBILT CYLINDRICAL TENDER.

AMERICAN LOCOMOTIVE COMPANY

NEW YORK

Class, 442 S 290 — Road Number, 4

BUILT FOR THE C. M. ST. P. & P. R. R.

GAUGE OF TRACK	CYLINDERS		DRIVING WHEEL DIAMETER	BOILER		FIRE BOX		TUBES		
	Diam.	Stroke		Inside Dia.	Pressure	Length	Width	Number	Diameter	Length
4′-8½″	19″	28″	84″	76 11/16″	300 lbs.	132 1/16″	75 3/16	160 43	2¼″ 5½″	19′-0″

WHEEL BASE			WEIGHT IN WORKING ORDER POUNDS				
Driving	Engine	Engine & Tender	Leading	Driving	Trailing	Engine	Tender
8′-6″	37′-7″	78′-10½″	72700	144300	73000	290000	266300

FUEL	EVAPORATING SURFACES, SQUARE FT.					SUPERHEATING SURFACE SQUARE FT.	GRATE AREA SQ. FT.	MAXIMUM TRACTIVE POWER	FACTOR OF ADHESION
Kind	Tubes	Flues	Fire Box	Syphon	Total				
Oil	1781	1170	254	40	3245	1029	69	30700 lbs.	4.70

Tender Type, 10-Wheeled — Capacity, Water, 13000 Gals — Fuel, 4000 Gals

ORDER No. S-1770

April, 1937

FIG. 9 MILWAUKEE ROAD HIAWATHA 4-4-2 ATLANTIC. THIS TOTALLY MODERN RENDITION OF A MUCH OLDER WHEEL ARRANGEMENT MAY WELL HAVE BEEN THE FASTEST LOCOMOTIVE EVER BUILT IN THE U.S. SPEED CONTESTS, HOWEVER, WERE NEVER CONDUCTED (AT LEAST OFFICIALLY) SO ONE CAN ONLY SPECULATE. THE DRIVING WHEEL DIAMETER WAS 84 IN. ITS ONLY PROBLEM WAS THAT THE POPULARITY OF THE MILWAUKEE'S HIGH SPEED TRAINS BECAME SO GREAT THAT IT COULD NOT HANDLE THE LONGER TRAINS NEEDED – THUS, THE 84 IN. DRIVERED HUDSON (4-6-4) WAS DEVELOPED (FIG 12).

AMERICAN LOCOMOTIVE COMPANY.
NEW YORK

Class, 462 S 303 — Road Number, 5000

BUILT FOR THE EXPERIMENTAL.

GAUGE OF TRACK	CYLINDERS		DRIVING WHEEL DIAMETER	BOILER		FIRE BOX		TUBES		
	Diam.	Stroke		Inside Dia.	Pressure	Length	Width	Number	Diameter	Length
4′-8½″	26″	28″	79″	82″	200 lbs.	108⅛″	90¼″	206 48	2¼″ 5½″	21′-0″

WHEEL BASE			WEIGHT IN WORKING ORDER—POUNDS				
Driving	Engine	Engine & Tender	Leading	Driving	Trailing	Engine	Tender
13′-8″	36′-10″	71′-10½″	58500	186000	58500	303000	203000

FUEL	HEATING SURFACES, SQUARE FT.						GRATE AREA SQ. FT.	MAXIMUM TRACTIVE POWER		FACTOR OF ADHESION	
Kind	Tubes	Flues	Fire Box	Arch Tubes	Total	Superheater		Engine	Booster	Drivers	Trailer
Soft Coal	2537	1445	218	35	4235	1245	67.8	40700 lbs.	9800 lbs.	4.57	5.97

Tender Type. 8-Wheeled. Capacity, Water, 10000 Gals Fuel, 15 Tons.

ORDER No. S-1411
January, 1924

FIG. 10 THE PACIFIC TYPE 4-6-2 WAS A VERY POPULAR PASSENGER LOCOMOTIVE BUILT IN LARGE QUANTITIES. THIS EXAMPLE REPRESENTS ONE OF THE LAST DESIGNS WITH A "BOOSTER" ENGINE ON THE TRAILING TRUCK FOR ADDITIONAL TRACTIVE EFFORT AT LOW SPEEDS. ALSO NOTE WATER SCOOP ON TENDER FOR TAKING ON WATER AT A TRACK PAN.

NEGATIVE NO. 7094

THE BALDWIN LOCOMOTIVE WORKS

Class 14-50-¼-E, 14 — Road No. 137

Built for United States Gov't for Chesapeake & Ohio R. R.

STANDARD LOCOMOTIVE, CLASS 4-8-2-B

Gauge	4′ 8½″
Cylinders	28″ x 30″
Valves	Piston, 14″ diam.

BOILER

Type	Conical wagon-top
Diameter	86″
Thickness of sheets	¾″ & 27/32″
Working pressure	200 lbs.
Fuel	Soft coal

Firebox

Material	Steel
Staying	Radial
Length	114⅛″
Width	96¼″
Depth, front	91″
Depth, back	69½″
Thickness of sheets, sides	⅜″
" " " back	⅜″
" " " crown	⅜″
" " " tube	½″

Water Space

Front	6″
Sides	5″
Back	5″

Tubes

Diameter	5½″	2¼″
Material	Steel	Steel
Thickness	No. 9 W. G.	No. 11 W. G.
Number	45	247
Length	20′ 6″	20′ 6″

Heating Surface

Firebox	220 sq. ft.
Combustion chamber	115 sq. ft.
Tubes	4,293 sq. ft.
Firebrick tubes	34 sq. ft.
Total	4,662 sq. ft.
Superheater	1,078 sq. ft.
Grate area	76.3 sq. ft.

DRIVING WHEELS

Diameter, outside	69″
" center	62″
Journals, main	12″ x 13″
" others	11″ x 13″

ENGINE TRUCK WHEELS

Diameter, front	33″
Journals	6½″ x 12″
Diameter, back	43″
Journals	9″ x 14″

WHEEL BASE

Driving	18′ 3″
Rigid	18′ 3″
Total engine	40′ 0″
Total engine & tender	75′ 8½″

WEIGHT

On driving wheels	243,000 lbs.
On truck, front	51,500 lbs.
On truck, back	57,500 lbs.
Total engine	352,000 lbs.
Total engine & tender	about 546,000 lbs.

TENDER

Wheels, number	8
" diameter	36″
Journals	6″ x 11″
Tank capacity	10,000 U. S. gals.
Fuel "	16 tons

Tractive force	58,000 lbs.
Service	Passenger

Construction Number 51881 — *Drawing Number 4*

FIG. 11 USRA HEAVY MOUNTAIN TYPE (4-8-2) LOCOMOTIVE. THIS IS AN EXAMPLE OF THE UNITED STATES RAILROAD ADMINISTRATION STANDARD DESIGN. AS SOON AS THE RAILROADS REVERTED BACK TO PRIVATE OPERATION AFTER WORLD WAR I, THEY LARGELY RETURNED TO INDIVIDUAL DESIGN STANDARDS. STANDARD LOCOMOTIVE DESIGNS DID NOT ARRIVE PERMANENTLY UNTIL THE ADVENT OF PRACTICAL DIESEL-ELECTRIC LOCOMOTIVES IN THE 1930'S AND THEREAFTER.

AMERICAN LOCOMOTIVE COMPANY

NEW YORK

Class, 464 S 415

Road Number, 100

BUILT FOR THE CHICAGO, MILWAUKEE, ST. PAUL & PACIFIC.

GAUGE OF TRACK	CYLINDERS		DRIVING WHEEL DIAMETER	BOILER		FIRE BOX		TUBES		
	Diam.	Stroke		Inside Dia.	Pressure	Length	Width	Number	Diameter	Length
4′-8½″	23½″	30″	84″	82½″	300 lbs.	144½″	96 3/16″	60 164	2¼″ 3¾″	19′-0″

WHEEL BASE			WEIGHT IN WORKING ORDER—POUNDS				
Driving	Engine	Engine & Tender	Leading	Driving	Trailing	Engine	Tender
14′-8″	42′-4″	89′-10″	82500	216000	116500	415000	375000

FUEL	EVAPORATING SURFACES, SQUARE FT.						SUPERHEATING SURFACE SQUARE FT.	GRATE AREA SQ. FT.	MAXIMUM TRACTIVE POWER	FACTOR OF ADHESION
Kind	Tubes	Flues	Fire Box	Arch Tubes	Syphons	Total				
Soft Coal	667	3041	348	19	91	4166	1695	96.5	50300 lbs.	4.29

Tender Type, 12 Wheeled Capacity, Water, 20000 Gals. Fuel, 25 Tons

ORDER No. S-1798

August, 1938

FIG. 12 MILWAUKEE ROAD 4-6-4 HUDSON WITH 84 IN. DRIVING WHEELS. THESE LOCOMOTIVES AND SIMILAR DESIGNS OWNED BY OTHER RAILROADS WERE THE HIGH SPEED PASSENGER LOCOMOTIVES OF THE LAST DAYS OF STEAM UNTIL REPLACED BY DIESELS. WHERE GRADES WERE ENCOUNTERED WITH HEAVY PASSENGER TRAINS THE HIGHER TRACTIVE EFFORT OF THE NORTHERN (4-8-4) AND OTHER FOUR-COUPLED TYPES WAS PREFERRED. THIS LOCOMOTIVE WAS BEAUTIFULLY STREAMLINED BY OTTO KUHLER FOR PUBLIC APPEAL, WHICH INTERFERRED TO SOME DEGREE WITH THE ACCESSIBILITY OF CERTAIN COMPONENTS.

LIMA LOCOMOTIVE WORKS, INCORPORATED

LIMA, OHIO

Class: 284 S 389 — Road No. 1400

Built for BOSTON & ALBANY R. R.

GAUGE OF TRACK	DRIVING WHEEL DIAMETER	FUEL KIND	CYLINDERS		BOILER		FIREBOX	
			DIAMETER	STROKE	DIAMETER	PRESSURE	LENGTH	WIDTH
4′-8½″	63″	SOFT COAL	28″	30″	88″	240 LBS.	150⅛″	96¼″

WHEEL BASE			TRACTIVE POWER		FACTOR OF ADHESION	TUBES AND FLUES		
DRIVING	ENGINE	ENGINE AND TENDER	MAIN CYLINDERS	WITH BOOSTER		NUMBER	DIAMETER	LENGTH
16′-6″	41′-8″	75′-8½″	69400 AT 60% CUT OFF	81400	3.6	90 204	2¼″ 3½″	20′-0″

AVERAGE WEIGHT IN WORKING ORDER, POUNDS						GRATE AREA SQ. FT.	HEATING SURFACES, SQUARE FEET				
ON DRIVERS	TRUCK	TRAILER FRONT AXLE	TRAILER REAR AXLE	TOTAL ENGINE	TENDER LOADED		TUBES AND FLUES	FIREBOX	ARCH TUBES	TOTAL	SUPER-HEATER
249500	36000	49500	54000	389000	201900	100	4773	284	53	5110	2111

Tender, Type 8 Wheel — Capacity, Water 10000 Gallons — Fuel, 16 Tons

Negative Order No. 1082

Equipped with THE LOCOMOTIVE BOOSTER

FIG. 13 BOSTON AND ALBANY 2-8-4 BERKSHIRE. THE FIRST PRODUCTION LIMA "SUPERPOWER" LOCOMOTIVE. THIS LOCOMOTIVE HAD HIGH HORSEPOWER OUTPUT MADE POSSIBLE BY THE LARGE FIREBOX (100 FT^2 GRATE AREA) CARRIED BY A 4-WHEEL TRAILING TRUCK. ITS DESIGN WAS TO INFLUENCE MOST HIGH CAPACITY LOCOMOTIVES BUILT THEREAFTER.

NEGATIVE NO. 12,200

THE BALDWIN LOCOMOTIVE WORKS

Class 4-8-4, 28 S, 100 — Road No. 2919

Built for **Atchison, Topeka & Santa Fe Railway**

Railroad Co.'s Class 2900

(Order covers 30 locomotives, Nos. 2900-2929)

Gauge	4′ 8½″
Cylinders	28″ x 32″
Valves	Piston, 15″ diam.

BOILER

Type	Conical
Diameter	F 90″; B 102″
Working pressure	300 lb.
Fuel	Oil

Firebox

Material	Steel
Staying	Radial
Length	$143\frac{31}{32}$″
Width	108″
C. C. Length	64″

Tubes

Diameter	3½″	2¼″
Number	220	52
Length	21′ 0″	21′ 0″

Heating Surface

Firebox	306 sq. ft.
Combustion chamber	124 sq. ft.
3½″ Flues	4210 sq. ft.
2¼″ Tubes	642 sq. ft.
Thermic syphon	30.5 sq. ft.
Total	5312.5 sq. ft.
Superheater	2366 sq. ft.
Grate area	108 sq. ft.
Firebox volume	579 cu. ft.
Min. Gas Area	10.5 sq. ft.

DRIVING WHEELS

Diameter, outside	80″
Diameter, center	72″
Journals, main—	diam. 14¼″ Roller bearing
Journals, others—	diam. $12\frac{9}{16}$″ Roller bearing

ENGINE TRUCK WHEELS

Diameter, front	42″
Journals—	diam. 8¼″ Roller bearing
Diameter, back	50″
Journals—	diam. 8″ Roller bearing

WHEEL BASE

Driving	21′ 3″
Rigid	13′ 10″
Total engine	50′ 0″
Total engine & tender	108′ 2″

WEIGHT
In Working Order

On driving wheels	293,860 lb.
On truck, front	93,820 lb.
On truck, back	122,470 lb.
Total engine	510,150 lb.
Total tender based on ⅔ oil & water capacity	369,690 lb.

TENDER

Wheels, number	Sixteen
Wheels, diameter	42″
Journals—	6½″ diam. Roller bearing
Tank capacity,	24,500 U.S. gals.
Fuel capacity	7,000 gals.

Tractive force (70%)	66,000 lb.
Service	Freight

Construction No. 69,807 *Built* March, 1944 *Drawing No.* 12

Equipped with Type "E" superheater, feed water heater, one thermic syphon, power reverse, one piece cast steel locomotive bed with integral cylinders but with separate cradle, and air brake on all driving, trailer truck and tender wheels, with two 8½″ cross-compound pumps.

1M 6-44 F. P.

FIG. 14 SANTA FE 4-8-4 IS THE HEAVIEST LOCOMOTIVE OF THE NORTHERN TYPE WHEEL ARRANGEMENT EVER BUILT. THIS OCCURRED AT LEAST IN PART AS A RESULT OF WARTIME MATERIALS RESTRICTION WHICH CAUSED THEM TO WEIGH ABOUT 8 TONS MORE THAN SIMILAR LOCOMOTIVES BUILT EARLIER. THE GREAT DIFFERENCES IN WEIGHT OF LOCOMOTIVES OF THE SAME WHEEL ARRANGEMENT IS SHOWN BY THE FACT THAT SANTA FE "2900" WEIGHED 510,150 LB WHILE THE TOLEDO, PEORIA AND WESTERN 4-8-4 WEIGHED ONLY 361,000 LB. THE TENDER OF THIS LOCOMOTIVE IS ALSO NOTABLE FOR ITS ENORMOUS SIZE.

MONTREAL LOCOMOTIVE WORKS, LTD.

MONTREAL, P. Q.

Class, 2104 S 447 — Road Number, 5922

BUILT FOR THE CANADIAN PACIFIC.

GAUGE OF TRACK	CYLINDERS		DRIVING WHEEL DIAMETER	BOILER		FIRE BOX		TUBES		
	Diam.	Stroke		Inside Dia.	Pressure	Length	Width	Number	Diameter	Length
4′-8½″	25″	32″	63″	82½″	285 lbs.	140 3/18″	96″	198 72	3½″ 2¼″	21′ 0″

WHEEL BASE			WEIGHT IN WORKING ORDER—POUNDS				
Driving	Engine	Engine & Tender	Leading	Driving	Trailing	Engine	Tender
22′-0″	46′-0½″	87′-5⅛″	35500	309900	101600	447000	281000

FUEL	EVAPORATING SURFACES, SQUARE FT.					SUPERHEATING SURFACE SQUARE FT.	GRATE AREA SQ. FT.	MAXIMUM TRACTIVE POWER		FACTOR OF ADHESION	
Kind	Tubes	Flues	Fire Box	Arch Tubes	Total			Engine	Booster	Drivers	Trailer
Oil	887	3755	367	45	5054	2032	93.5	76905 lbs.	12000 lbs.	4.03	4.79

Tender Type, 12-Wheeled — Capacity, Water, 12000 Imp. Gal's — Fuel, 4500 Imp. Gal's

ORDER No. Q-386

November, 1938

FIG. 15 THE LARGEST STEAM LOCOMOTIVE USED IN CANADA – THE CANADIAN PACIFIC "SELKIRK" 2-10-4. CANADIAN LOCOMOTIVE DESIGN WAS SIMILAR TO U.S. PRACTICE, EXCEPT THAT DUE TO A DESIRE TO MAKE THE LOCOMOTIVES OPERABLE ON DISTRICTS WITH LIGHTER RAIL, THE CANADIANS DID NOT OPT FOR THE EXTREMELY HEAVY DESIGNS. ARTICULATED LOCOMOTIVES WERE ABANDONED VERY EARLY WITH ONLY A FEW BUILT EXPERIMENTALLY. NOTE THE VESTIBULE "ALL WEATHER" CAB WHICH REFLECTS CANADIAN WINTER WEATHER CONDITIONS.

NEGATIVE NO. 12,285

THE BALDWIN LOCOMOTIVE WORKS

Class 4-4-4-4, 19¾ SR, 1 — Road No. 6110

Built for **The Pennsylvania Railroad**

Railway Co.'s Class: Engine "T-1"; Tender 180P76
(Order covers two locomotives, Nos. 6110-6111)

Gauge	4′ 8½″
Cylinders (4)	19¾″ x 26″
Valves	Poppet type

BOILER

Type	Modified Belpaire
Diameter	F. 91½″ B. 100″
Working pressure	300 lb.
Fuel	Bituminous coal

Firebox

Material	Steel
Staying	Radial
Length	138″
Width	96″
Comb. Cham. Length	96″

Tubes

Diameter	5½″	2¼″
Number	69	184
Length	18′ 0″	18′ 0″

Heating Surface

Firebox	269 sq. ft.
Combustion chamber	150 sq. ft.
Tubes 2¼″	1940 sq. ft.
Tubes 5½″	1779 sq. ft.
Circulators	71 sq. ft.
Total	4209 sq. ft.
Superheater	1680 sq. ft.
Grate area	92 sq. ft.
Firebox volume	641 cu. ft.

DRIVING WHEELS

Diameter, outside	80″
Diameter, center	72″
Journals, main— 11½″ diam. roller bearing	
Journals, others— 11½″ diam. roller bearing	

ENGINE TRUCK WHEELS

Diameter, front	36″
Journals— 7″ diam. roller bearing	
Diameter, back	42″
Journals— 8″ diam. roller bearing	

WHEEL BASE

Driving	25′ 4″
Rigid	18′ 5″
Total engine	51′ 11″
Total engine & tender	107′ 0″

WEIGHT
In Working Order

On driving wheels	268,200 lb.
On truck, front	100,200 lb.
On truck, back	128,800 lb.
Total engine	497,200 lb.
Total tender, fully loaded	433,000 lb.
(Average weight per pr. drivers	67,050 lb.)

TENDER

Wheels, number	Sixteen
Wheels, diameter	36″
Journals— 6½″ diam. roller bearing	
Tank capacity	19,500 U. S. gal.
Fuel capacity	41 tons

Tractive force	65,000 lb.
Service	Passenger

Construction No. 62,451 — *Built* 1942 — *Drawing No.* 1

Equipped with Type "A-S" superheater, feed water heater, stoker, five circulators, power reverse, one piece cast steel locomotive bed with integral cylinders, and air brake on all front truck, trailer truck, driving and tender wheels, with two 8½″ cross-compound air pumps.

Booster (estimated to develop 13,500 lb. tractive force), on engine 6111 only.

FIG. 16 PENNSYLVANIA RAILROAD'S 4-4-4-4 DUPLEX LOCOMOTIVE. WHILE THIS LOCOMOTIVE APPEARS TO BE AN ARTICULATED DESIGN, IT HAS A ONE-PIECE RIGID CAST STEEL ENGINE BED. THE PURPOSE OF THE DUPLEX DRIVE WAS TO REDUCE DYNAMIC AUGMENT, BUT IN REALITY IT PRODUCED A VERY "SLIPPERY" LOCOMOTIVE WITH HIGHER MAINTENANCE COSTS. THIS LOCOMOTIVE IS EQUIPPED WITH FRANKLIN TYPE 'A' POPPET VALVE STEAM DISTRIBUTION WHICH WAS A PROMISING DEVELOPMENT "CUT SHORT" BY TOTAL DIESELIZATION.

AMERICAN LOCOMOTIVE COMPANY
NEW YORK

Class, 060 - 181 | Road Number, 83

BUILT FOR THE UNION R. R.

GAUGE OF TRACK	CYLINDERS		DRIVING WHEEL DIAMETER	BOILER		FIRE BOX		TUBES		
	Diam.	Stroke		Inside Dia.	Pressure	Length	Width	Number	Diameter	Length
4′-8½″	22″	28″	51″	70½″	200 lbs	77⅝″	72¼″	320	2″	15′-0″

WHEEL BASE			WEIGHT IN WORKING ORDER—POUNDS		
Driving	Engine	Engine & Tender	Driving	Engine	Tender
11′-0″	11′-0″	45′-10″	180500	180500	114600

FUEL	EVAPORATING SURFACES, SQUARE FT.			GRATE AREA SQ. FT.	MAXIMUM TRACTIVE POWER	FACTOR OF ADHESION
Kind	Tubes	Fire Box	Total			
Soft Coal	2499	143	2642	38.9	45200 lbs.	3.99

Tender Type, 8 Wheeled | Capacity, Water, 6000 Gals. | Fuel, 8 Tons

ORDER No. S- 1906

February, 1944

FIG. 17 A TYPICAL 0-6-0 SWITCHER WITH ALL OF THE ENGINE WEIGHT ON SMALL DRIVING WHEELS FOR MAXIMUM TRACTIVE EFFORT. THE SPEED AND HORSEPOWER CAPABILITIES ARE VERY MODEST DUE TO LOW YARD SPEED REQUIREMENTS. NOTE THAT THE TENDER IS SMALL AND HAS A RECESSED COAL BUNKER FOR GOOD REARWARD VISABILITY AS A SWITCHER HAS TO OPERATE EXTENSIVELY IN BOTH DIRECTIONS.

AMERICAN LOCOMOTIVE COMPANY,
NEW YORK.

Class 2662 C S 354 — Road Number, 1307

BUILT FOR THE BOSTON & ALBANY

GAUGE OF TRACK	CYLINDERS			DRIVING WHEEL DIAMETER	BOILER		FIRE BOX		TUBES.		
	DIAM.		STROKE		Diameter	Pressure	Length	Width	Number	Diameter	Length
	H. P.	L. P.									
4′-8½″	21½″	34″	32″	57″	82″	200 lbs.	108⅛″	75¼″	232 40	2¼″ 5½″	22′-0″

WHEEL BASE			WEIGHT IN WORKING ORDER—POUNDS.				
Driving	Engine	Engine & Tender	Leading	Driving	Trailing	Engine	Tender
10′-0″ & 10′-0″	46′-9″	75′-8½″	25500	302500	26000	354000	154700

FUEL	HEATING SURFACES, SQUARE FT.						GRATE AREA SQ. FT.	MAXIMUM TRACTIVE POWER	FACTOR OF ADHESION
Kind	Tubes	Super. Flues	Fire Box	Arch Tubes	Total	Superheater			
Soft Coal	2994	1262	198	27	4481	1082	56.5	67500 lbs.	4.48

Tender, Type 8-Wheeled. Capacity, Water, 8000 Gals. Fuel, 12 Tons

ORDER No. S-1139

FIG. 18 BOSTON AND ALBANY 2-6-6-2 MALLET TYPE ARTICULATED LOCOMOTIVE. THIS PARTICULAR WHEEL ARRANGEMENT WAS BUILT IN THE LARGEST QUANTITIES OF ALL ARTICULATED TYPES. IT WAS A COMPOUND LOCOMOTIVE AS EVIDENCED BY THE LARGE LOW PRESSURE FRONT CYLINDERS. THE LOW PRESSURE CYLINDERS UTILIZED SLIDE VALVES WHILE THE HIGH PRESSURE CYLINDERS WHICH HAD TO HANDLE HIGH PRESSURE SUPERHEATED STEAM HAD PISTON VALVES.

AMERICAN LOCOMOTIVE COMPANY

NEW YORK

Class, "Four Cylinder Simple" 4884 S 762 Road Number, 4002

BUILT FOR THE UNION PACIFIC.

GAUGE OF TRACK	CYLINDERS		DRIVING WHEEL DIAMETER	BOILER		FIRE BOX		TUBES		
	Diam.	Stroke		Inside Dia.	Pressure	Length	Width	Number	Diameter	Length
4′-8½″	23¾″	32″	68″	95″	300 lbs.	235$\frac{1}{32}$″	96$\frac{3}{16}$″	75 184	2¼″ 4″	22′-0″

WHEEL BASE			WEIGHT IN WORKING ORDER—POUNDS				
Driving	Engine	Engine & Tender	Leading	Driving	Trailing	Engine	Tender (⅔ Load)
18′-3″ & 18′-3″	72′-5½″	117′-7″	97000	540000	125000	762000	342200

FUEL	EVAPORATING SURFACES, SQUARE FT.					SUPERHEATING SURFACE SQUARE FT.	GRATE AREA SQ.FT.	MAXIMUM TRACTIVE POWER	FACTOR OF ADHESION
Kind	Tubes	Flues	Fire Box	Circulators	Total		225 x 96$\frac{3}{16}$		
Soft Coal	967	4218	593	111	5889	2466	150.3	135375 lbs.	3.99

Tender Type 14-Wheeled Capacity, Water, 24000 Gals. Fuel, 28 Tons

ORDER No. S-1844

September, 1941

FIG. 19 THE UNION PACIFIC "BIG BOY" 4-8-8-4. THIS LOCOMOTIVE REPRESENTED THE MAXIMUM-SIZED STEAM LOCOMOTIVE EVER BUILT, YET IT WAS EASILY CAPABLE OF 60 MPH OPERATION WHERE TRACK CONDITIONS PERMITTED. NOTE THE HIGH CAPACITY PEDESTAL-BED TENDER WHICH IN ITSELF WAS AN OUTSTANDING DESIGN. WHILE FEW RAILROADS COULD USE SUCH A LARGE LOCOMOTIVE, IN ITS OWN ENVIRONMENT, IT WAS A SUPERB PERFORMER WITH FEW PROBLEMS.

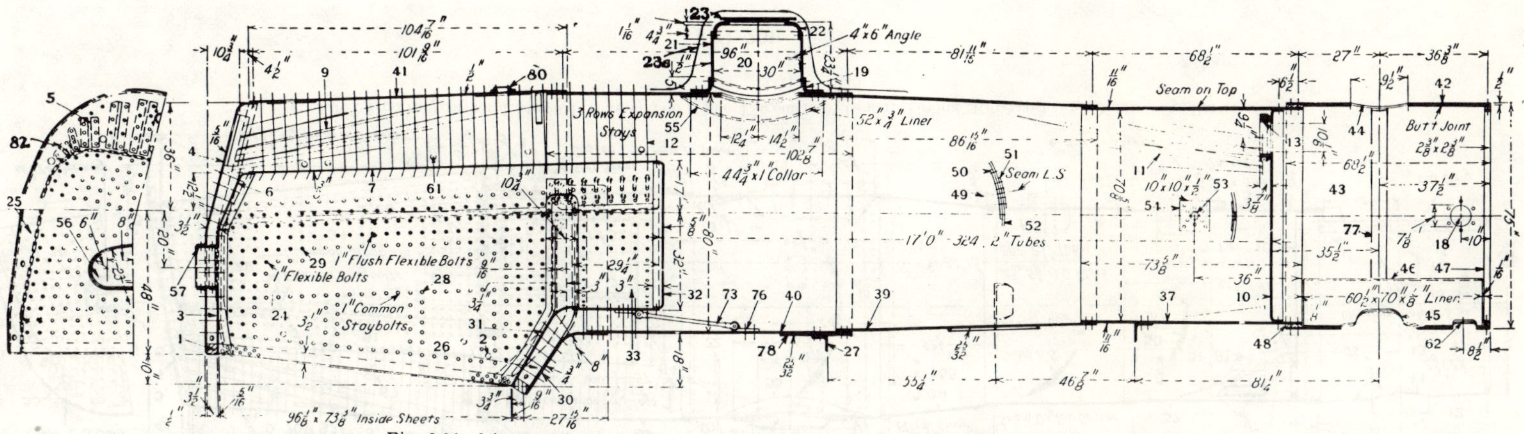

Fig. 3.01 Diagram of a Conical Type Locomotive Boiler, Giving Reference to Names of Parts.
Note: These boilers are not of recent design but still serve for names of parts.

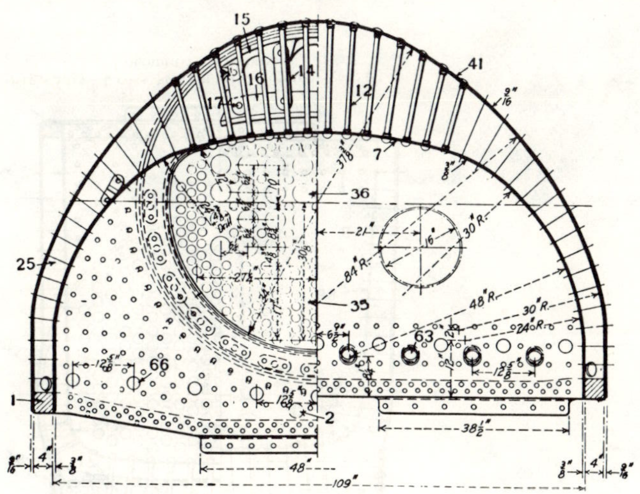

Fig. 3.02 Cross section of a Wootten Type Locomotive Firebox.

Names of Parts of Locomotive Boilers. Figs. 3.01-3.04

1 *Foundation or Mudring*
2 *Blowoff Cock Hole*
3 *Back Water Space*
4 *Back-Head*
5 *Back-Head Brace Angle*
6 *Back or Door Sheet of Firebox*
7 *Crownsheet*
8 *Outside Firebox Sheet*
9 *Back-Head Brace*
10 *Front Tube-Sheet*
11 *Front Tube-Sheet Brace*
12 *Radial Stay*
13 *Stiffening or Tube-Sheet Ring*
14 *Sling Stay*
15 *Sling Stay Tee*
16 *Crownbar*
17 *Crownbar Bolt*
18 *Smokebox Cleaning Hole*
19 *Dome Base, Flange or Collar*
20 *Dome*
21 *Dome Shell*
22 *Dome Cap*
23 *Dome Cover*
23a *Dome Casing*
24 *Firebox*
25 *Side Water Space*
26 *Front Water Space*
27 *Waist-Sheet Angle*
28 *Common Short Staybolt*
29 *Flexible Staybolt*
30 *Throat-Sheet*
31 *Inside Firebox Throat-Sheet*
32 *Back Tube-Sheet*
33 *Combustion-Chamber*
34 *Throat or Belly-Brace*
35 *Tube (Small)*
36 *Flue (Large) for Superheater Unit*
37 *First Course*
33 *Second Course*
39 *Conical or Gusset Course*
40 *Dome Course*
41 *Roof-Sheet*

(Continued on Page 251)

FIG. 20a

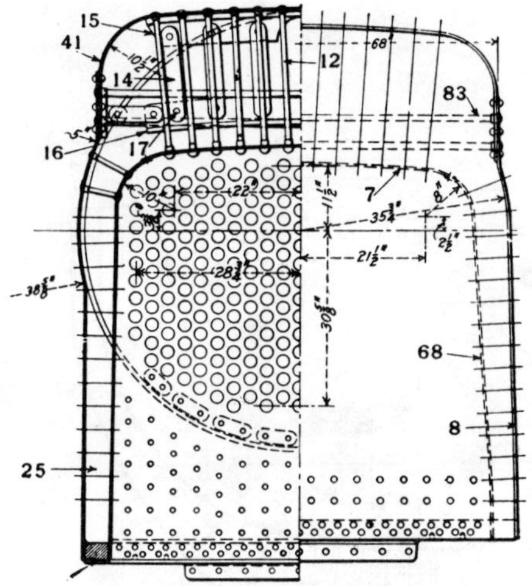

Fig. 3.03 Cross section of a Belpaire Type Locomotive Firebox.

Names of Parts, Figs. 3.01-3.04 (Cont.)

42 *Smokebox Shell*
43 *Smokebox*
44 *Smokestack Opening*
45 *Steampipe Opening*
46 *Smokebox Bottom Liner*
47 *Smokebox Front Ring*
48 *Smokebox and Shell Ring*
49 *Longitudinal Seam*
50 *Inside Welt*
51 *Outside Welt*
52 *Shell-Plate*
53 *Check-Valve Hole*
54 *Check-Valve Hole Pad or Liner*
55 *Dome Liner*
56 *Firedoor Hole*
57 *Firedoor Hole Ring*
58 *Arch-Tube*
59 *Arch-Tube Plug*
60 *Brick Arch*
61 *Washout Plug Hole*
62 *Cinder Valve Hole*
63 *Grate-Bar Thimble*
64 *Dash-Plate*
65 *Drypipe Hanger*
66 *Grate Water-Tubes*
67 *Washout Plug Hole Flange*
68 *Firebox Side-Sheet*
69 *Combustion-Chamber Course*
70 *Safety-Valve Dome or Turret*
71 *Safety-Valve Turret Swash-Plate*
72 *Injector Delivery-Pipe Hanger*
73 *Combustion-Chamber Brace*
74 *Belly Washout Hole*
75 *Belly Washout Hole Flange*
76 *Boiler Brace Foot*
77 *Smokebox Joint Ring*
78 *Waist-Sheet Angle Line*
79 *Water-Gage Cock Hole*
80 *Whistle Flange*
81 *Gusset Stay*
82 *Crowfoot*
83 *Boiler Cross Brace*

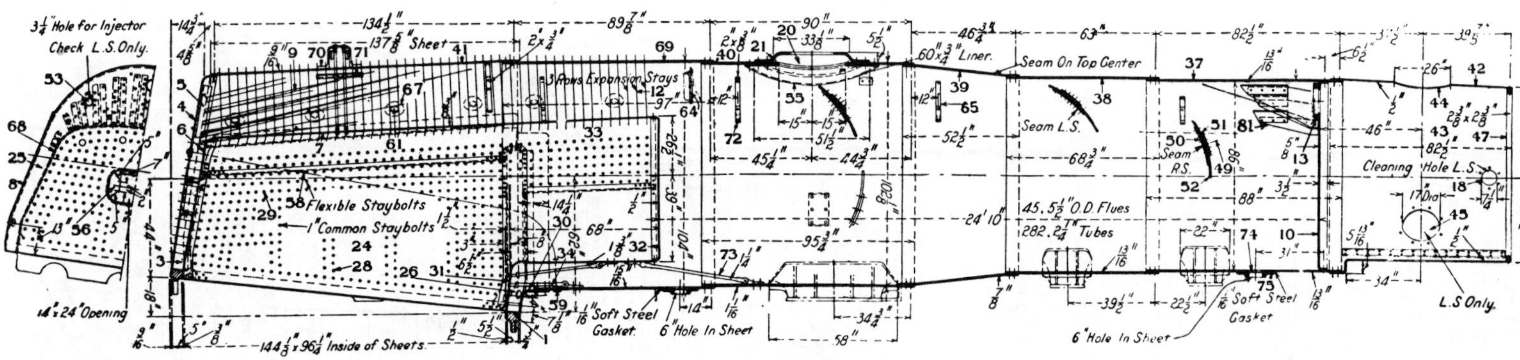

Fig. 3.04 Diagram of an Extended Wagon-Top Type Locomotive Boiler, Giving References to Names of Parts.

FIG. 20b

Parts Per Stoker Name of Part
1 Main Drive Shaft Key
2 Main Drive Shaft Thrust Button
1 Oil Box, Complete
1 Oil Box Lid
1 Oil Box Lid Hinge Pin
2 Rear Drive Shaft Collar
1 Engine Bracket
12 1" x 2¾" Hex. Hd. Bolt and Nut
6 ¾" x 4⅞" Hex. Hd. Bolt and Nut
1 Front Conveyor Screw
1 Intermediate Conveyor Screw
1 Tender Conveyor Screw
1 Tender Trough
1 Inside Rear Bowl
1 Screw Thrust Washer
1 Housing Thrust Washer
1 Main Drive Gear Housing
1 Main Drive Gear Housing Cover
1 Main Drive Gear
1 Main Drive Pinion and Shaft, Complete
1 Main Drive Shaft
1 Front Conveyor Trough Cover Guide
Stoker Steam Gauge
are not required and neither are
arts 1062, 1064, 1100 and 1101 re-
t.
se parts see part catalog No. 52
d B Stokers.

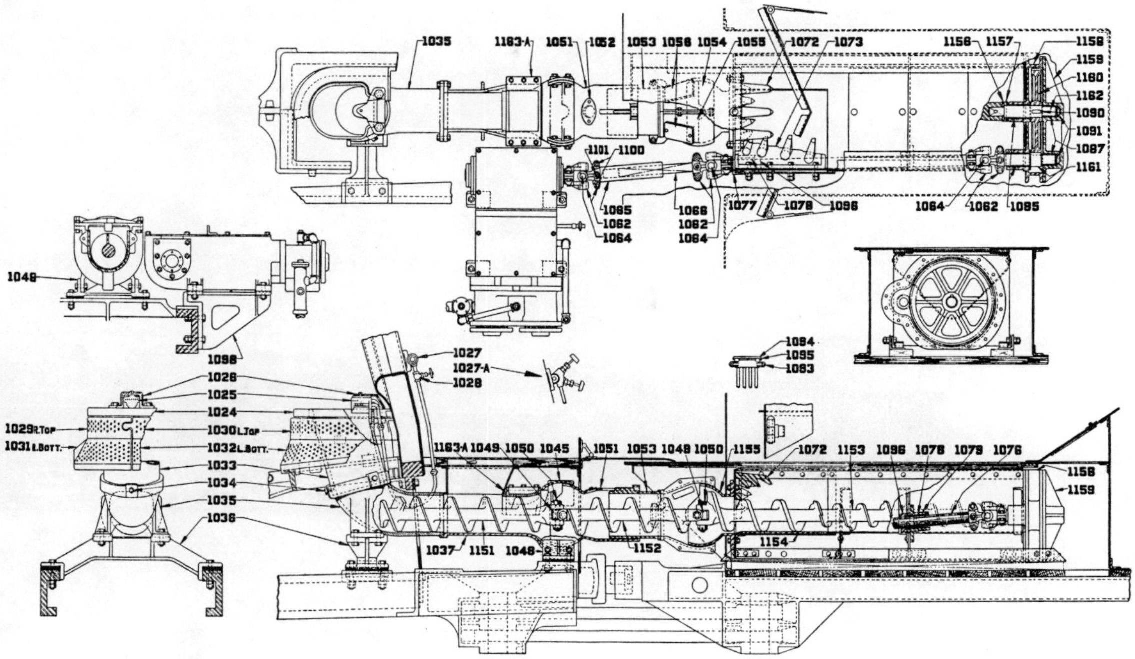

Part No.	Parts Per Stoker	Name of Part
1024(v)	1	Firing Table
1025(v)	1	Distributor Jet
1026(v)	1	Distributor Jet Hood
1027(v)	1	Distributor Jet Manifold
1027-A(v)	1	Distributor Jet Manifold, Complete
1028	5	Distributor Jet Control Valves
1029(v)	1	Protecting Grate—Right Top
1030(v)	1	Protecting Grate—Left Top
1031(v)	1	Protecting Grate—Right Bottom
1032(v)	1	Protecting Grate—Left Bottom
1033(v)	1	Upper Vertical Housing
1034	3	Upper Vertical Housing Bolt
1035(v)	1	Front Elbow
1036(v)	1	Front Elbow Support
1037(v)	1	Front Conveyor Trough
1045	1	Front Conveyor Trough Bowl Cover
1048(v)	1	Front Conveyor Trough Support
1049	2	Universal Joint Block
1050	4	Universal Joint Block Bolt
1051(v)	1	Intermediate Conveyor Shell—Front Half
1052	1	Intermediate Conveyor Shell Hand Hole Cover
1053(v)	1	Intermediate Conveyor Shell—Rear Half
1054	1	Outside Rear Bowl—Right Half
1055	1	Outside Rear Bowl—Left Half
1056	2	Outside Rear Bowl Cover
1062	12	Universal Joint Clip
1064	3	Universal Joint Block
1065(v)	1	Intermediate Drive Shaft—Hollow Section
1066(v)	1	Intermediate Drive Shaft—Solid Section
1072	1	Main Crusher
1073	1	Auxiliary Crusher
1076(v)	1	Rear Drive Shaft
1077(v)	1	Rear Drive Shaft Bearing—Front
1078	2	Rear Drive Shaft Bearing Bushing
1079(v)	1	Rear Drive Shaft Bearing—Rear
1085	2	Main Drive Gear Housing Bushing
1087	2	Main Drive Gear Housing Cover Bushing
1090	1	Main Drive Shaft Key
1091	2	Main Drive Shaft Thrust Button
1093(v)	1	Oil Box, Complete
1094	1	Oil Box Lid
1095	1	Oil Box Lid Hinge Pin
1096	2	Rear Drive Shaft Collar
1098(v)	1	Engine Bracket
1100	12	1" x 2¾" Hex. Hd. Bolt and Nut
1101	6	¾" x 4⅞" Hex. Hd. Bolt and Nut
1151(v)	1	Front Conveyor Screw
1152(v)	1	Intermediate Conveyor Screw
1153(v)	1	Tender Conveyor Screw
1154(v)	1	Tender Trough
1155	1	Inside Rear Bowl
1156	1	Screw Thrust Washer
1157	1	Housing Thrust Washer
1158(v)	1	Main Drive Gear Housing
1159(v)	1	Main Drive Gear Housing Cover
1160(v)	1	Main Drive Gear
1161(v)	1	Main Drive Pinion and Shaft, Complete
1162	1	Main Drive Shaft
1163-A	1	Front Conveyor Trough Cover Guide
912		Stoker Steam Gauge

When the stoker engine is mounted on the locomotive tender, part numbers 1065 and 1066 are not required and neither are parts 1077, 1078, 1079 and 1096 unless the tender unit is 123" or longer. The total number of parts 1062, 1064, 1100 and 1101 required will also vary according to the position of the stoker engine and length of the tender unit.

Numbers followed by (v) represent parts that vary in different applications. In ordering these parts see part catalog No. 52 and variable sheet.

Part numbers shown in bold face type are used on both Type B and Modified B Stokers.

FIG. 21 STANDARD MODIFIED TYPE "B" STOKER

AMERICAN LOCOMOTIVE COMPANY

NEW YORK

Class "Four-Cylinder Simple," 2884 S 717 — **Road Number, 5000**

BUILT FOR THE NORTHERN PACIFIC.

GAUGE OF TRACK	CYLINDERS		DRIVING WHEEL DIAMETER	BOILER		FIRE BOX GAINES ARCH		TUBES		
	Diam.	Sfroke		Inside Dia.	Pressure	Length	Width	Number	Diameter	Length
4′-8½″	26″	32″	63″	103¼″ Inside 110¼″ Outside	250 lbs.	266⅛	114¼″	92 280	2¼″ 3½″	22′-0″

WHEEL BASE			WEIGHT IN WORKING ORDER—POUNDS					
Driving	Engine	Engine & Tender	Leading	Driving	Trailing FRONT	Trailing REAR	Engine	Tender
16′-9″ & 16′-9″	66′-8″	111′-11″	48500	553000	55000	60500	717000	401000

FUEL	EVAPORATING SURFACES, SQUARE FT.					SUPERHEATING SURFACE SQUARE FT.	GRATE		MAXIMUM TRACTIVE POWER		FACTOR OF ADHESION	
Kind	Tubes	Flues	Fire Box	Syphons	Total		LENGTH	WIDTH	Engine 70% Cut-off	BOOSTER 50% Cut off	Drivers	Trailer
							229¼″	114¼″				
Rose Bud Coal	1186	5615	610	262	7673	3219	Area 182 Sq. Ft		140000 lbs.	13400 lbs.	3.95	4.51

Tender Type, 12 Wheeled — **Capacity, Water, 21200 Gals.** — **Fuel, 27 Tons**

ORDER No. S-1617

December, 1928

FIG. 22 THE NORTHERN PACIFIC "YELLOWSTONE" 2-8-8-4 WHICH HAD THE LARGEST FIREBOX EVER USED ON A LOCOMOTIVE (182 FT²). THE NP USED SEMIBITUMINOUS "ROSE BUD" COAL FROM ITS CAPTIVE MINE AT COLSTRIP, MONTANA. SINCE THE HEAT CONTENT OF THIS COAL WAS RELATIVELY LOW, MODERN NP LOCOMOTIVES USED EXCEPTIONALLY LARGE FIREBOXES. THE COST OF THE COAL, HOWEVER, WAS REPORTED TO BE AS LOW AS $0.65 PER TON. NOTE THE AUXILLARY DOOR IN THE WATER LEG OF THE FIREBOX TO FACILITATE FIRE CLEANING. THIS LOCOMOTIVE WAS RESTRICTED TO SPEEDS WELL BELOW ITS CAPABILITY BECAUSE OF COUNTERBALANCE PROBLEMS WHICH RESULTED IN TRACK DAMAGE IF OPERATED TOO FAST.

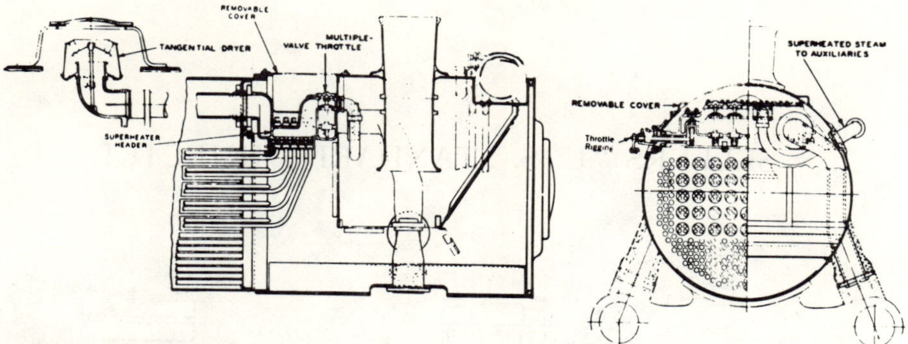

Fig. 3.299—Typical Application of American Multiple-Valve Throttle with Elesco Type A Superheater.

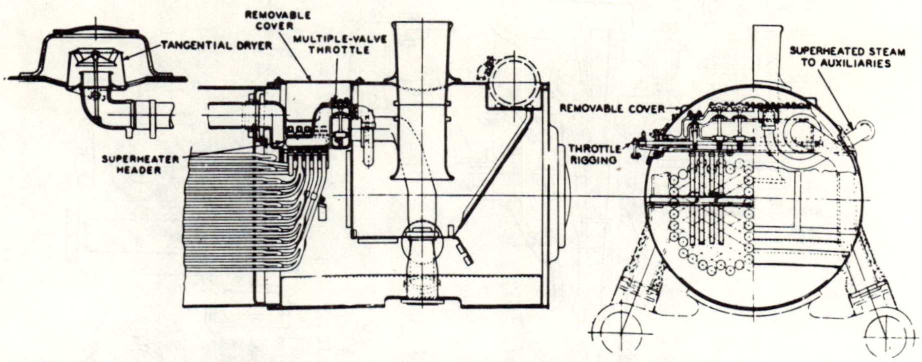

Fig. 3.300—Typical Application of American Multiple-Valve Throttle with Elesco Type E Superheater.

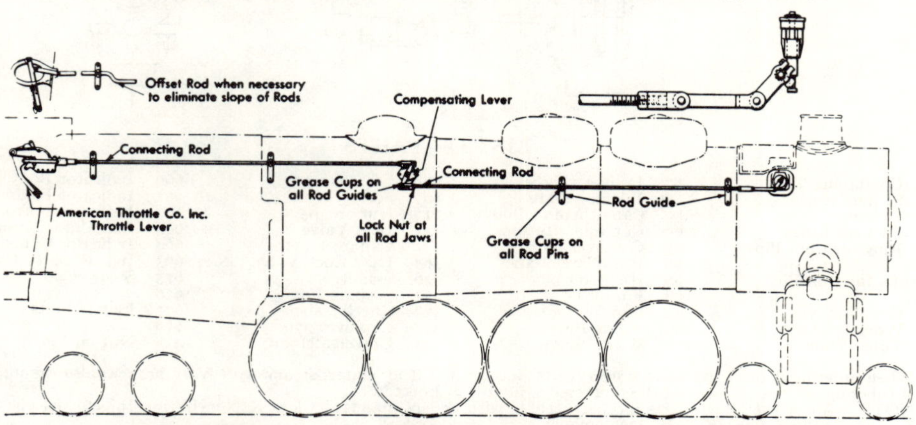

Fig. 3.301—Throttle Lever and Outside Connecting Rods for American Multiple-Valve Throttle.

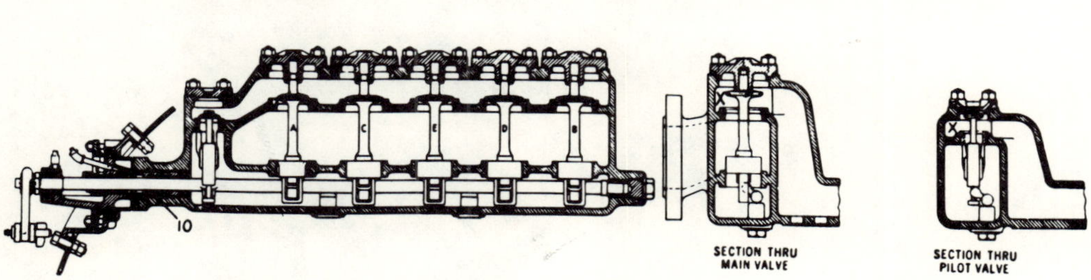

Fig. 3.302—American Multiple Throttle in Combination with Locomotive Superheater Header.

American Throttle Company, Inc.

(*See also Page 367*)

FIG. 23

SELLERS "U. S. STANDARD" INJECTOR

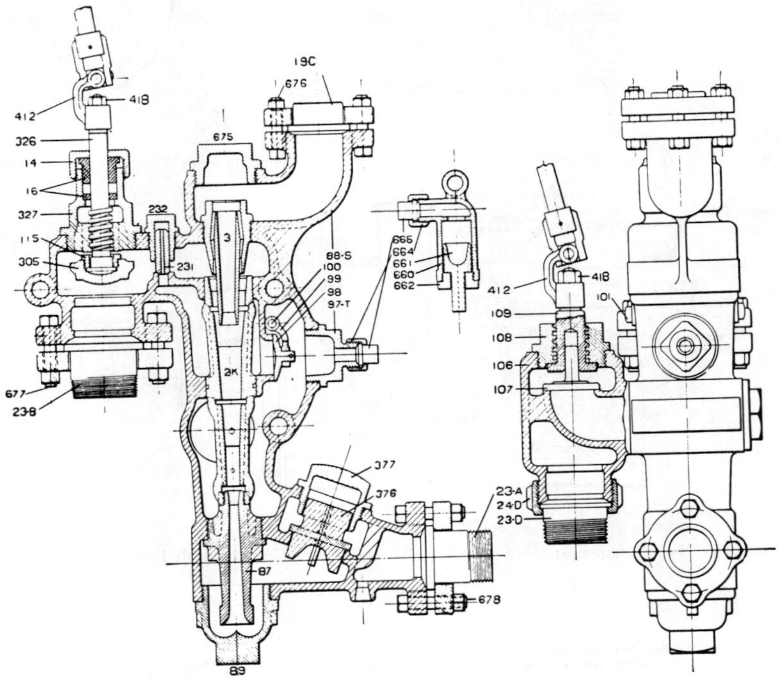

LIST OF PARTS

2k Combining Tube
3 Steam Nozzle
14 Follower
16 Packing Rings
19c Ring for Cop. Pipe
23a, 23b, 23d } do. for Iron Fipe
24d Coupling Nut
27d Wrench for 24
60 Tube Wrench

87 Delivery Tube
88s Injector Body
89 Cap at end of Body
96 Cap opposite 106
97t Cap over 98
98 Hot Water Valve
99 Hinge for 98
100 Pin for 99
101 Cap for 100
106 Waste Pipe
107 Waste Valve

108 Cap over 107
109 Screw Stem
115 Nut on 305
231 Inlet Valve
232 Cap over 231
305 Lazy Cock Valve
326 Spindle
327 Stuffing Box
376 Check Valve
377 Cap over 376
412 Universal Joint

660 Indicator Body
661 Indicator Piston
662 Indicator Bottom Cap
663 Ind. Union, Iron Pipe
664 Indicator Coupling Nut
665 Ind. Ring for Cop. Pipe
675 Steam Cap
676, 677, 678 } Flange Bolts
418 Nut on 326

When ordering flanges and bolts state type of injector and if for water, steam or delivery branch; also the number and size of tubes or parts required, and the progressive number of injector.

NOTE. All of the above parts interchangeable with Sellers Standard Class K Non-Lifting Injector except tubes for No. 8½, 2500 gallons and No. 10½, 4000 gallons.

FIG. 24

Fig. 4.46—Worthington Feedwater Heating Equipment, Type SA, applied to a locomotive.

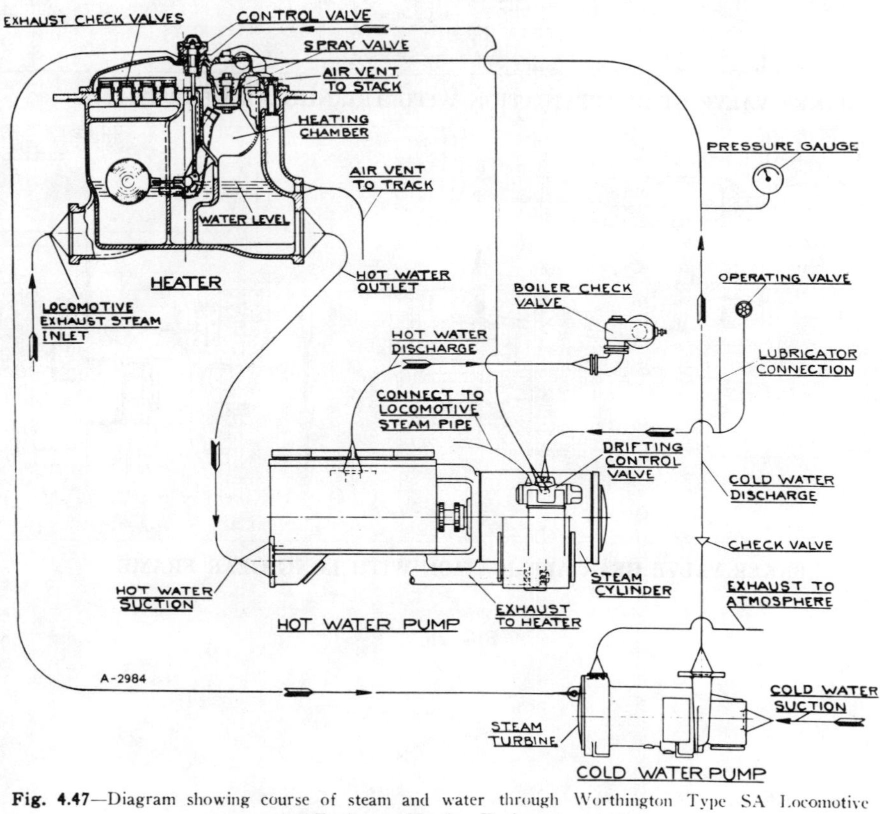

Fig. 4.47—Diagram showing course of steam and water through Worthington Type SA Locomotive Feedwater Heating Equipment.

FIG. 25

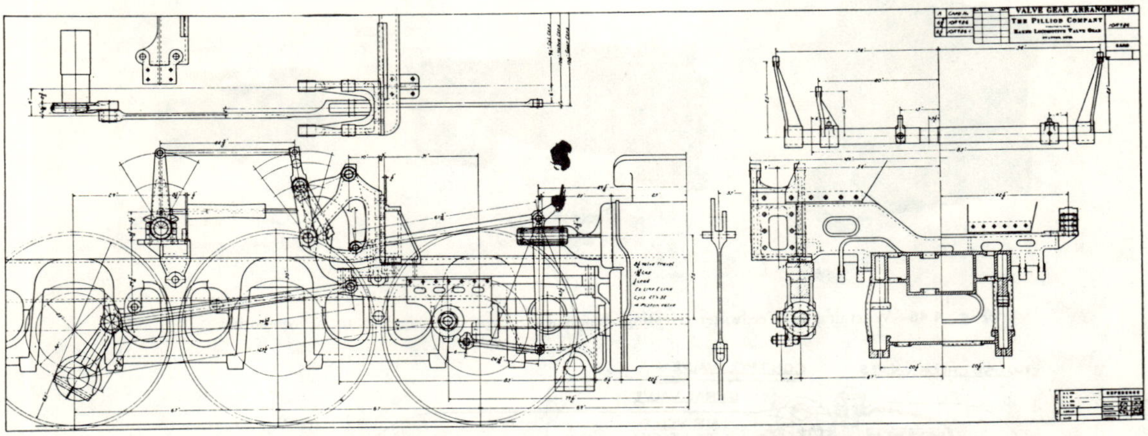

BAKER VALVE GEAR APPLICATION WITH TRIANGULAR GEAR FRAME

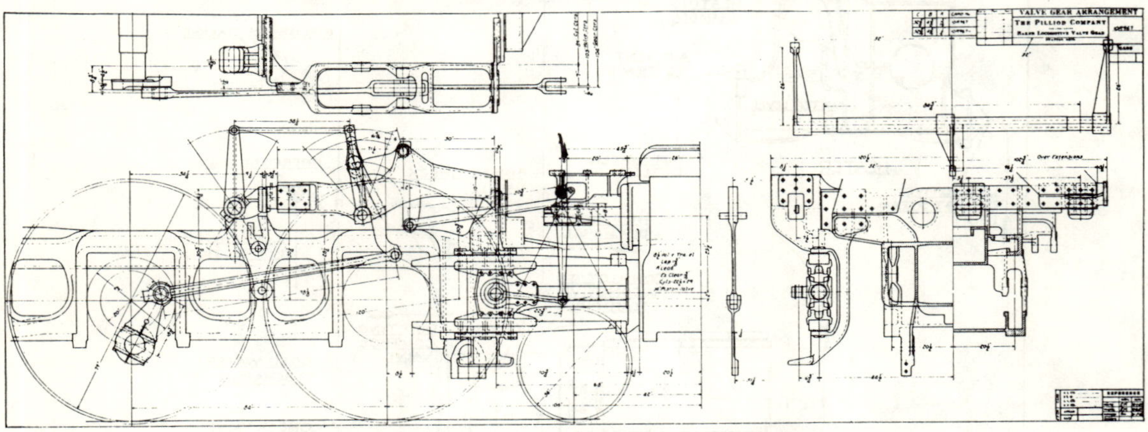

BAKER VALVE GEAR APPLICATION WITH LONG GEAR FRAME

FIG. 26

AMERICAN LOCOMOTIVE COMPANY
NEW YORK

Class, 4 12 2 S 515, "Three-Cylinder," **Road Number, 9085**

BUILT FOR THE UNION PACIFIC.

GAUGE OF TRACK	CYLINDERS		DRIVING WHEEL DIAMETER	BOILER		FIRE BOX GAINES ARCH		TUBES		
	Diam.	Stroke		Inside Dia.	Pressure	Length	Width	Number	Diameter	Length
4′-8½″	Inside 27″ Outside 27″	31″ 32″	67″	90″	220 lbs.	184½″	108¼″	40 222	3½″	22′ 0″

WHEEL BASE			WEIGHT IN WORKING ORDER—POUNDS				
Driving	Engine	Engine & Tender	Leading	Driving	Trailing	Engine	Tender
30′-8″	52′-4″	91′-6½″	81000	372000	62000	515000	308800

FUEL	EVAPORATING SURFACES, SQUARE FT.					SUPERHEATING SURFACE SQUARE FT.	GRATE		MAXIMUM TRACTIVE POWER	FACTOR OF ADHESION
Kind	Tubes	Flues	Fire Box	Arch Tubes	Total		Length 144″	Width 108¼″		
Semi-Bit. Coal	803	4459	529	62	5853	2560	Area 108.25 Sq.Ft.		96650 lbs.	3.84

Tender Type, 12 Wheeled **Capacity, Water, 18000 Gals.** **Fuel, 22 Tons**

ORDER No. S-1701
August, 1930

FIG. 27 THE THREE-CYLINDER UNION PACIFIC 4-12-2. THIS DESIGN WAS OF INTEREST IN THAT IT REPRESENTED THE LIMITS OF RIGID DRIVING WHEELBASE (30 FT 8 IN.) AND THE MAXIMUM SIZE OF THREE-CYLINDER LOCOMOTIVE DESIGNS. THE CENTER CYLINDER WAS CONTROLLED BY GRESLEY CONJUGATED VALVE MOTION LOCATED OVER THE PILOT DECK. SINCE THE CENTER CYLINDER HAD TO DRIVE A CRANKED AXLE, THE ACCESSIBILITY OF THESE COMPONENTS WAS POOR. THESE LOCOMOTIVES COULD NOT BE OPERATED ON A HEAVILY CURVED RAILROAD DUE TO THE LONG RIGID WHEELBASE.

NEGATIVE NO. 11,611

THE BALDWIN LOCOMOTIVE WORKS

2-10-4 TYPE LOCOMOTIVE

Class 16-54-¾-F, 11 Road No. 5010

Built for **Atchison, Topeka & Santa Fe Ry.**

Railway Co.'s Class 5001

(Order covers 10 locomotives, Nos. 5001 to 5010)

Gauge	4′ 8½″
Cylinders	30″ x 34″
Valves	Piston, 15″ diam.

BOILER

Type	Conical
Diameter	94″
Working pressure	310 lb.
Fuel	Oil*

Firebox

Material	Steel
Staying	Radial
Length	162″
Width	108″
Depth, front	94⅞″
Depth, back	76½″

Tubes

Diameter	3½″	2¼″
Number	249	56
Length	21′ 0″	21′ 0″

Heating Surface

Firebox	319 sq. ft.
Combustion chamber	145 sq. ft.
Tubes	5443 sq. ft.
Thermic syphons	146 sq. ft.
Total	6053 sq. ft.
Superheater	2675 sq. ft.
Grate area	121.5 sq. ft.

DRIVING WHEELS

Diameter, outside	74″
Diameter, center	66″
Journals, main	15″ x 13″
Journals, front	12″ x 14″
Journals, others	12″ x 13″

ENGINE TRUCK WHEELS

Diameter, front	37″
Journals	8″ x 14″
Diameter, back	40″
Journals	9″ x 14″

WHEEL BASE

Driving	26′ 2″
Rigid	19′ 3″
Total engine	50′ 2″
Total engine & tender	100′ 3¾″

WEIGHT
In Working Order

On driving wheels	371,990 lb
On truck, front	51,120 lb.
On truck, back	115,410 lb.
Total engine	538,520 lb.
Total tender	393,200 lb.

TENDER

Wheels, number	Twelve
Wheels, diameter	37″
Journals	7″ x 13″
Tank capacity—	
Water	21,000 U.S. gal.
Oil	7,000 U.S. gal.

Tractive force	93,000 lb.
Service	Freight

Construction No. 62,162 *Built* August 1938 *Drawing No.* 2

Equipped with Type E superheater, feed water heater, three thermic syphons, Baldwin power reverse, Baldwin Disc driving wheel centers, one piece cast steel locomotive bed with integral cylinders, limited cut-off, and air brake on all driving and tender wheels, with two 8½″ cross-compound pumps.

*Engines 5001-5005 burn coal.

FIG. 28 THE SANTA FE 2-10-4 REPRESENTED THE MAXIMUM OUTPUT AVAILABLE FROM TWO CYLINDERS. THE PISTON THRUST OF 219,000 LB WAS THE HIGHEST EVER DEVELOPED AND REQUIRED 15 × 13 IN. JOURNAL BEARINGS ON THE MAIN DRIVING AXLE TO WITHSTAND THIS FORCE. THE DESIGN WAS OUTSTANDING IN THAT IT UTILIZED 74 IN. DIAMETER DRIVING WHEELS WHICH GAVE IT HIGH SPEED CAPABILITY ALONG WITH HIGH TRACTIVE EFFORT.

LOCOMOTIVE BED ARRANGED FOR POPPET VALVES
PENNSYLVANIA RAILROAD 4-4-4-4 TYPE, CLASS T-1 LOCOMOTIVES
2010-B
PATTERN 54190-1
12-13-45
NICKEL STEEL

FIG. 29

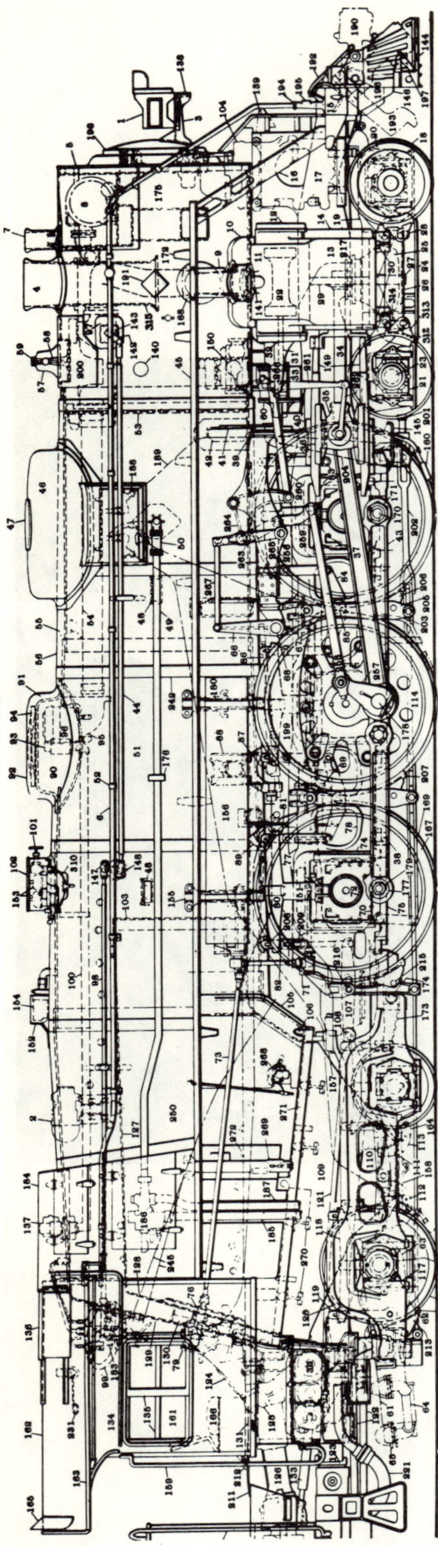

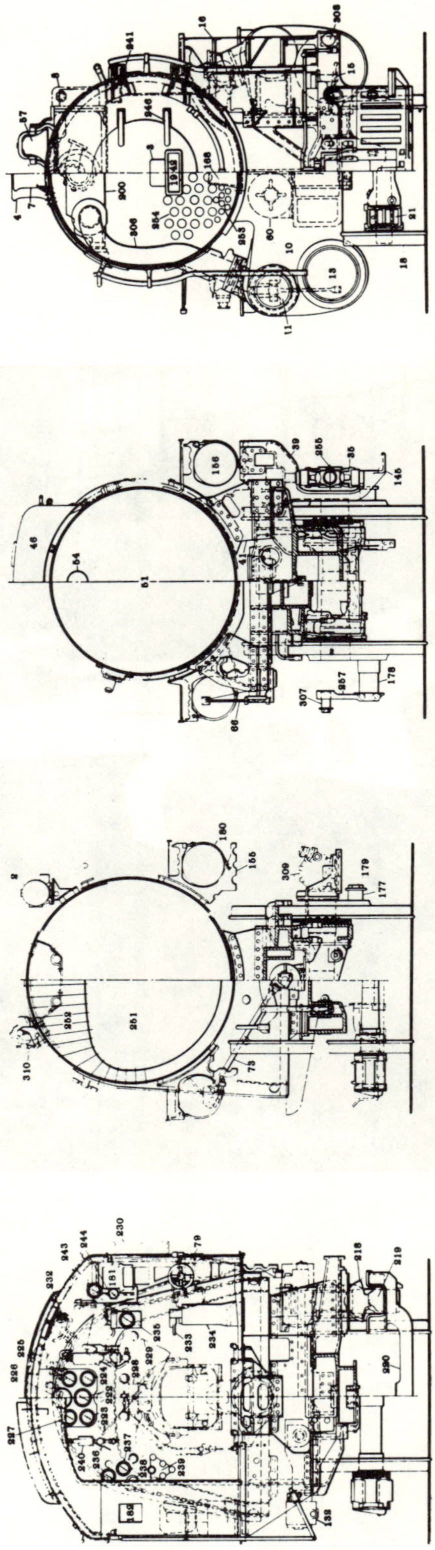

2.01- Diagram of a 4-6-4 or Hudson Type Passenger Locomotive with Principal Parts Numbered.
(*For Tender see Page 118*)

FIG. 30a

Names of Principal Parts of a Steam Locomotive

1. Headlight
2. Headlight Turbo-Generator
3. Headlight Support
4. Smokestack
5. Smokebox Front
6. Handrail
7. Booster Exhaust Stack
8. Feedwater Heater
9. Steam Pipe Casing
10. Cylinder Saddle
11. Valve Chamber
12. Valve Chamber Head Cover
13. Cylinder
14. Cylinder Head Cover
15. Bumper
16. Front Steps
17. Front of Frame or Bed
18. Front Truck Wheel
19. Front Truck Frame
20. Front Truck Pedestal
21. Front Truck Journal Box
22. Piston Valve
23. Front Truck Axle
24. Front Truck Pedestal Tie Bar
25. Front Truck Transom
26. Front Truck Equalizer
27. Front Truck Spring
28. Front Truck Spring Hanger
29. Piston
30. Cylinder Cock
31. Valve Stem
32. Valve Stem Gland
33. Valve Stem Pin
34. Piston Rod
35. Crosshead
36. Crosshead Shoe
37. Main Rod
38. Side Rod Rear
39. Guide Yoke
40. Guide Yoke Brace
41. Waist Sheet
42. Waist Sheet Connection
43. Side Rod Front
44. Throttle Rod
45. Running Board
46. Sandbox
47. Sandbox Cap
48. Sandbox Step
49. Injector Delivery Pipe
50. Boiler Check Valve
51. Boiler
52. Handrail Bracket
53. Front Tubesheet
54. Drypipe
55. Boiler Lagging
56. Boiler Jacket
57. Bell
58. Bell Stand
59. Bell Yoke
60. Driver Brake Cylinder
61. Booster Engine
62. Booster Crank Case
63. Booster Gear Case
64. Booster Steam Pipe
65. Booster Steam Pipe Ball and Socket Joint
66. Reverse Shaft Arm
67. Reverse Shaft Lever
68. Reverse Shaft
69. Power Reverse Cylinder
70. Driving Box
71. Power Reverse Shaft Gear Box
72. Driving Axle
73. Power Reverse Operating Shaft
74. Frame Pedestal
75. Pedestal Binder
76. Power Reverse Operating Shaft Worm
77. Top Frame Rail (if separate frame)
78. Bottom Frame Rail (if separate frame)
79. Power Reverse Operating Handwheel
80. Driving Spring
81. Driving Spring Clip
82. Driving Spring Hanger
83. Driving Spring Hanger Pin
84. Driving Equalizer
85. Equalizer Fulcrum Pin—Driving
86. Valve Pilot Connecting Shaft
87. Valve Pilot Cam Box
88. Valve Pilot Cam Box Bracket
89. Valve Pilot Speed Recorder Wheel
90. Dome
91. Dome Lagging
92. Dome Casing
93. Turret Supply Steam Dryer
94. Dome Cover
95. Drypipe Elbow
96. Steam Dryer
97. Throttle Valve-Inside of Smokebox
98. Throttle Rod
99. Throttle Lever Quadrant
100. Turret Drypipe
101. Syphon Valve
102. Safety Valve Casing
103. Back Tubesheet
104. Air Compressor
105. Throat Sheet
106. Washout Plug
107. Driving and Trailing Truck Upper Equalizer
108. Driving and Trailing Truck Equalizer Fulcrum
109. Ashpan
110. Ashpan Hopper
111. Ashpan Hopper Slide Pull Link
112. Ashpan Hopper Dump Shaft
113. Ashpan Hopper Dump Shaft Bearing
114. Driving Wheel
115. Trailing Truck Wheel
116. Trailing Truck Journal Box
117. Trailing Truck Pedestal
118. Trailing Truck Spring
119. Trailing Truck Spring Hanger
120. Trailing Truck Frame
121. Rear Frame or Cradle
122. Drawbar
123. Chafing Plate
124. Stoker Elevator Conveyor
125. Stoker Intermediate Conveyor Ball Joint
126. Stoker Intermediate Conveyor
127. Crownsheet
128. Backsheet
129. Backhead of Boiler
130. Firedoor Opening
131. Foot Board or Deck
132. Stoker Engine
133. Stoker Drive Shaft
134. Cab
135. Cab Window Sash
136. Cab Ventilator
137. Turret
138. Number Plate
139. Headlight Step
140. Inspection Hole Cover
141. Steam Chest Relief Valve Base
142. Throttle Cam Shaft Arm
143. Throttle Stuffing Box
144. Pilot
145. Crosshead Step
146. Pilot Bracket
147. Throttle Clevis
148. Throttle Rod Equalizer
149. Guide
150. Mechanical Lubricator
151. Driving Box Spring Saddle
152. Whistle Shield
153. Safety Valve
154. Whistle
155. Main Reservoir Bracket
156. Main Air Reservoir
157. Expansion Bracket or Pad
158. Ashpan Hopper Slide
159. Cab Handhole
160. Sand Pipe Outlet
161. Cab Window Opening
162. Cab Roof
163. Cab Roof Water Gutter
164. Ashpan Dump Shaft Arm
165. Cab Smoke Deflector
166. Cab Seat
167. Driving Wheel Tire
168. Exhaust Nozzle
169. Driving Wheel Flange
170. Front Crank Pin
171. Front Crank Pin Nut
172. Running Board Bracket
173. Driving and Trailing Truck Lower Equalizer
174. Driving and Trailing Truck Equalizer Pin
175. Smokebox
176. Injector Delivery Pipe Bracket
177. Rear Crank Pin
178. Main Crank Pin
179. Rear Crank Pin Nut
180. Main Reservoir Hanger
181. Cab Front Window
182. Cab Rear Window
183. Throttle Lever
184. Turret Cover
185. Suction or Feed Pipe
186. Injector
187. Injector Overflow Pipe
188. Sander
189. Sand Pipe
190. Pilot Drop Coupler
191. Builder's Name Plate
192. Uncoupling Rod
193. Coupler Counterweight
194. Flag Staff
195. Signal Lamp Bracket
196. Drop Coupler Pivot
197. Pilot Step
198. Smokebox Door
199. Counterbalance
200. Superheater Header and Throttle
201. Driver Brake Rod Adjuster
202. Driver Brake Pull Rod
203. Driver Brake Beam
204. Driver Brake Lever
205. Driver Brake Hanger
206. Driver Brake Shoe Head
207. Driver Brake Shoe
208. Driving Spring Hanger Gib
209. Transverse Equalizer
211. Cab Apron
212. Cab Apron Hinge
213. Trailing Truck Brake Shoe Head
214. Trailing Truck Brake Shoe
215. Brake Head Pin
216. Brake Hanger Pin
217. Front Truck Center Plate
218. Trailing Truck Centering Device Rocker
219. Trailing Truck Centering Device Base
220. Trailing Truck Transom
221. Tank Water Hose
222. Main Steam Gage
223. Water Pump Steam Gage
224. Tender Water Level Gage
225. Booster Steam Gage
226. Pyrometer for Superheated Steam
227. Train Steamheat Gage
228. Firebox Door Operating Cylinder
229. Firebox Door
230. Windshield
231. Whistle Lever
232. Whistle Lever Support
233. Brake Valve
234. Brake Valve Pedestal
235. Valve Pilot Speed Recorder and Cutoff Gage
236. Stoker Engine Steam Gage
237. Stoker Single Pointer Jet Gage
238. Stoker Double Pointer Jet Gage
239. Stoker Distributor Jet Manifold and Control Valves
240. Water Gage Glass
241. Front-End Hinge
242. Boiler Jacket Band
243. Duplex Air Brake Gage
244. Independent Air Brake Gage
245. Arch Tubes
246. Smokebox Door Hinge
250. Firebox
251. Combustion Chamber
252. Staybolts
253. Boiler Tubes
254. Superheater Flues
255. Wrist Pin
256. Valve Gear Frame
257. Eccentric Crank
258. Eccentric Rod
259. Valve Gear Arm
260. Valve Rod
261. Combination Lever
262. Union Link
263. Reverse Yoke
264. Bell Crank
265. Radius Bar
266. Valve Crosshead
267. Valve Gear Reach Rod
268. Blowoff Cock
269. Pipe Bracket
270. Ashpan Side Support
271. Mud Ring
272. Combustion Holes
306. Steam Pipe
307. Eccentric Crank Pin
308. Push Pole Pocket
309. Boiler Feedwater Pump
310. Low Water Alarm
311. Equalizing Reservoir
312. Front Truck Brake Shoe
313. Front Truck Brake Head
314. Front Truck Brake Hanger
315. Smokestack Extension

FIG. 30b

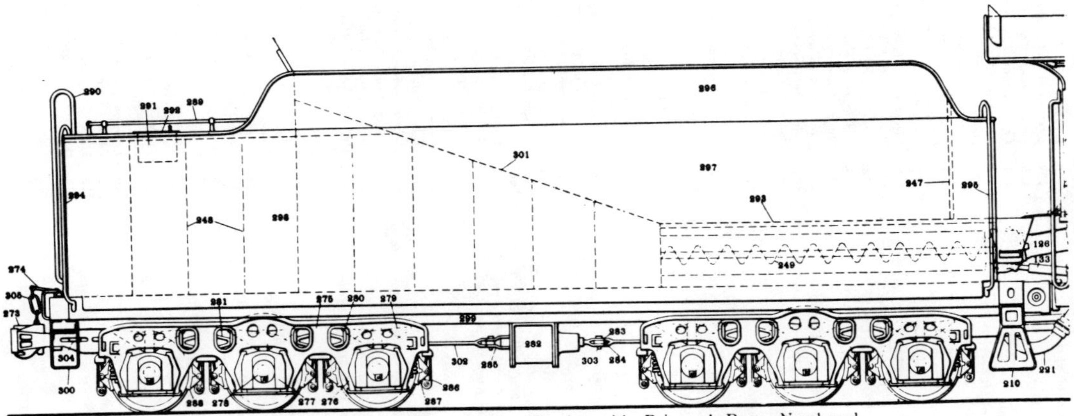

Fig. 2.02—Diagram of a Locomotive Tender with Principal Parts Numbered.

Names of Principal Parts of a Locomotive Tender

- **126.** *Stoker Intermediate Conveyor*
- **133.** *Stoker Drive Shaft*
- **210.** *Tender Gangway Step*
- **221.** *Tank Water Hose*
- **247.** *Coal Gate*
- **248.** *Tender Bracing and Splash Plates*
- **249.** *Tender Coal Conveyor Screw to Stoker*
- **273.** *Tender Coupler*
- **274.** *Tender Uncoupling Rod*
- **275.** *Tender Truck Frame*
- **276.** *Tender Truck Pedestal*
- **277.** *Tender Journal Box*
- **278.** *Tender Journal Bearing*
- **279.** *Tender Truck Spring*
- **280.** *Tender Truck Spring Hanger*
- **281.** *Tender Truck Equalizer*
- **282.** *Tender Brake Cylinder*
- **283.** *Tender Brake Cylinder Push Rod*
- **284.** *Tender Brake Cylinder Lever*
- **285.** *Tender Brake Lever Fulcrum*
- **286.** *Tender Brake Hanger*
- **287.** *Tender Brake Shoe Head*
- **288.** *Tender Brake Shoe*
- **289.** *Tender Hand Rail*
- **290.** *End Ladder*
- **291.** *Tender Filling Hole*
- **292.** *Tender Filling Hole Cover*
- **293.** *Tender Deck*
- **294.** *Tender Rear Handhold*
- **295.** *Tender Front Handhold*
- **296.** *Tender Collar*
- **297.** *Fuel Space*
- **298.** *Water Trank or Cistern*
- **299.** *Tender Frame*
- **300.** *Tender Rear Step*
- **301.** *Tender Slope Sheet*
- **302.** *Tender Brake Rod*
- **303.** *Tender Brake Pin*
- **304.** *Tender Coupler Draft Key*
- **305.** *Uncoupling Chain*

FIG. 30c

NEGATIVE NO. 12,250

THE BALDWIN LOCOMOTIVE WORKS

SINGLE EXPANSION ARTICULATED LOCOMOTIVE

Class 4-8-8-2-24 SA, 112 — Road No. 4211

Built for **Southern Pacific Lines**

Railway Co.'s Class AC-10

(Order covers 40 locomotives, Nos. 106 to 145)

Gauge	4′ 8½″
Cylinders (4)	24″ x 32″
Valves	Piston, 11″ diam.

BOILER

Type	Conical
Diameter	94$\frac{1}{16}$″
Working pressure	250 lb.
Fuel	Oil

Firebox

Material	Steel
Staying	Radial
Length	196⅛″
Width	102¼″
Volume	899 cu. ft.
C. C. length	68″

Tubes

Diameter	2¼″	3½″
Number	91	240
Length	22′ 0″	22′ 0″

Heating Surface

Firebox	337 sq. ft.
Combustion chamber	141 sq. ft.
2¼″ Tubes	1174 sq. ft.
3½″ Tubes	4818 sq.ft.
Total	6470 sq. ft.
Superheater	2616 sq. ft.
Grate area	139 sq. ft.

DRIVING WHEELS

Diameter, outside	63½″
Diameter, center	56″
Journals, main	12″ x 15″
Journals, others	11″ x 15″

ENGINE TRUCK WHEELS

Diameter, front	33″
Journals	7″ x 14″
Diameter, back	36″
Journals	8″ x 14″

WHEEL BASE

Driving	44′ 7″
Rigid	11′ 4″
Total engine	67′ 3″
Total engine & tender	112′ 2″

WEIGHT
In Working Order

On driving wheels	531,700 lb.
On truck, front	76,400 lb.
On truck, back	49,800 lb.
Total engine	657,900 lb.
Total tender (fully loaded)	393,300 lb.

TENDER

Wheels, number	Twelve
Wheels, diameter	36″
Journals	7″ x 14″
Tank capacity— Water	22,000 U.S. gal.
Oil	6,100 U.S. gal.

Tractive force	124,300 lb.
Service	Passenger & freight

Construction No. 64,293 *Built* February, 1942 *Drawing No.* 4

Equipped with Type E superheater, feed water heater, power reverse, one-piece cast steel locomotive bed with integral cylinders for each unit, and air brake on all driving and tender wheels, with two 8½″ cross-compound pumps.

FIG. 31 THE SOUTHERN PACIFIC "CAB-IN-FRONT" 4-8-8-2 ARTICULATED LOCOMOTIVE. THIS DESIGN WAS MADE POSSIBLE BY THE USE OF OIL AS FUEL AS THERE WOULD HAVE BEEN NO PRACTICAL WAY TO CONVEY COAL THE FULL LENGTH OF THE ENGINE. THE VISIBILITY ON THESE LOCOMOTIVES WAS SUPERB AS THE OPERATING CREW WAS UP FRONT AND SMOKE WAS EMITTED BEHIND THEM.

Loco. Type	Wheel Arrangement	Generally Accepted Name	Approximate Production
4-2-0	ooO	Jervis	Unknown
4-4-0	ooOO	American	25,600
4-6-0	ooOOO	Ten Wheeler	17,100
4-8-0	ooOOOO	Mastodon	600
2-6-0	oOOO	Mogul	10,900
0-8-0 (Road)	OOOO	8-Wheel Connected	400
2-8-0	oOOOO	Consolidation	33,400
2-10-0	oOOOOO	Decapod	4,100
4-4-2	ooOOo	Atlantic	1,900
2-6-2	oOOOo	Prairie	1,700
4-6-2	ooOOOo	Pacific	6,800
2-8-2	oOOOOo	Mikado	14,100
4-8-2	ooOOOOo	Mountain	2,400
2-10-2	oOOOOOo	Santa Fe	2,200
4-6-4	ooOOOoo	Hudson	500
2-8-4	oOOOOoo	Berkshire	750
4-8-4	ooOOOOoo	Northern	950
2-10-4	oOOOOOoo	Texas	450
4-10-2	ooOOOOOo	Southern Pacific	60
4-12-2	ooOOOOOOo	Union Pacific	90
Articulared	Various	Articulated (Mallet, Simple)	3,100
Nonarticulated (4-Cyl.)	Various	Duplex	80
0-4-0	OO	4-Wheel Switcher	1,500
0-6-0	OOO	6-Wheel Switcher	15,500
0-8-0	OOOO	8-Wheel Switcher	2,800
Geared Engines	Various	Shay, Climax, Heisler, etc.	3,800
Miscl.	Various	Tank, etc.	24,220
		TOTAL –	175,000

FIG. 32 U.S. STEAM LOCOMOTIVE PRODUCTION 1831-1950 INCLUDING EXPORT AND RAILROAD COMPANY SHOPS

STEAM PASSENGER LOCOMOTIVE

For the lighter passenger service.

Atlantic (4-4-2) Type
Cylinders, 23½-inch diameter, 26-inch stroke
Steam Pressure, 205 pounds per square inch
Driving Wheel Diameter . . . . 80 inches
Weight on Driving Wheels . 136,000 pounds

Class E-6s
Total Weight of Locomotive and Tender in Working Order, 411,250 pounds
Tractive Effort . . . . 31,275 pounds

STEAM PASSENGER LOCOMOTIVE

For local passenger service.

Ten-Wheel (4-6-0) Type
Cylinders, 24-inch diameter, 28-inch stroke
Steam Pressure, 205 pounds per square inch
Driving Wheel Diameter . . . . . . 68 inches
Weight on Driving Wheels . 178,000 pounds

Class G-5s
Total Weight of Locomotive and Tender in Working Order, 409,900 pounds
Tractive Effort . . . . 41,328 pounds

STEAM PASSENGER LOCOMOTIVE

For high-speed through passenger service.

Pacific (4-6-2) Type
Cylinders, 27-inch diameter, 28-inch stroke
Steam Pressure . . 205 pounds per square inch
Driving Wheel Diameter . . . . . . 80 inches
Weight on Driving Wheels, 209,300 pounds

Class K-4s
Total Weight of Locomotive and Tender in Working Order, 541,150 pounds
Tractive Effort . . . . 44,460 pounds

STEAM PASSENGER LOCOMOTIVE

For high-speed through passenger service.

Pacific (4-6-2) Type
Cylinders, 27-inch diameter, 28-inch stroke
Steam Pressure, 205 pounds per square inch
Driving Wheel Diameter . . . . 80 inches
Weight on Driving Wheels, 223,000 pounds

Class K-4s Streamlined
Total Weight of Locomotive and Tender in Working Order, 630,000 pounds
Tractive Effort . . . . . 44,460 pounds

FIG. 33a PENNSYLVANIA RAILROAD "STANDARD" LOCOMOTIVES. WHILE THESE DESIGNS WERE UNIQUE TO THE P.R.R., THEY WERE BUILT IN QUANTITIES AS LARGE AS SEVERAL HUNDRED TO INTERNAL R.R. STANDARDS. ALSO NOTE THAT MANY OF THE TENDERS ARE EQUIPPED WITH WATER SCOOPS FOR TRACK PAN WATER SUPPLY.

STEAM PASSENGER OR FREIGHT LOCOMOTIVE

For heavy through passenger service or fast freight service.

Mountain (4-8-2) Type
Cylinders: 27-inch diameter, 30-inch stroke
Steam Pressure . . 250 pounds per square inch
Driving Wheel Diameter . . . . 72 inches
Weight on Driving Wheels, 271,000 pounds

Class M-1a
Total Weight of Locomotive and Tender in Working Order, 768,360 pounds
Tractive Effort . . . . 64,550 pounds

STEAM FREIGHT LOCOMOTIVE

For the heaviest freight service.

Decapod (2-10-0) Type
Cylinders: 30½-inch diameter, 32-inch stroke
Steam Pressure . . 250 pounds per square inch
Driving Wheel Diameter . . . . 62 inches
Weight on Driving Wheels . 352,500 pounds

Class I-1s
Total Weight of Locomotive and Tender in Working Order, 590,800 pounds
Tractive Effort . . . . 90,000 pounds

STEAM PASSENGER LOCOMOTIVE

For high-speed through passenger service.

Pennsylvania (6-4-4-6) Type Class S-1
Cylinders—4 . . . . . 22-inch diameter, 26-inch stroke
Steam Pressure . . . . . . 300 pounds per square inch
Driving Wheel Diameter . . . . 84 inches
Weight on Driving Wheels . . . . 270,000 pounds
Total Weight of Locomotive and Tender in Working Order . . . . 1,052,200 pounds
Tractive Effort . . . . 76,400 pounds

Designed to combine power, speed and economy of operation to a degree never before achieved, the fully streamlined new Class S-1 is expected to anticipate steam locomotive development for years to come. In distinction from more conventional types, it is equipped with four cylinders instead of two, each pair of cylinders providing power for two pairs of driving wheels. This locomotive is the outcome of extensive studies of modern trends in railroad operation and motive power and the requirements of the future. It is capable of sustained speeds of more than 100 miles per hour.

FIG. 33b THE S-1 WAS A ONE-OF-A-KIND EXPERIMENT. IT WAS THE LARGEST *PASSENGER* STEAM LOCOMOTIVE EVER BUILT.

STEAM FREIGHT LOCOMOTIVE
Largely used in local freight and branch line service.

Consolidation (2-8-0) Type
Cylinders: 26-inch diameter, 28-inch stroke
Steam Pressure . . 205 pounds per square inch
Driving Wheel Diameter . . . . . 62 inches
Weight on Driving Wheels . 223,000 pounds

Class H-10s
Total Weight of Locomotive and Tender in Working Order, 424,050 pounds
Tractive Effort . . . . . 53,197 pounds

STEAM FREIGHT LOCOMOTIVE
For heavy freight service.

Mikado (2-8-2) Type
Cylinders: 27-inch diameter, 30-inch stroke
Steam Pressure . . 205 pounds per square inch
Driving Wheel Diameter . . . . 62 inches
Weight on Driving Wheels . 240,200 pounds

Class L-1s
Total Weight of Locomotive and Tender in Working Order, 497,050 pounds
Tractive Effort . . . . . 61,465 pounds

STEAM SWITCHING LOCOMOTIVE
For general switching service.

Six-Wheel (0-6-0) Type
Cylinders: 22-inch diameter, 24-inch stroke
Steam Pressure . . 205 pounds per square inch
Driving Wheel Diameter . . . . 56 inches
Weight on Driving Wheels . 180,300 pounds

Class B-6sb
Total Weight of Locomotive and Tender in Working Order, 305,300 pounds
Tractive Effort . . . . 36,144 pounds

STEAM SWITCHING LOCOMOTIVE
For heavy switching and hump service.

Eight-Wheel (0-8-0) Type
Cylinders . . 27-inch diameter, 30-inch stroke
Steam Pressure, 250 pounds per square inch
Driving Wheel Diameter . . . . . . 56 inches
Weight on Driving Wheels, 278,000 pounds

Class C-1
Total Weight of Locomotive and Tender in Working Order, 435,250 pounds
Tractive Effort . . . . 76,154 pounds

FIG. 33c

4 The History of the Diesel-Electric Locomotive in the United States

A. N. ADDIE

INTRODUCTION

Diesel-electric locomotives are the dominant form of motive power for railroads in the United States. The development of a successful diesel-electric locomotive was predicated on the availability of a number of important components, amongst which were a light weight, reliable engine and a torque conversion device to exploit the maximum engine power over a wide vehicle speed range. In addition, there was a need for an organization with sufficient engineering and mass production manufacturing capabilities to integrate the various components of the complex system into an overall locomotive design and manufacture it at a competitive price.

The principles embodied in the diesel-electric locomotive were developed in the latter part of the nineteenth century, but successful, wide-spread commercial application of these principles was not achieved until about 50 years had elapsed. The invention of the diesel engine and its subsequent development into a light weight traction prime mover, the development of the series traction motor, the modification of the dc generator to make it compatible with traction requirements, the invention of the axle-hung traction motor arrangement, the mastery of the art of control and the systems engineering necessary to combine these devices into a reliable traction vehicle were all necessary steps to be taken before the mass production techniques of American industry could be used to replace the existing accepted form of railroad motive power – the steam locomotive.

This paper will attempt to trace the evolution of the major components of the diesel-electric locomotive which made the change of traction mode which occurred in the United States in the 1940's and 1950's a reality.

EARLY DEVELOPMENTS

EARLY DIESEL ENGINE DEVELOPMENT

The work of Dr. Rudolph Diesel at the MAN Plant at Augsburg, Germany, in 1892, resulted in a patent for a diesel engine. The first engine (Fig. 1) was fueled with coal and served to confirm the compression ignition principle. Practical success was achieved when a second engine ran on partly refined oil, establishing the basic principles upon which a locomotive power plant could be based. In fact, Dr. Diesel himself foresaw the application of the diesel engine to railroads. In Dr. Diesel's book, entitled "The Theory and Construction of a Rational Heat Motor," [1] published in 1893, the author, referring to application of his new en-

FIG. 1 DR. RUDOLPH DIESEL'S FIRST ENGINE

gine, said: "We consider the new motor specially applicable on railways, to replace the ordinary steam locomotive, not only on account of its great economy of fuel, but because there is no boiler. In fact, the day may possibly come when it may completely change our present system of steam locomotive on existing lines of rails."

Dr. Diesel collaborated with the firms of Klose and Sulzer to design the first experimental diesel locomotive in 1909 (Fig. 2). The locomotive used a 1000 hp, two-stroke cycle, four-cylinder engine mounted perpendicular to the longitudinal axis of the locomotive and drove the wheels through a mechanical linkage. It was tried out on the Hessian-Prussian State Railroad in 1913, pulling passenger trains with limited success.

The first diesel engines were slow running, heavy machines employing high pressure air to inject the fuel into the working cylinder. Solid fuel injection was not possible until Robert Bosch developed a practical fuel injection pump and nozzle system about 1925. The early engines found extensive application in the marine and stationary power generation fields, but because of high specific weight and high first cost were not considered suitable for vehicle propulsion.

World War I saw extensive application of the lighter weight diesel engines in Germany Navy submarines. These engines were four-stroke cycle machines of 1200 hp output at 450 rpm, designed and manufactured by MAN in 1912 [2]. Typical specific weights of 47 lb per brake horsepower were attained, which compare with a value of 10 lb per brake horsepower attained in railroad engines today. Application of the diesel engine to vehicle propulsion in the United States awaited the solution of three major problems: 1. the reduction of weight and bulk of the engine; 2. the development of an efficient method for transmitting the engine power which was developed at constant torque to a decreasing torque at the wheel with increasing vehicle speed; and 3. a means for starting the vehicle without stalling the engine.

TRANSMISSION DEVELOPMENT

The basic problem of the transmission in an internal combustion engine-powered vehicle is one of converting efficiently a constant torque produced by the prime mover operating at its rated

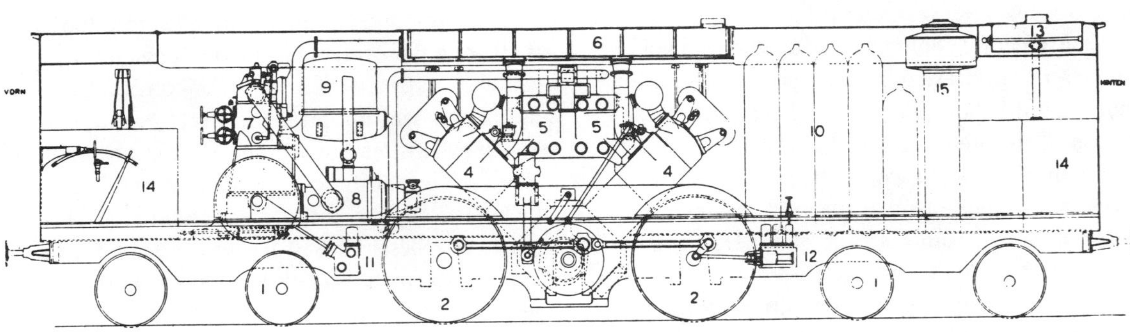

FIG. 2 FIRST EXPERIMENTAL 1000 HP DIESEL LOCOMOTIVE, KLOSE AND SULZER, 1909

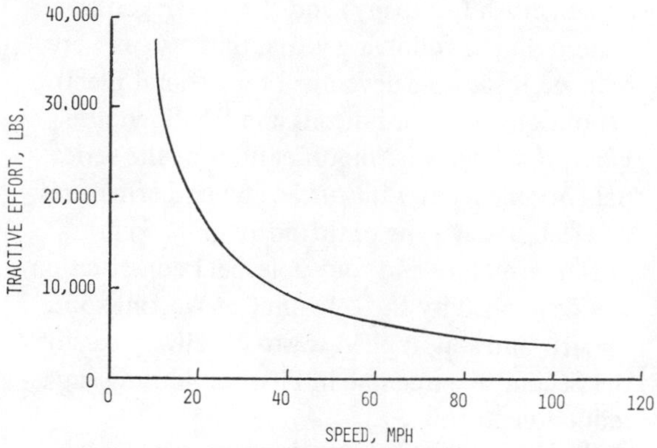

FIG. 3 IDEAL TRACTIVE EFFORT-SPEED CURVE FOR LOCOMOTIVE PROPULSION

power and speed to a variable torque on the wheels of the vehicle so that the power delivered to the wheels is maximized. The ideal relationship of torque to wheel speed or tractive force to vehicle speed is the positive branch of a rectangular hyperbola (Fig. 3). Multiratio geared transmissions approach this ideal rapidly as the number of ratio steps increase. However, the coupling of an internal combustion engine through a relatively rigid drive train to a relatively inflexible steel wheel on a steel rail is a system fraught with difficulties. The torsional vibrations of the prime mover and the dynamic loading of the steel wheel on the steel rail give rise to resonances and shock loadings which are particularly difficult to sustain without the flexibility of the pneumatic tire available to the highway vehicle.

In the absence of electric transmission and the hydraulic torque converter (the Föttinger torque converter was invented in 1905, but lay dormant for some 28 years), there were a number of courageous attempts. Noteworthy amongst these attempts to solve the problems for rail vehicles were the gasoline engine-powered rail cars used on short and branch line railroads as early as 1905. Typical of these vehicles was the J. G. Brill Company Model 55 gasoline car (Fig. 4), which made use of a four-cylinder, four-cycle, 68 hp gasoline engine, driving through a multiple disk clutch, three-speed geared transmission with propeller shaft drive to bevel gear boxes on the axle of one truck. There were also self-propelled cars [3] built by National Steel Car Corporation, Ltd., A Meister Sons Company, The Sykes Company, Four-Wheel Drive Auto Company, International Motor Company and the Edwards Railway Motor Car Company. The McKeen car designed by William R. McKeen, Superintendent of Motive Power, Union Pacific Railroad, introduced on the Chicago, Rock Island and Pacific Railway Company in 1905, employed a 68 hp gasoline engine, friction clutch and chain drive. About 138 cars were built and operated in the United States and foreign countries. A rail car using a hydrostatic transmission built by Waterbury Gear Company was also developed.

While each of the gasoline-driven, mechanical-drive rail cars had some measure of success in light passenger vehicle application on branch line railroads, neither the drive system nor the prime movers employed were adequate for consideration on the main line, and steam locomotives were the undisputed method for hauling both passenger and freight trains on the main line.

A major step forward toward the use of the internal combustion engine and electric transmission

FIG. 4 J. G. BRILL COMPANY MODEL 55 GASOLINE RAIL CAR INCORPORATING A FOUR-CYLINDER, FOUR-STROKE, 68 HP ENGINE AND MECHANICAL TRANSMISSION

on rail vehicles was found in the technology developed for street railways.

DC MOTOR DEVELOPMENT

Both the development of the axle-hung dc motor and gear drive to the axle are attributable to Frank J. Sprague [4], a leading figure in the development of electric transportation. He patented the "Sprague drive" in 1866 while conducting experiments on the Manhattan Electrified Railroad. This arrangement, which is used universally in the United States today, solves the problem of maintaining the alignment between the motor pinion mounted on the armature shaft and the bull gear mounted on the axle by fixing the motor frame to the axle in support bearings and suspending the other side of the motor frame through springs to the truck frame.

All early dc railroad motors which were commercially successful were of the double reduction type to accommodate the low torque, high speed dc motors available. Spur gearing of 10/1 ratio was employed. There were four principle motor types [5] in 1890: The Van Depoele (Thomson Houston), the Sprague (Edison Company), the Short (Brush Company) and the Westinghouse system. In the following years, these types were reduced to designs developed by General Electric (Thomson-Houston-Edison) and Westinghouse Electric, which were modifications of the series field arrangement with surface wound armatures. In 1891, a new type of dc motor (Fig. 5) using slotted armature and four-pole field construction was developed by B. G. Lamme at Westinghouse Electric and was applied to street railway cars of the Second Avenue line in Pittsburgh using single-reduction gearing.

In 1891, the axle-hung drive arrangement (Fig. 6) was used with single-reduction spur gearing. This motor (Westinghouse No. 38) represented the arrangement of dc traction motor drive that has persisted with success through the years on practically all of railroad motive power in the United States. Competitive motors were developed by the General Electric Company, culminating in a general design similar to the Westinghouse motor, both incorporating four radially-bolted field poles, drum-wound armature, axle bearings incorporated in the motor frame, and adjustable spring tension brush holders. In 1907, General Electric Company introduced series dc motors for street railway ap-

FIG. 5 EARLY WESTINGHOUSE ELECTRIC COMPANY DC TRACTION MOTOR, 1891. [5]

FIG. 6 AXLE DRIVE SYSTEM USING SINGLE-REDUCTION GEARING AND WESTINGHOUSE ELECTRIC TRACTION MOTOR [5]

plication with interpoles (commutating poles), greatly improving brush and commutator life.

DC GENERATOR DEVELOPMENT

The electrically-propelled street car received its current through an overhead trolley wire system fed initially by steam engine-driven dc generators. As in the case of street railway dc motors, the development of the dc generator was the product of the same four competitors, Edison-Sprague, Thomson-Houston, Westinghouse and Short (Brush Company). Each competitor began dc railway generator development to supply power for street railways and the designs progressed from large, bipolar types to the multipolar design by Thomson-Houston in 1890 [reference 5, page 651]. These machines were huge, slow-speed, belted-drive arrangements.

In 1893, large, direct-coupled, steam engine-driven dc generators were exhibited in the Chicago World's Fair for lighting systems and street railways. The desire for nearly constant voltage as load current varied made it necessary to use self-excited shunt or compound-wound machines. These machines were soon to be displaced by rotary converters operating from polyphase ac power systems after the first successful application of 25-cycle ac power at Niagara Falls in 1893.

EARLY RAILCARS AND DIESEL-ELECTRIC LOCOMOTIVES

With the development of the dc generator, dc traction motor, and a practical axle drive system, it was now possible to overcome the shortcomings of the mechanical transmission and build a rail car employing these components together with a gasoline engine. The General Electric Company, an early advocate of the electric drive, began building rail cars in 1909 powered by an eight-cylinder, four-stroke, 45-degree "V" type gasoline engine rated at 200 brake horsepower at 550 rpm built by General Electric. It drove a dc generator which was separately excited and required the operator to select the proper field excitation to load the engine to its torque capacity at each throttle setting, a rather demanding task requiring considerable skill.

The first reference to the combination of the diesel engine and an electric transmission to propel a rail vehicle is the diesel-electric rail car built in 1913 by ASEA (Fig. 7) in conjunction with the Atlas Diesel Engine Company in Sweden [6]. This vehicle used a 60 hp diesel engine coupled to a dc generator supplying current to one axle-hung motor. About 30 cars of this general type, but using 150 hp engines and two axle-hung motors, were built for service in Sweden, Denmark, France

FIG. 7 ASEA-ATLAS DIESEL ENGINE COMPANY DIESEL-ELECTRIC RAIL CAR, 1913

FIG. 8 GENERAL ELECTRIC COMPANY DIESEL-ELECTRIC 60-TON SWITCHING LOCOMOTIVE, 1923

and Tunis by the Diesel Electric Car Company (DEVA), a new company formed by ASEA and Atlas Diesel.

Diesel-electric rail cars were built soon afterward by Sulzer Brothers (Switzerland) in cooperation with Brown Boveri Company (Mannheim, Germany) for operation on the Saxon and Prussian Railways using a 300 V, 140 kW dc generator with a jackshaft drive to a side connecting rod from six-pole series motors mounted on a four-wheel truck. A Ward Leonard system was used to control generator excitation [7]. In this system, the engine and generator ran at constant speed with the field of the generator separately controlled. The generator fed current to a series traction motor. Efficiency of the system was poor at low loads.

EARLY DIESEL-ELECTRIC LOCOMOTIVES

After applying eight-cylinder, four-stroke gasoline engines of 200 brake horsepower of General Electric design with electric transmissions to rail cars in 1909, the General Electric Company decided to concentrate on building electrical apparatus and left the manufacture of engines to others. A 180 hp gasoline electric switching locomotive was built at Erie using a Stirling Dolphin six-cylinder gasoline engine [8], but the first diesel-electric locomotive appeared in the United States in 1923. This was a 300 hp, 60-ton switcher which used a General Electric-designed, Ingersoll Rand-built, four-stroke diesel using solid injection and having a rated power of 300 brake horsepower at 600 rpm (Fig. 8). The engine weighed 17,000 lb (56.7 lb/hp), the dc generator 9750 lb. One of the most significant features of the electric transmission system was the use of separately-excited and differential series fields in the dc generator and an engine governor which permitted engine speed to be varied with load to improve the system efficiency. The system was granted a patent in 1917 [9]. This generator control system converted the usual dc generator shunt characteristic to a constant kilowatt characteristic, which is much more suitable to a traction vehicle. The locomotive was assembled by the American Locomotive Company.

Following the successful demonstration of the 300 hp General Electric experimental switcher on fourteen railroads, the first commercially-produced locomotive, the Jersey Central Lines No. 1000 (Fig. 9) was delivered by American Locomotive Company on October 20, 1925 [10]. This locomotive was similar to the original experimental unit. Some time later, a 600-hp, 106-ton diesel-electric locomotive was delivered to the Red River Lumber Company. This unit utilized two 300 hp diesel engines and dc generators to power the four axle-hung traction motors [Fig. 10].

FIG. 9 JERSEY CENTRAL LINES NO. 1000 DIESEL-ELECTRIC SWITCHER BUILT BY GENERAL ELECTRIC COMPANY AND AMERICAN LOCOMOTIVE COMPANY, 1925

FIG. 10 RED RIVER LUMBER COMPANY 600 HORSEPOWER, 100-TON DIESEL-ELECTRIC LOCOMOTIVE BUILT BY GENERAL ELECTRIC AND AMERICAN LOCOMOTIVE COMPANY

FIG. 11 FIRST ROAD DIESEL-ELECTRIC LOCOMOTIVE BUILT BY CANADIAN LOCOMOTIVE COMPANY, 2660 HP, 334 TONS, 1928

About the same time, the Baldwin Locomotive Works completed a 1000 hp, diesel-electric locomotive using a 12-cylinder Knudsen engine and Westinghouse Electric Company transmission. The dc generator was separately-excited and used Ward Leonard speed control with separately-excited dc traction motors [7].

Based on theoretical work done by V. I. Grinovetsky, Ja. M. Gakkel and Ja. V. Lomonossoff in the U.S.S.R., a diesel-electric locomotive was built by the Baltijisky Plant in Leningrad for test on August 5, 1924 [11]. A second diesel-electric locomotive built to U.S.S.R. designs was completed at the Esslingen Machinenfabrik in Germany shortly afterward and employed a crude means for control of the excitation of the main generator to control tractive effort with speed.

The first road diesel locomotive to appear in North America was a design of the Canadian Locomotive Company engineered by Sir Henry W. Thornton in 1928 [10]. This 334-ton locomotive (Fig. 11) was rated at 2660 hp and actually consisted of two separate carbodies, each supported on a two-axle lead truck, four driving axles, and a trailing idler axle in the rear.

PASSENGER DIESEL-ELECTRIC LOCOMOTIVES

During the period 1909-1917, about 86 gas-electric rail cars were built by General Electric Company [8] and in addition several manufacturers produced gasoline mechanical-drive cars which had limited success in branch line operations. These cars were generally underpowered and were plagued with service problems. The General Electric Company discontinued building the gasoline-electric car in 1917 and began to concentrate efforts on railroad electrification.

In 1922, H. L. Hamilton, a former employe of the White Motor Company in Cleveland, Ohio, recognized the need for an improved performance rail car vehicle and in August, 1922, formed the Electro-Motive Engineering Company in Cleveland, Ohio, with the intent of designing a gasoline-electric rail car powered by a 175 brake horsepower marine engine, a General Electric dc generator feeding two 105 hp axle-hung General Electric series dc motors in a powered truck located under the engine compartment. The dc generator control was that developed by Herman Lemp at General Electric's Erie Works. The St. Louis Car Company, a streetcar and bus manufacturer, supplied the carbody for the rail car. Unfortunately, the 175 brake horsepower marine engine failed to produce its rating and a new engine was designed for the application by Carl Salisbury, Chief Engineer of the Winton Engine Company of Cleveland, Ohio, (Fig. 12). This gasoline engine embodied a number of principles which bear mentioning since they remain, in modified form, features of current General Motors en-

FIG. 12 WINTON 175 BRAKE HORSEPOWER MODEL 106A (7 × 8) GASOLINE ENGINE USED IN ELECTRIC-MOTIVE ENGINEERING COMPANY DESIGNED RAIL CAR

FIG. 13 ELECTRIC-MOTIVE ENGINEERING COMPANY DESIGNED GASOLINE ELECTRIC RAIL CAR, M300, 150 HP, 1924

gines. They include individual cylinder heads, removable without disturbing the exhaust manifolds; sleeved cylinder bores; and handhole doors for inspection and maintenance of crankshaft and bearings.

Hamilton was successful in obtaining tentative orders from the Chicago Great Western and the Northern Pacific for two rail cars based on a guaranteed performance of 50 mph, pulling a 35 ton trailer, and 30 day service without major failure. The 57 ft long, 35 ton M300 rail car (Fig. 13), was delivered to the Chicago Great Western and the B-3 to the Northern Pacific in 1924. In 1925, 36 additional cars were delivered.

A number of the early McKeen cars were also repowered with Electric-Motive Engineering Company-designed propulsion systems.

A significant contributor to the development of the gas-electric rail cars and one who was later to have a major influence in the design of the diesel-electric locomotive, was Richard Dilworth, a former associate of Herman Lemp at General Electric Company. Dick Dilworth became Chief Engineer of the Electro-Motive Corporation on January 1, 1926, and remained an active contributor until retirement in 1952.

In 1927, two six-cylinder, 275 hp distillate engines were installed in a 40 ft mail car by the Horton Shop of the Rock Island Railroad under direction of Electro-Motive Company engineers and the first distillate-electric locomotive for passenger service came into being.

Gasoline and distillate fuel prices were escalating in 1930 and there was a definite need for an internal combustion engine capable of burning lower cost fuel. Mr. C. F. Kettering and the General Motors Corporation were successful in supplying this demand through the development of the light weight, two-cycle diesel engine.

Although his first engines operated on the four-stroke cycle, Dr. Diesel recognized that his thermodynamic cycle could also be carried out in a two-stroke cycle. Both Sulzer Brothers and MAN developed engines operating on the two-stroke prin-

FIG. 14 C. F. KETTERING'S TWO-CYCLE (8 × 10), SINGLE-CYLINDER DIESEL ENGINE, 1928

ciple, the former using a separately-driven scavenge pump and the latter a piston-driven scavenge pump [12]. The two-cycle engine became the favored design in marine surface applications because it could easily be made direct reversing. Sulzer Brothers introduced a two-cycle, four-cylinder reversible marine engine in 1905 and unsuccessful attempts were made to adapt this type engine to automobiles in 1909.

The Busch-Sulzer Brothers Diesel Engine Company of St. Louis, Missouri, was the sole licensee of Dr. Diesel's patients in the United States and Canada, a license which remained valid until 1927.

Mr. Kettering began his research and development work on the two-cycle diesel engine in 1928 with the specific objective of developing a light weight design with improved response and lower cost per horsepower than available marine diesel engines. "Boss Ket's" original two-cycle, single-cylinder engine is shown in Fig. 14. In June, 1930, General Motors Corporation purchased the Winton Engine Company, the Electro-Motive Company's supplier of gasoline and distillate engines for rail car application, and in December, 1930, the Electro-Motive Company was also purchased.

A number of very significant engineering contributions were made during this period by the combined efforts of General Motors Research and Winton Engine Company engineers including: 1. the unit injector, incorporating the pumping and fuel metering function into a single device so as to avoid troublesome high pressure lines; 2. the application of the engine-driven positive displacement (Roots type) blower for scavenging; and 3. the welded steel crankcase, making significant reduction in overall weight over cast steel versions. This development program came to a major turning point when the 8-201 engine, rated at 600 hp, with a specific weight of 20 lb per brake horsepower, was used to supply power to the Chevrolet model assembly plant at the Chicago World's Fair in 1933 (Fig. 15). Mr. Ralph Budd of Chicago, Burlington and Quincy Railroad Company, on viewing the display, decided on the use of this

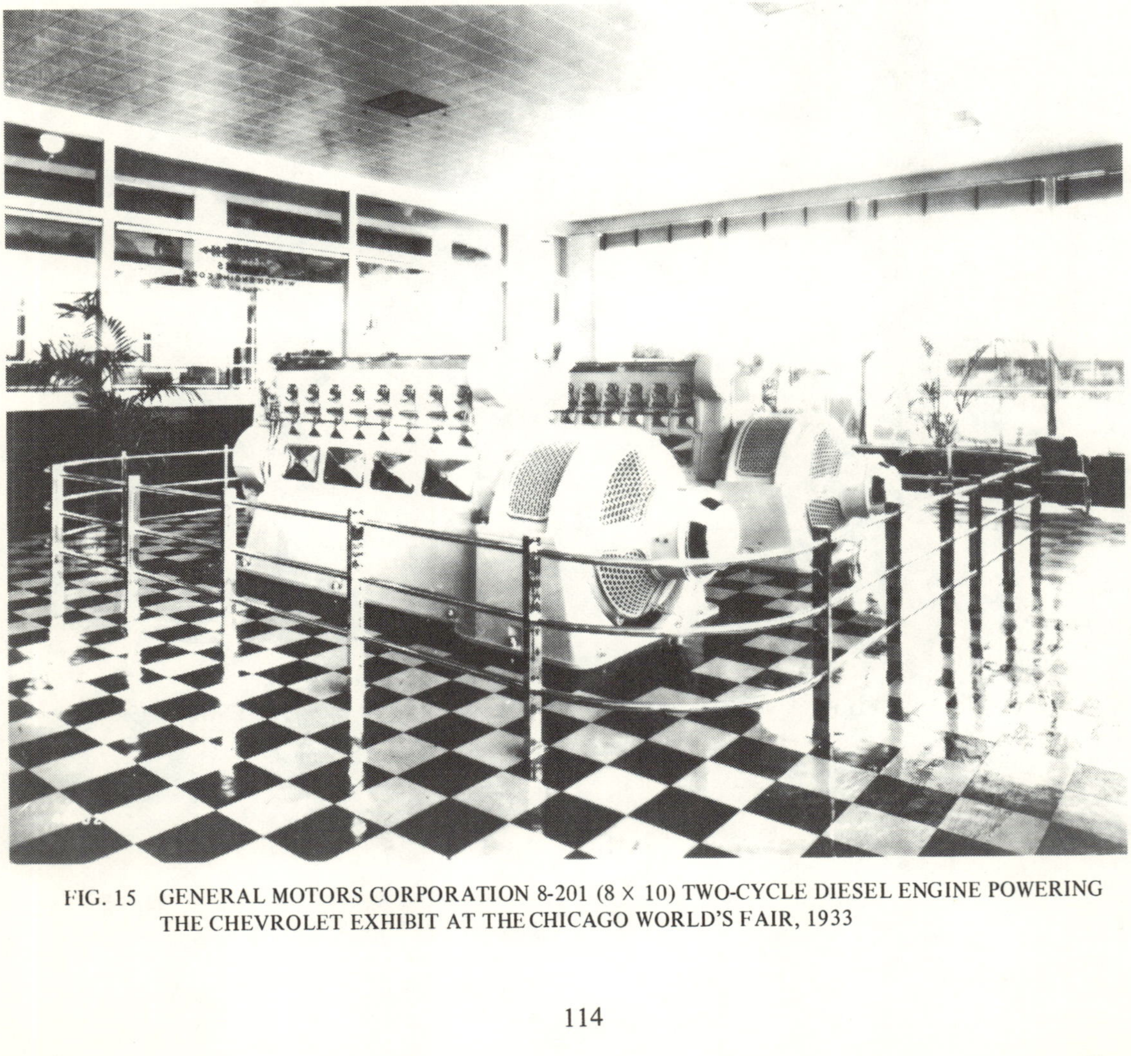

FIG. 15 GENERAL MOTORS CORPORATION 8-201 (8 × 10) TWO-CYCLE DIESEL ENGINE POWERING THE CHEVROLET EXHIBIT AT THE CHICAGO WORLD'S FAIR, 1933

FIG. 16 CHICAGO, BURLINGTON & QUINCY PIONEER ZEPHYR, 1934, POWERED BY GENERAL MOTORS (8 × 10) DIESEL ENGINE

type of engine in the three-car Pioneer Zephyr streamlined passenger train (Fig. 16). A redesigned version of the engine, designated the 8-201A, was installed, together with a General Electric Company dc generator, in the stainless steel carbody manufactured by the Edward G. Budd Company in Philadelphia and was delivered to the Chicago, Burlington and Quincy Railroad on April 18, 1934. The articulated three-car train weighed 97-1/2 tons and was powered by two dc motors mounted in the lead unit General Steel Casting truck. The Pioneer Zephyr was immediately successful in capturing the public's attention and is generally credited with ushering in the diesel-electric locomotive into main line passenger service. On May 26, 1934, the Pioneer Zephyr made its historic trip from Denver to Chicago, a distance of 1015.4 miles in 13 hr, 4 min and 58 sec (average speed 76.61 mph), exceeding 100 mph along the way [13].

Almost simultaneously with the building of the Pioneer Zephyr, the Union Pacific Railroad, under direction of its President, Averell Harriman, ordered the City of Salina, a streamlined, articulated, three-car train, weighting 85 tons total. The cars were built by Pullman Standard Car Company in Chicago using aluminum and powered by a 12-cylinder, "V"-type, Winton distillate engine in the lead unit. A second train of six articulated cars followed shortly afterward (Fig. 17). It was powered by a 12-cylinder version of the Winton 201A diesel engine rated at 900 brake horsepower and developed for United States Navy submarines. The light weight, articulated passenger train principle, though inflexible in regard to the make up of the train consist, was popular in the middle 1930's. Nevertheless, the train was merely an extension of the early rail car design pulling several trailers. In 1935, Electro-Motive introduced the Nos. 511 and 512 prototypes of two-unit passenger locomotives rated at 3600 hp total (Fig. 18). Each unit, weighing 240,000 lb, had two 12-cylinder "V" 201A engines rated at 900 hp each. The locomotives incorporated General Electric generators and traction motors and were built in the Erie, Pennsylvania, Works of General Electric. The

FIG. 17 UNION PACIFIC RAILROAD M-10001 POWERED BY GENERAL MOTORS 12-201A DIESEL ENGINE

FIG. 18 BOX TYPE PASSENGER LOCOMOTIVE 511-512 PROTOTYPES POWERED BY FOUR GENERAL MOTORS 12-CYLINDER 201A DIESEL ENGINES

Atchison, Topeka and Santa Fe used this passenger locomotive to pull its Super Chief passenger trains in transcontinental service.

Following the successful operation of the two-unit "box cab" passenger locomotive prototypes, Electro-Motive introduced a 3600 hp two-unit, non-articulated, passenger locomotive on the Baltimore and Ohio on August 22, 1935, thereby initiating the concept of multiple unit, diesel-electric passenger locomotive, which was to become a standard operation in future years.

In March, 1935, General Motors Corporation began the construction of its locomotive manufacturing plant in LaGrange, Illinois, with capabilities for building the locomotive carbody on a cast underframe and installing General Motors 201A diesel engines and General Electric electrical equipment. The first product to emerge from the new plant was a 600 hp switching locomotive in 1936 (Fig. 19). There followed the beginning of mass production and product standardization to minimize price and operating cost, a principle of which General Motors is a major proponent. Innovations in locomotive design introduced at LaGrange included the use of submerged arc, automatic welding techniques to join 5-in. steel sections to produce switcher underframes, which are less expensive than the cast structures previously utilized.

One of the most significant steps in the history of the success of the diesel-electric locomotive was taken in 1937 when General Motors expanded the LaGrange manufacturing facilities to make ready for fully-integrated manufacture of the major locomotive components, including the diesel engine and the electric transmission.

The 201A diesel engines, originally designed for submarine application and applied to the early streamliners and switchers, were not optimum for railroad application because of a lack of power, complex crankcase, short life of power assembly components, and complex valve gear. Accordingly, General Motors initiated the design of a relatively narrow, 45-degree "V", two-cycle engine (Fig. 20) using an 8-1/2 by 10 in. bore and stroke which at 800 rpm achieved a rating of 91.6 brake horsepower per cylinder (80 BMEP). It was initially built in 8 and 12-cylinder versions at 733 and 1100 brake horsepower, permitting a traction rating (horsepower to main generator) of 600 and 2000 hp respectively for switchers and two-engine passenger locomotives. The 16-cylinder version of the 567 engine developed 1465 brake horsepower (1350 hp for traction).

The first Electro-Motive main line passenger locomotives to be produced at LaGrange in 1937 were the E-type units (Fig. 21) built in cab and

FIG. 19 FIRST 600 HP SWITCHING LOCOMOTIVE TO BE BUILT AT ELECTRO-MOTIVE DIVISION'S LAGRANGE PLANT, 1936

FIG. 20 GENERAL MOTORS CORPORATION 12-567 ENGINE, 8-1/2 × 10 IN. BORE AND STROKE, RATED 1100 BRAKE HORSEPOWER AT 800 RPM, 1938

FIG. 21 E-3 2000 HP STREAMLINED PASSENGER LOCOMOTIVE BUILT BY ELECTRO-MOTIVE DIVISION AND POWERED BY TWO 12-567 DIESEL ENGINES AND ELECTRO-MOTIVE BUILT GENERATOR AND TRACTION MOTORS

noncab configurations. Incorporating two 12-201A engines, each rated at 900 hp for traction and employing General Electric or Westinghouse designed and manufactured dc generators and traction motors, these locomotives utilized a characteristic streamlined nose on a truss-type carbody mounted on two A-1-A trucks. These were the first of the now famous E series locomotives, the last of which was the Model E-9 rated at 2250 hp for traction.

The design of electrical transmission components for manufacture at the new Electro-Motive Plant was begun at the Delco Products Division of General Motors Corporation in Dayton, Ohio, in 1936 and moved to LaGrange in 1937, and by 1938, Electro-Motive Division was producing its own generators and traction motors.

In 1938, the E-4 streamlined passenger locomotive was delivered to the Seaboard Coast Line to pull the Orange Blossom Special. This locomotive was the first to have the new 12-567 engine and Electro-Motive designed and manufactured electrical transmission and marked the beginning of integrated manufacture at LaGrange.

FREIGHT DIESEL-ELECTRIC LOCOMOTIVES

In 1939, Electro-Motive built the first production freight locomotives, the Model FT (Fig. 22), which consisted of a lead unit permanently coupled to a trailing unit. Two of these sets of units were coupled together to form a four-unit consist with each lead unit incorporating an engineman's cab. The complete consist was 193 ft long and weighed 456 short tons. The four Model 16-567 engines delivered a total of 5400 hp to the main generator for traction. The introduction of the Model FT locomotive marked the beginning of dieselization of freight operations on the United States railroads.

1940-1957: DIESELIZATION OF THE UNITED STATES RAILROADS

American Locomotive Company had produced mainly yard switching locomotives up to 1940 using McIntosh and Seymour Corporation 600 hp, four-cycle engines. In 1940, Alco and General Electric entered the road locomotive field with a 2000 hp passenger locomotive using two M&S engines.

During the war years, 1940-1944, the production of passenger locomotives was discontinued and only freight locomotives and yard switchers were manufactured. A total of 1097 Model FT freight locomotives were built between the years 1940 and 1947 by Electro-Motive Division.

FIG. 22 FIRST FREIGHT DIESEL-ELECTRIC LOCOMOTIVE, GENERAL MOTORS CORPORATION MODEL FT – 5400 HP

A new type of locomotive was introduced by Alco-General Electric in 1941 featuring an offset cab and a 1000 hp engine. This unit was a forerunner of the popular road switcher which was built in large numbers in later years and renamed general purpose locomotive because of its multi-duty capabilities.

In 1944, Alco began the production of a 1500 hp, V12, and 2000 hp, V16, four-cycle, turbocharged engine which permitted increased penetration into the road locomotive field.

The Baldwin Locomotive Works, though it produced an experimental 1000 hp locomotive powered by a Knudsen-type 12-cylinder, two-cycle engine in 1925 and a 1000 hp locomotive using a four-cycle Krupp, six-cylinder engine in 1929, was not a serious diesel locomotive competitor until 1937 when 900 hp yard switchers were produced. These switchers used the De La Vergne eight-cylinder, four-cycle engine and Westinghouse Electric transmissions. In 1956, Baldwin-Lima-Hamilton Company ceased the manufacture of diesel-electric locomotives.

Fairbanks Morse & Company entered the locomotive business in 1944 with eight and ten-cylinder, opposed-piston, two-cycle engines previously developed for marine application rated at 1000 and 1500 hp respectively. A later version of the engine design at 2000 horsepower used 12 cylinders. General Electric Company electrical transmission equipment was utilized rated at 1200 hp and road switchers rated at 1600 and 2400 hp were produced.

In the United States, the movement of freight by rail must contend with a wide range of operating conditions. Railroad routes in the United States vary from relatively flat terrain with ruling grades up to 1 percent, to mountainous profiles with 20-mile long grades of over 2.0 percent. In the western part of the country, altitudes of 8000 ft or more are encountered and distances between the starting point and the destination may be over

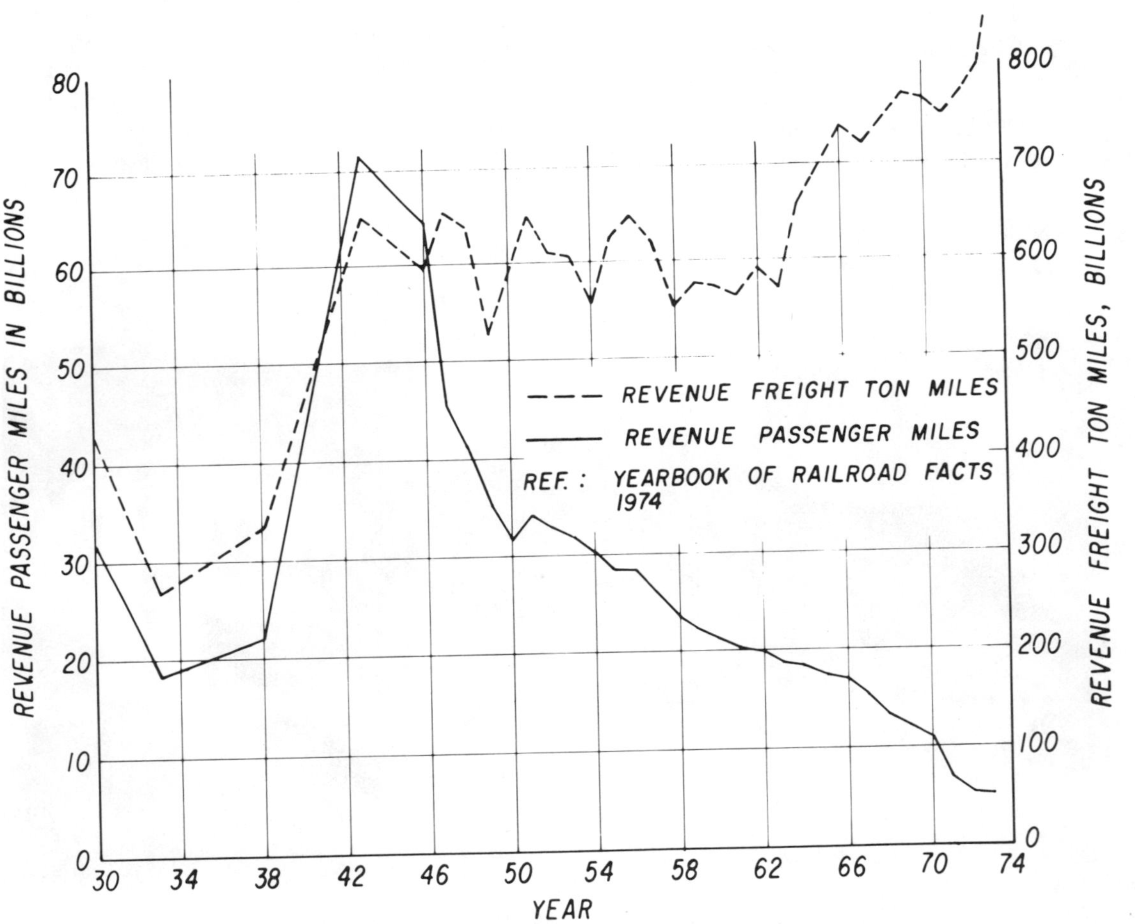

FIG. 23 PASSENGER AND FREIGHT TRAFFIC TRENDS IN THE UNITED STATES

2000 miles.

As railroads developed in the United States, the cost of labor to operate trains has been historically high. The introduction of the air brake, Janney coupler, improved brake shoes, and the improvement in control of starting tractive effort made possible with the diesel-electric locomotive, permitted railroads to increase the length of freight trains and thereby minimize operating costs. The limit in train length was set by the working load capability of the Standard AAR Type D freight car coupler of 175,000 lb, which was used on new freight cars until 1937, when the AAR Type E coupler was adopted with its working load capability of 225,000 lb, more recently uprated to 250,000 lb. As a result of these factors, the drag operation became the accepted practice for United States railroad freight trains and as much as 6000 trailing tons became common.

An axle load limit of 60,000 lb and a usable adhesion under adverse conditions of 18 percent for then existing two-axle trucks and available control systems, resulted in use of four-unit diesel-electric locomotive consists capable of 173,800 lb of tractive effort at minimum continuous speed. This level of tractive effort closely approached the working load limit of early Type D couplers, many of which were still in use in 1939. At the time of the introduction of the first FT freight locomotive, the power capability of the 16-cylinder diesel engines was 1350 traction horsepower, and a four-unit consist provided 5400 hp for traction at the head end of the train. The minimum continuous speed on a grade was 13 mph.

As shown in Fig. 23, the passenger traffic in the United States, which had peaked during the period of World War II (1940-1944), dropped rapidly after 1946, whereas freight traffic remained at a relatively high level. This caused diesel locomotive manufacturers to concentrate heavily on the production of freight diesel-electric locomotives. Electro-Motive Division introduced the Model F-3 locomotive (Fig. 24) in 1946, utilizing the Model 16-567B engine supplying 1500 hp for traction. This locomotive had the exterior appearance of previous passenger locomotives, using a full-width, stressed side-frame carbody and two-axle trucks. Many improvements over the early Model FT freight units were made to reduce maintenance and improve reliability. A total of 5875 of the

FIG. 24 DIESEL-ELECTRIC FREIGHT LOCOMOTIVE, GENERAL MOTORS CORPORATION MODEL F-3 – 1500 HP

Model F-3 and F-7 locomotive units were manufactured by General Motors between 1946 and 1957.

GENERAL PURPOSE FREIGHT LOCOMOTIVES

In 1949, a four-axle 1500 hp GP-7 locomotive (Fig. 25), was introduced to the United States railroads as a general purpose unit, originally intended for application on railroad secondary lines. This carbody design proved to be well adapted to the visibility requirements of yard operations and the accessibility of propulsion system components was enhanced by the absence of carbody superstructure. The purely functional bidirectional design provided cost advantages over comparable full-width carbody types. Furthermore, the relatively low speeds of freight trains in the United States made the reduction of profile drag obtained with the full-width carbody and streamlined nose unimportant. As a result of these factors, the road switcher form became accepted as the standard for United States freight operations.

A total of 2619 GP-7 locomotives were built by Electro-Motive between the years 1949 and 1954. Of this total, 2448, or 93.5 percent, were still in operation on the United States railroads after some 20 to 25 years' service. The GP-7 was succeeded by the 1750 hp GP-9 locomotive in 1954 and 3480 of this model were built in the next six years.

The success of the general purpose type locomotive introduced in the United States in 1949 is largely attributable to several factors:

1. The integration of the complete locomotive design under a single manufacturer who exercised close control over the design and manufacture of each of the important components of the propulsion system; i.e., engine, generator, traction motor, trucks, cooling system, and controls.

2. The establishment of component designs for mass production.

FIG. 25 DIESEL-ELECTRIC GENERAL PURPOSE FREIGHT LOCOMOTIVE, GENERAL MOTORS CORPORATION MODEL GP-7 – 1500 HP

FIG. 26. SPECIAL-DUTY ROAD SWITCHER LOCOMOTIVE, GENERAL MOTORS CORPORATION MODEL SD-7 – 1500 HP

3. The provision for availability to the customer of parts with a maximum of interchangeability.

4. The development of a reliable, high-performance diesel engine.

5. The development of electrical components; e.g., traction motors, generators, and controls to keep pace with engine developments.

In 1950, a six-axle special-duty, road switcher locomotive, Model SD-7 (Fig. 26), was introduced by the Electro-Motive Division. This locomotive provided 1500 hp for traction using the same basic components as the four-axle general purpose locomotive. It was originally intended for low-speed, high-tractive effort applications, but as engine development resulted in increased power, the six-axle locomotive was used in a wider range of applications. Of major importance in the C-C locomotive design was the introduction by Electro Motive Division of a six-wheel Flexi-coil truck which eliminated the loading pads used in other six-wheel truck designs and made it suitable for high-speed operation. This use of a spring to permit lateral flexibility was an innovation by General Motors and is now used extensively throughout the world. In 1954, the 1750 hp Model SD-9 was introduced on the American railroads and this locomotive superseded the SD-7 model.

ENGINE DEVELOPMENT FOR FREIGHT LOCOMOTIVES

The single most important component of the diesel-electric locomotive is the prime mover. The evolution of the design of the Series 567 diesel engines used in General Motors freight locomotives from 1939 to 1957 has been documented in reference 14.

The 16-cylinder version of this two-stroke, normally-aspirated engine (Fig. 27), delivered 1500 hp at 835 rpm into the main generator for traction in the first general purpose locomotive, the Model GP-7, and special-duty road switcher, Model SD-7. It is sufficient here to mention that this engine, during its evolution, has had the greatest application to railroad motive power of any known diesel engine in the world. A total of 23,000 locomotives have been powered by the Series 567 engine in 6, 8, 12 and 16-cylinder con-

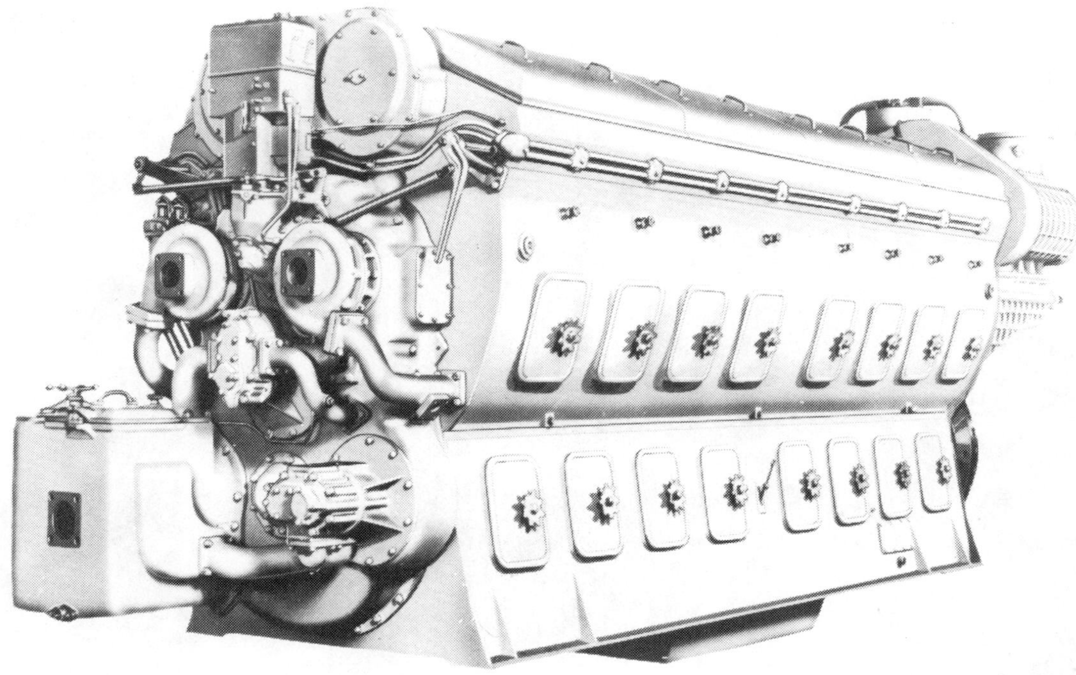

FIG. 27 GENERAL MOTORS CORPORATION MODEL 16-567B TWO-STROKE CYCLE DIESEL ENGINE – 1500 BRAKE HP

figurations. The basic principles upon which the original 201A engines were built were retained, including the two-stroke cycle, unit injector, replaceable power assemblies (piston, rod, cylinder liner and cylinder head), and welded low carbon steel cylinder block and oil pan.

1957-1970: DEVELOPMENTS FOR HIGH LOCOMOTIVE POWER

Development of the normally-aspirated Series 567 diesel engines continued in the 1940's and 1950's to increase the rating of this engine. By 1959, the Electro-Motive Division Model 16-567 engine was producing 1800 hp for traction in general purpose locomotives. Further increases in power with the same 8½ in. bore by 10 in. stroke cylinder became available in 1959 by application of the turbocharging principle. Although four-stroke cycle diesel engines had been turbocharged prior to this time and were in general use on United States railroads in the 1940's using Buchi designs, the two-stroke cycle railroad diesel engine lagged in this development because of the requirements for a positive air supply for starting and acceleration.

Early attempts at two-stroke cycle turbocharging used one or more turbochargers arranged in series or in parallel with the positive displacement blowers, making a cumbersome installation with poor engine response characteristics. Electro-Motive Division solved these problems with a gear-driven overrunning turbocharger installed in place of the positive displacement blowers (Fig. 28). This design, described in reference 15, together with appropriate engine modifications, permitted an initial increase in engine rating by about 25 percent and ultimately by 50 percent with accompanying improvement in specific fuel consumption of nominally 10 percent. The unique design of the turbocharger permits the two-cycle engine to receive the required positive flow of air for starting and low-speed running while allowing the turbocharger rotor to run free from the gear drive at high engine power. Thus, excellent engine acceleration capabilities are obtained and the turbocharger is free at high engine load to compensate for reduced density air at altitude. The ability to accelerate and run under a wide range of power and atmospheric conditions without visible emissions has become more important with the current concern for environment in the United States.

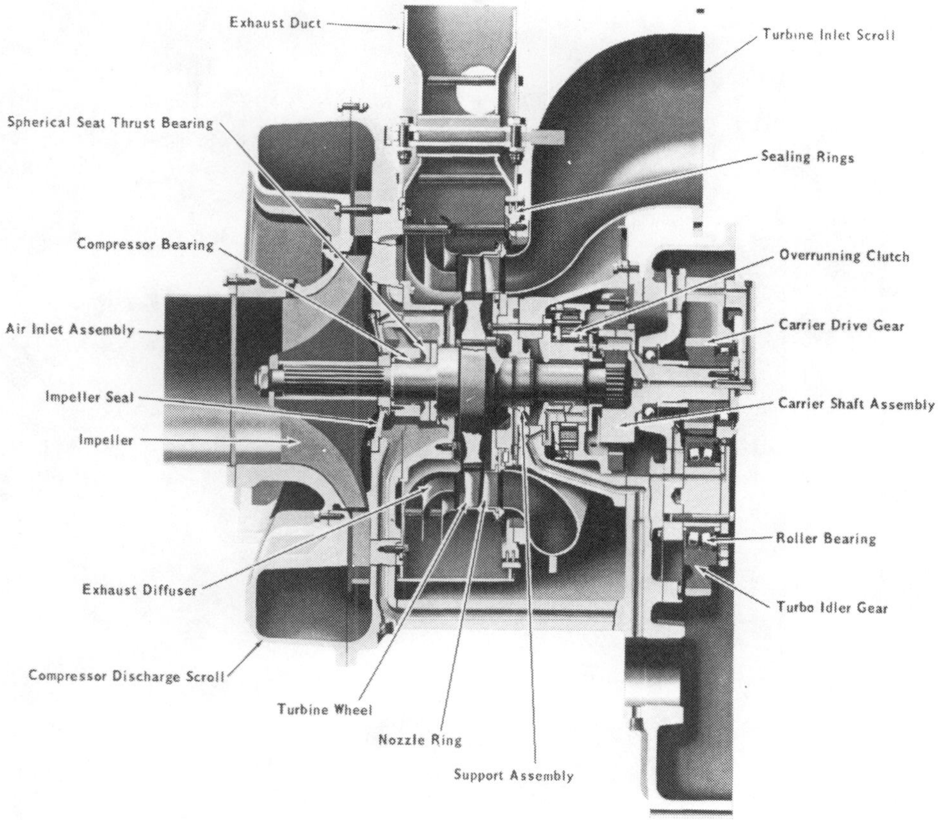

FIG. 28 ELECTRO-MOTIVE DIVISION GEAR-DRIVEN OVERRUNNING TURBOCHARGER

FIG. 29 GENERAL ELECTRIC GENERAL PURPOSE DIESEL-ELECTRIC LOCOMOTIVE MODEL U25C – 2500 HP

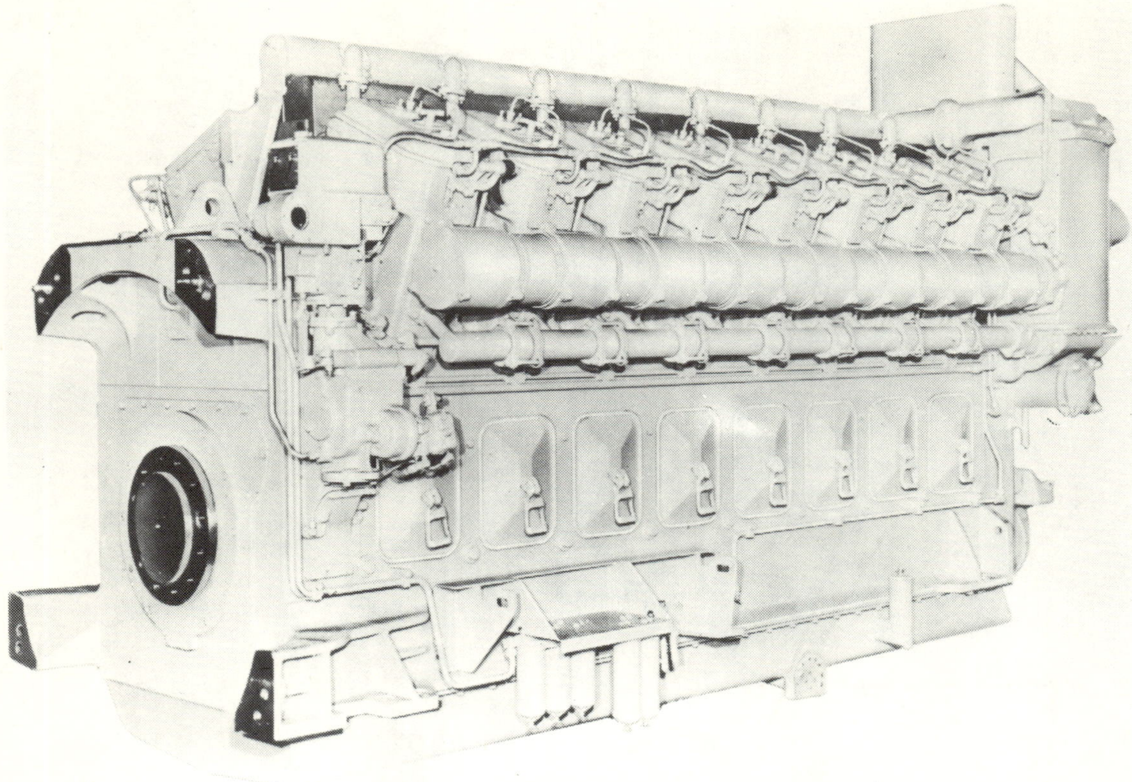

FIG. 30 COOPER BESSEMER MODEL 7VDL-16-T DIESEL ENGINE USED IN GENERAL ELECTRIC U25C LOCOMOTIVE – 2500 HP

With the introduction of turbocharged two-stroke cycle engines by General Motors in the six-axle SD-24 locomotive (2400 hp for traction) in 1959, the trend toward increased power per locomotive unit was accelerated. The General Electric Company entered the United States road locomotive field in 1960 with the four-axle Model U25C (2500 hp for traction) (Fig. 29). This locomotive was equipped with a Cooper Bessemer Corporation Model 7VDL-16-T (Fig. 30) 16-cylinder, four-stroke, 9 in. by 10-1/2 in. turbocharged engine. Previous to this time, the General Electric Company had a long history of locomotive manufacture concentrated in the non-United States market and confined to small industrial switching locomotives in the United States. In 1964, General Electric Company began manufacture of a modification of the Cooper Bessemer engine at their own facility, making them the second fully-integrated supplier of diesel-electric locomotives in North America.

From the earliest railroad application of the General Motors 16-567B normally-aspirated engine at 1350 traction horsepower to the final application of the 16-567D3A turbocharged version of this engine for 2500 traction horsepower in the Model GP-35 locomotive, interchangeability, reliability, and performance have been the most important goals in the development of the engine. In 1965, an uprated design version of the General Motors railroad diesel engine, known as the Series 645 engines (Fig. 31), was introduced. This engine (described in reference 16) embodied the proven features of the earlier design, but enabled increased ratings to be obtained by virtue of an increase in cylinder bore, 9-1/16 in. and operating speed to 900 rpm. Development of the Model 16-645E3 turbocharged engine made available a power plant for a locomotive of 3000 tractive horsepower rating.

The major technological breakthrough permitting the design of a 3000 hp locomotive was the development of high capacity silicon diodes for rectifying the output of a three-phase alternating current generator (Fig. 32) to replace the engine-driven dc generator previously used. The increased

FIG. 31 GENERAL MOTORS CORPORATION 16-645E3 ENGINE – 3300 BRAKE HORSEPOWER

power absorption and current capacity of the rectified alternator eliminated the restrictions previously imposed by the dc generator.

Accompanying the diesel engine development and traction generator development programs, have been simultaneous efforts in all other components of the propulsion system. The traction motor used in General Motors domestic locomotives was originally developed for application to the early passenger locomotives equipped with 36-in. wheels. This axle-hung, nose-suspended dc series motor evolved from a continuous current rating of 700 A to its present 1050 A rating in basically the same frame size (Fig. 33). The 50 percent growth in current capacity has been made possible by continual research and development efforts, particularly in the field of insulation which has made it possible to provide more copper and greatly improved, reduced thickness insulation. The high degree of interchangeability during traction motor development makes it possible for railroads to constantly upgrade their existing motors and take advantage of all of the latest technological improvements.

Application of the Model 16-645E3 engine, the AR10 alternating current generator with its silicon diode rectifier bridge, and the upgraded dc traction motor made it possible to introduce the four-axle 3000 hp Model GP-40 in 1965 (Fig. 34). This locomotive, while retaining the road switcher configuration of the first general purpose locomotive, the GP-7, has exactly twice the power rating. As of January, 1974, 11,624 of the general purpose, four-axle locomotives with rating between 1500 and 3000 traction horsepower have been produced by General Motors since the introduction of this type in 1949.

While a significant portion of railroad freight operations in the United States have been adequately served by application of multiple units of the four-axle general purpose locomotives, there

FIG. 32 GENERAL MOTORS CORPORATION MODEL AR10 TRACTION ALTERNATOR WITH SILICON DIODE RECTIFIERS

FIG. 33 GENERAL MOTORS CORPORATION MODEL D77 DC TRACTION MOTOR

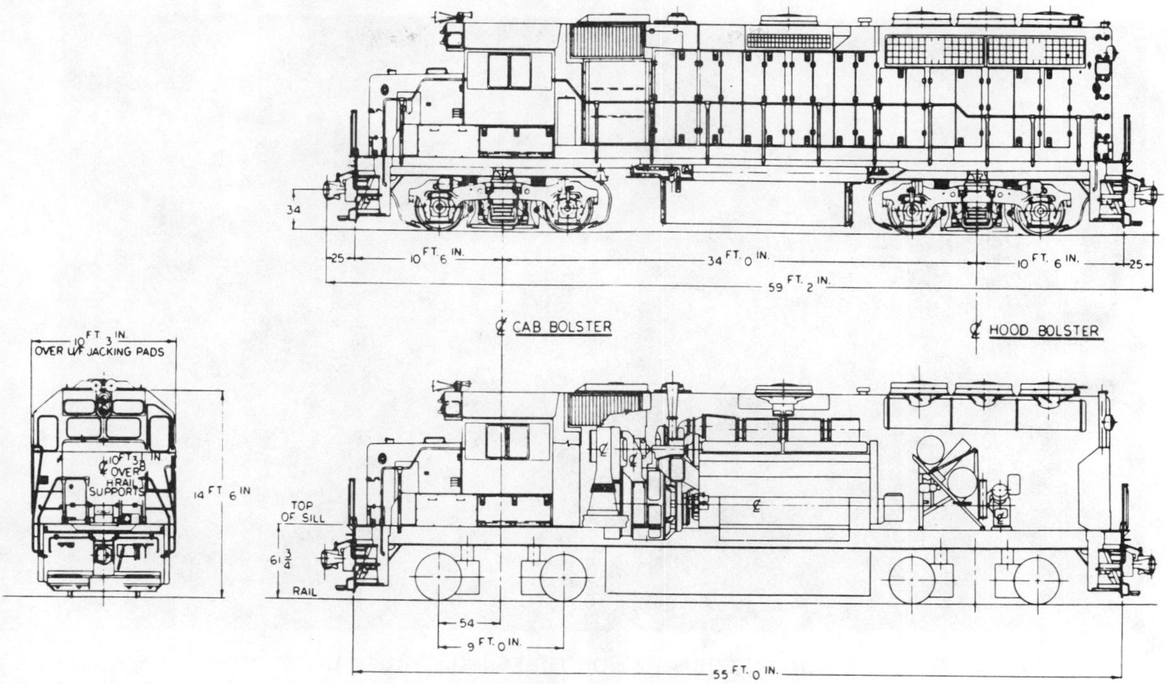

FIG. 34 GENERAL MOTORS CORPORATION GENERAL PURPOSE DIESEL-ELECTRIC LOCOMOTIVE MODEL GP-40 – 3000 HP

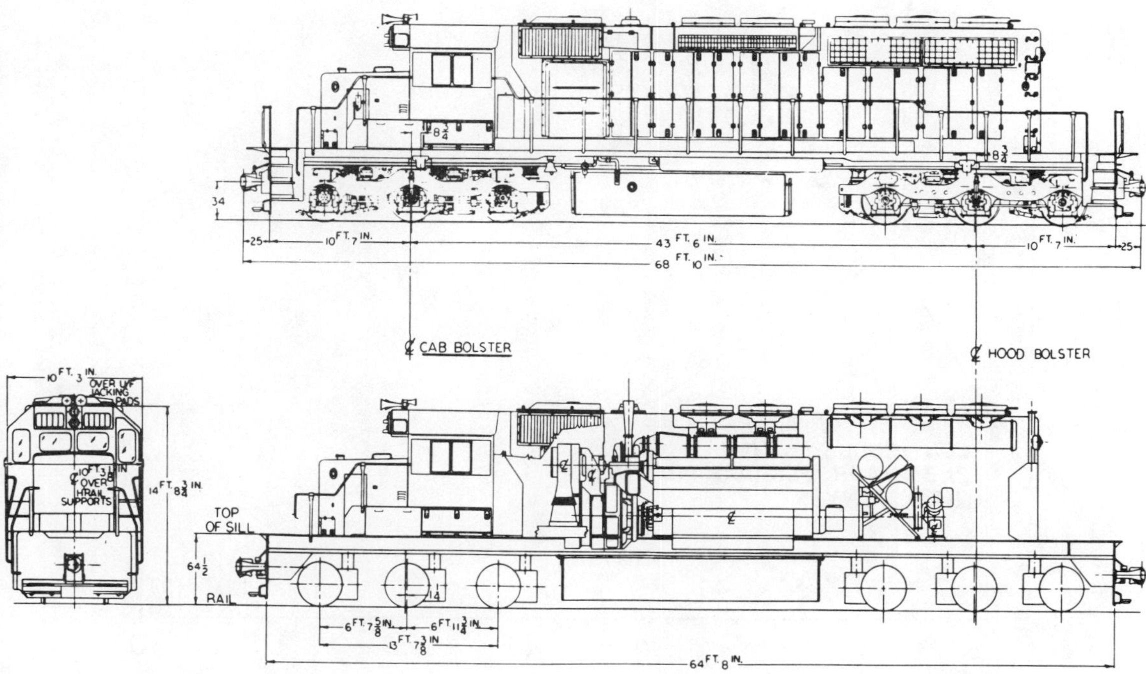

FIG. 35 GENERAL MOTORS CORPORATION GENERAL PURPOSE DIESEL-ELECTRIC LOCOMOTIVE MODEL SD-40 – 3000 HP

FIG. 36 GENERAL MOTORS CORPORATION THREE-AXLE TRUCK MODEL HT-C

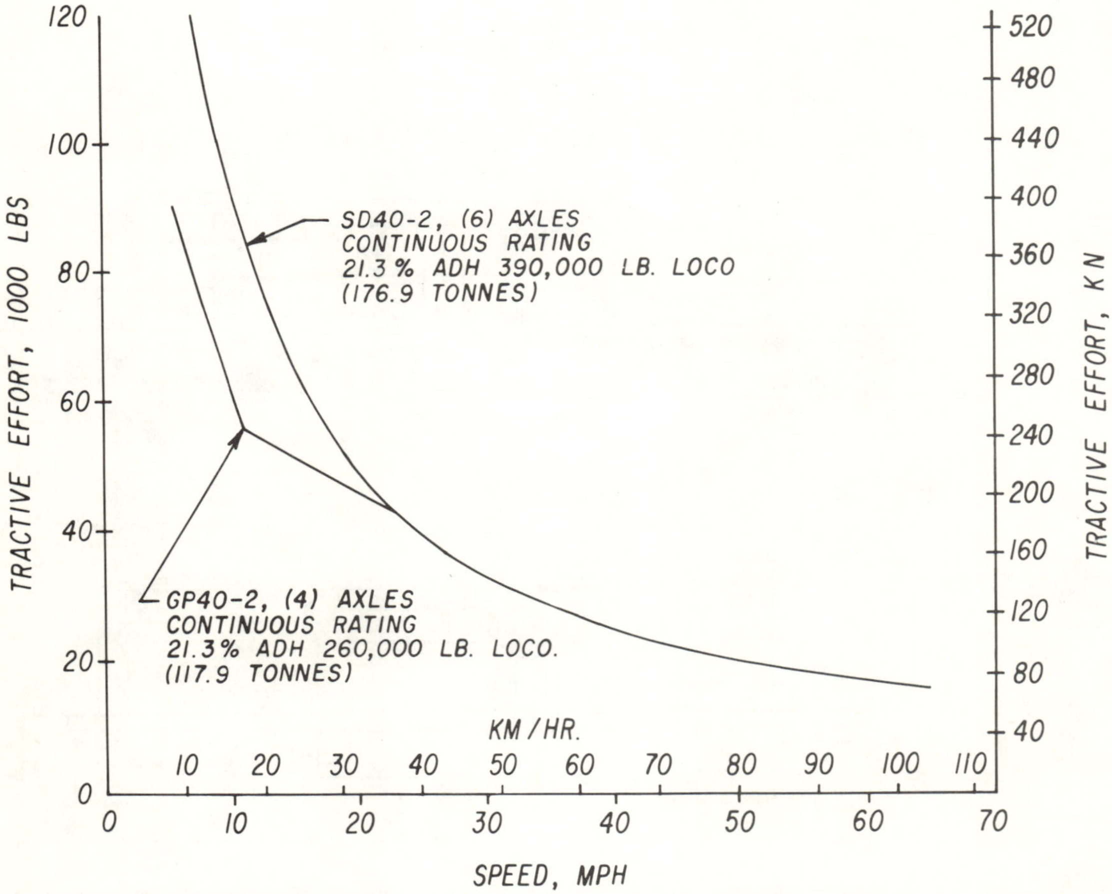

FIG. 37 COMPARISON OF PERFORMANCE OF GENERAL MOTORS CORPORATION SIX-AXLE AND FOUR-AXLE 3000 HP LOCOMOTIVES

remain requirements for heavy drag operations which demand a high level of tractive effort in a single unit. These requirements can be met with a six-axle, six-motor locomotive unit which utilizes the same basic propulsion systems provided in the four-axle designs. The General Motors SD-40 (Fig. 35) is typical of this type of locomotive.

An outstanding feature of this design is the HT-C three-axle truck described in reference 17 and shown in Fig. 36. This truck design produces a 10-20 percent improvement in adhesion efficiency over previous three-axle truck designs used in the United States. The unidirectional orientation of the traction motors, large diameter center plate, and lower driving force between the truck frame and bolster make possible this improvement in weight transfer while retaining the simplicity and ease of maintenance necessary for North American railroad practice. The truck is equipped with a low rate metal spring primary suspension and rubber springs for the secondary suspension.

The broader range of tractive effort at rated power (Fig. 37) of the six-axle SD-40 locomotive as compared with the four-axle GP-40 has caused many railroads in recent years to apply the SD-40 locomotive for general purpose freight applications.

Both General Motors and General Electric have produced locomotives having traction ratings of 3600 hp on six axles designed for high-speed, high-tonnage, main-line freight operations. Some of the western railroads in the United States have deviated from the low-speed, long-train, heavy-drag principle which has been characteristic of American railroad freight practice in the past. In particular, the Atchison, Topeka and Santa Fe and the Union Pacific Railroads are utilizing the high speed or manifest freight principle wherein trains are sized to reduce transit time at the sacrifice of tonnage.

The "Super C" trains operated by the Atchison, Topeka and Santa Fe in this service average 1500 trailing tons and reach peak speeds of 80 mph. These trains are powered by four six-axle, 3600 traction horsepower locomotives of the SD-45 type design (Fig. 38), which results in 9.6 traction horsepower per ton. The six-axle SD-45 locomotive is powered by a 20-cylinder version (20-645E3) of the General Motors Series 645, two-stroke cycle diesel engine, which develops 3900 gross horsepower at 900 rpm. This engine is one of the few 20-cylinder "V" engine designs in existence [18].

In order to minimize the number of locomotive units required to pull 5000-ton freight trains up to 70 mph, the Union Pacific has made use of a special 6600 hp, eight-axle locomotive powered by two 3300 traction horsepower diesel engines. This

FIG. 38 GENERAL MOTORS CORPORATION HEAVY TONNAGE HIGH SPEED FREIGHT LOCOMOTIVE MODEL SD-45

FIG. 39 GENERAL MOTORS CORPORATION EIGHT-AXLE SPECIAL PURPOSE LOCOMOTIVE MODEL DDA-40-X – 6600 HP

locomotive, known as the DDA-40-X (Fig. 39), is used in consists of three units, making a total of 19,800 traction horsepower available, or 3.96 hp per ton.

Although the high-power freight locomotive using turbocharged diesel engines satisfied a major demand by railroads in the 1960's to increase speed of high-tonnage, main-line freight operations and to reduce the number of locomotive units necessary to pull specific tonnage trains, there remained a large segment of secondary-line operations where train size and speeds are lower. For these operations, the normally-aspirated version of the General Motors 16-645E engine, installed in a general purpose, four-axle type locomotive having a rating of 2000 hp for traction is an optimum application.

1970 TO 1975 – NEW DEVELOPMENTS FOR HIGH RELIABILITY AND LOW MAINTENANCE

The increase in power per locomotive unit, which characterized the trend in North American freight locomotives in the 1960's, was arrested in 1970 because of increasing emphasis on locomotive reliability. High power diesel-electric locomotives employed in freight service in the United States accumulate on the average between 150,000 and 200,000 miles per year, with some units reaching 250,000 miles per year. Locomotive lifetimes exceed 20 years, making it possible for a unit to accumulate in excess of 5,000,000 miles. During this period, the major components of the locomotive can be replaced as required with spare components. For example, the diesel engine normally operates between 6 and 12 years before removal for rebuild. Combustion components of the diesel engine, however, such as cylinder heads, pistons and cylinder liners require more frequent replacement, but can be exchanged in the locomotive during scheduled maintenance periods without removal of the engine. The same philosophy of parts replacement without extensive overhaul is applicable to the traction motor and its reduction gearing, the main generator, cooling system components, auxiliaries, electrical controls and truck running gear.

In January, 1972, General Motors introduced a new series of high reliability locomotives in North America. Prior to the introduction of these models, all new component designs to be included were subject to extensive in-plant testing, followed by multiple applications in full-scale locomotives on different railroads. These field tests, conducted in cooperation with the railroads, together with the in-plant tests on the components and continuous statistical monitoring of field failures, have made it possible to introduce a new diesel-electric model line with unprecedented success in regard to reliability. These locomotives retained the power levels established in the late 1960's, but provided major improvements to eliminate maintenance,

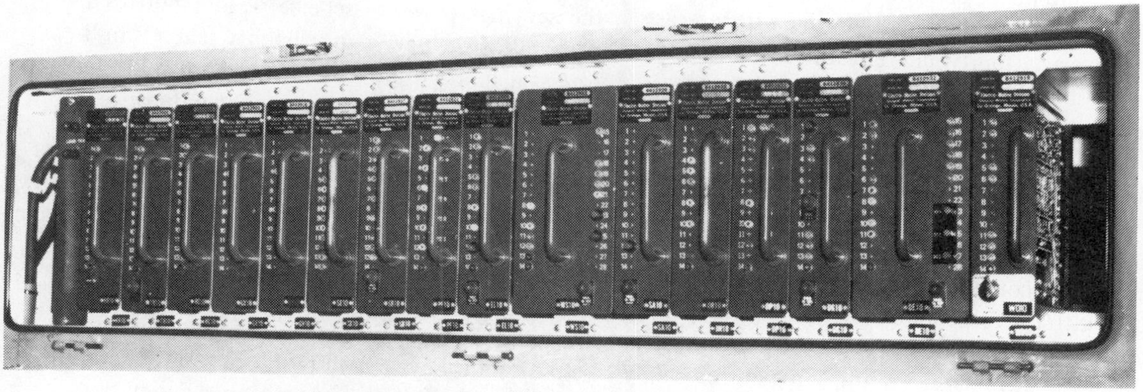

FIG. 40 CONTROL MODULES USED IN GENERAL MOTORS DASH-2 LOCOMOTIVES

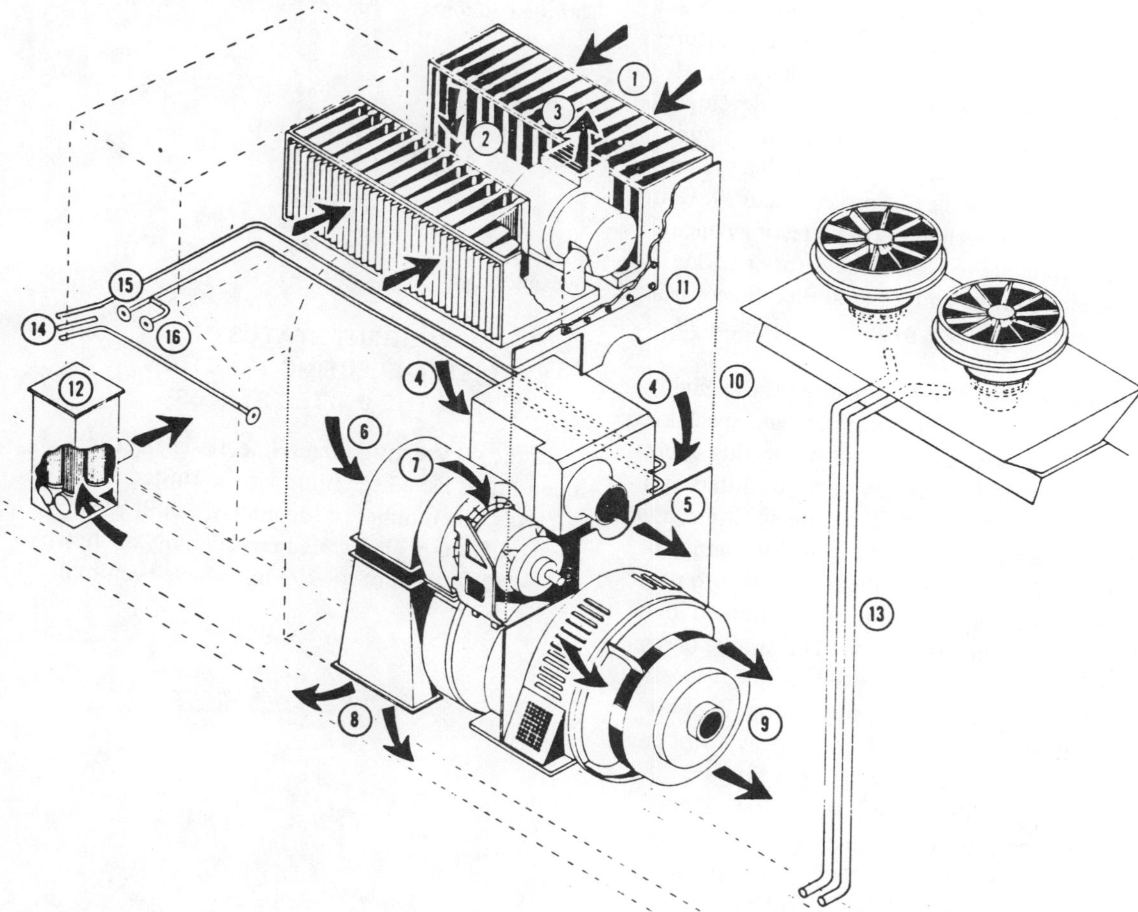

1. Outside Air Intake To Inertial Filters
2. Clean Air Into Sealed Compartment
3. Blower Driven Air Carrying Dirt
4. Intake For Engine Air Filter
5. Clean Air To Engine
6. Intake To Traction Motor Blower
7. Intake To Generator Blower
8. Cooling Air Ducted To Traction Motors
9. Generator Cooling Air Pressurizes Engine Compartment
10. Outline Of Sealed Central Air Compartment
11. Access Panel
12. Electrical Cabinet Air Filter
13. Cooling Air Piped To Dynamic Brake Blower Motor Bearing
14. Hose Stems For Manometer Connection
15. Filter Vacuum Switch FVS
16. Engine Filter Switch EFS

FIG. 41 INERTIAL FILTERS USED IN GENERAL MOTORS DASH-2 LOCOMOTIVES

extend maintenance periods, simplify diagnosis of control circuit malfunctions and otherwise increase locomotive availability and reliability. An account of the methods used by General Motors to achieve current levels of reliability is presented in reference 19.

As locomotive power ratings increased, the degree of complexity of electrical control necessary to provide for optimum performance increased. To offset this complexity, functions were incorporated into discrete modules in numbers ranging from 12 to 16 per locomotive. Use of the control module (Fig. 40) provides a simple means for railroad maintenance personnel to isolate trouble and replace defective control components rapidly to prevent "out-of-service" delays. In addition, many of the control modules have self-checking features and redundant circuits to improve reliability. At the same time that these modules were developed, General Motors designed and built testing equipment to prove the concept of a simplified system for checking the complex electrical circuitry. Commercial equipment using this basic testing concept was made available to the railroads for checking both General Motors and General Electric control modules simultaneously with the production of these new controls.

The overall locomotive design requires a highly integrated and coordinated program and specialized equipment was developed for all supporting equipment such as water-cooling systems, oil filters and cooling systems, carbody filters, engine filters, and other appurtenances. Wherever possible, monitoring systems are included in the locomotive to protect the components and to insure continued locomotive over-the-road operation. The unique design of the locomotive insures that the engine, generator, and traction motors are given a supply of air at ambient temperature and free of heavy dust concentrations, snow and water. Inertial filters (Fig. 41) are provided for primary filtration and they have been proven over many years of service and are highly favored since they greatly reduce routine maintenance. Highly efficient paper and glass filters have been developed as secondary filters for the engine that meet "throw-away" element requirement, which is well suited to the American railroads' maintenance practices. One interesting point pertains to the main electric cabinets which are pressurized using two-stage filtered air to prevent the ingress of dirt.

As an indication of the level of reliability of the modern diesel-electric locomotive in North America, statistics from a substantial number of the new high-power diesel-electric locomotives incorporating the new high reliability features indicate availability of 90-95 percent. Availability takes into account lost time for routine maintenance, inspections required by United States laws, as well as time lost as a result of parts replacements.

Reliability of diesel-electric locomotives is closely associated with the duty cycle imposed on the locomotive during its assigned freight duty. The duty cycle varies depending on the profile of the railroad route and the class of train. A representative duty cycle used by Electro-Motive Division for high horsepower locomotive design purposes for North American operations is as follows:

Throttle Position	Percent Power	Percent Time
8	100	30
7	86	3
6	66	3
5	51	3
4	35	3
3	23	3
2	12	3
1	5	3
Idle	0	41
Dynamic brake	–	8

CURRENT STATUS OF DIESEL-ELECTRIC TRACTION IN THE UNITED STATES

In the 40 years that main line diesel-electric locomotives have been built for the United States railroads, the number of locomotives built each by all manufacturers has varied considerably as shown in Fig. 42. A total of 48,209 units have been built

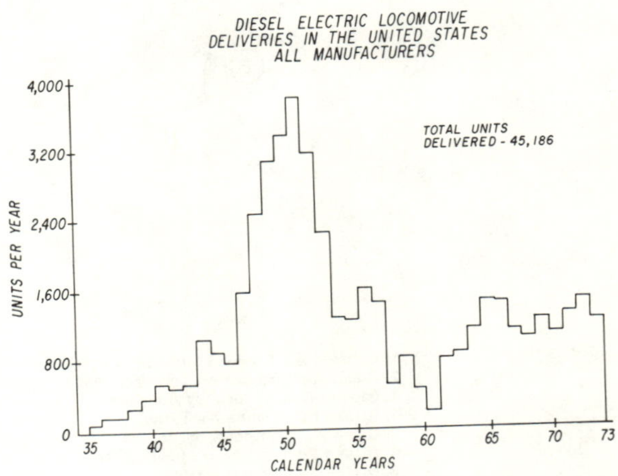

FIG. 42 DOMESTIC DIESEL-ELECTRIC LOCOMOTIVE DELIVERIES IN THE UNITED STATES – ALL MANUFACTURERS

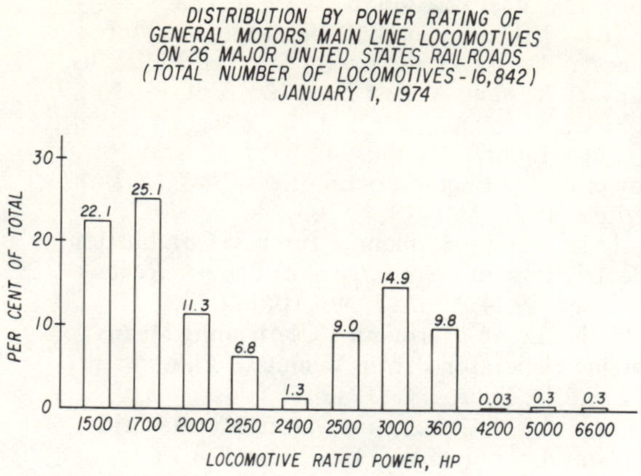

FIG. 43 DISTRIBUTION OF GENERAL MOTORS LOCOMOTIVES BY HORSEPOWER RATING ON 26 MAJOR UNITED STATES RAILROADS

as of January, 1977. The distribution of General Motors locomotives by horsepower rating on 26 major railroads, accounting for some 65 percent of the current locomotive fleet, is shown in Fig. 43. The large concentration, approximately 47 percent, of locomotives of the 1500-1750 hp class reflects the high rate of production between 1949 and 1953. The second concentration at the 2500-3600 hp level reflects the more recent turbocharged power. The distribution by age of locomotive types built by the Electro-Motive Division of General Motors Corporation is shown in Fig. 44.

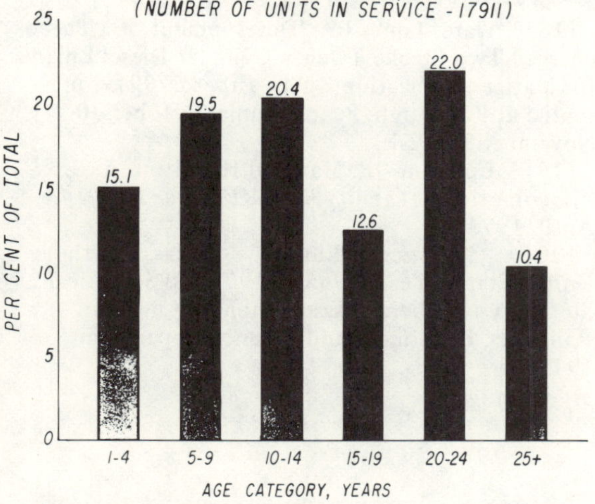

FIG. 44 DISTRIBUTION BY AGE OF GENERAL MOTORS LOCOMOTIVES ON 26 MAJOR UNITED STATES RAILROADS

The potential life of a diesel-electric locomotive is still being developed, since large numbers of General Motors' locomotives built 20-24 years ago are still in service. Technological improvements in locomotive components have resulted in large rebuilding programs where old locomotives are exchanged for "new" locomotives in which some of the original components are upgraded to modern standards. It is the continuing research and development on diesel locomotives which has influenced United States railroads to replace older locomotives with these "new" rebuilt units and, therefore, the life of a diesel locomotive can be said to be based on technical obsolescence rather than physical deterioration.

CURRENT COSTS OF DIESEL TRACTION IN THE UNITED STATES

The Association of American Railroads tabulates the annual railroad statistics covering costs of repair and fuel for all locomotives operated. This report covers shunters, branch line, main line, freight and passenger locomotives in aggregate. A summary of the A.A.R. information on operating costs for 1975 is as follows:

No. of locomotives (close of year.)	27,667
Total repair costs, 10^6 U.S. $	$791.8
Total fuel used, 10^6 U.S. gal	3,314.0
Total cost of fuel, 10^6 U.S. $	$994.2
Total trailing 1000 gross ton miles, MGTM	1,694,213,150
Fuel used per MGTM, U.S. gal/MGTM	1.956
Repair cost per MGTM, U.S. $	$0.467
Fuel cost per MGTM, U.S. $	$0.586
Total operating cost per MGTM (Excluding crew)	$1.053

Of the total locomotive unit usage represented in this table, the following is the distribution of types of service:

Freight road	74.9
Yard switching	21.9
Passenger	3.2
Total	100.0 percent

The data presented are the latest available from the Association of American Railroads [20] and are for the year 1975.

CONCLUSION

The development of the diesel-electric locomotive in the United States is one of the important milestones in the application of technology. The locomotive vehicle embraces a number of highly complex systems which must perform their function cooperatively and reliably under restrictions of limited space and weight in a less than ideal environment. It is clear that refinement in designs which has taken place over the 88 years which have elapsed since Dr. Diesel ran his first engine is the product of the efforts of many individuals with the backing of numerous companies.

The unexcelled efficiency and durability of the diesel prime mover and the highly flexible electric transmission makes this combination an extremely difficult system to obsolete. There have been several relatively unsuccessful attempts by competing systems involving other types of prime movers and transmissions in the last 30 years to displace the diesel-electric, but the inherent advantages of efficiency, durability, flexibility and cost of this system have not been superseded. At the present time, there exists only one viable challenger to the diesel-electric locomotive; i.e., the electrified railroad. It seems probable that at some future time electric locomotives will be substituted for diesel motive power on high traffic density lines. The introduction of this system, however, is not likely to take place as a result of economics, but more likely because of institutional factors beyond the control of individual railroads.

REFERENCES

[1] Diesel, Rudolf, "Theory and Construction of a Rational Heat Motor," Translated from German by Bryan Donkin, E. & F.N. Spon, London, 1894.

[2] Gibson, Holbrook C., "The German Submarine Diesel Engine," *Transactions of SAE*, Part 1, 1921, p. 641.

[3] "Car Builders Cyclopedia," 11th Edition, 1925, pp. 337, 359.

[4] Middleton, William D., "When the Steam Railroads Electrified," Kalmbach Publishing Company.

[5] Lamme, B.G., "Electrical Engineering Papers," Westinghouse Electric & Manufacturing Company, East Pittsburgh, 1919, p. 723.

[6] Nilsson, Enoch, "Diesel Electric Motor Cars Are Used Extensively on Swedish Railroads," *Railway Review*, August 29, 1925, Vol. 77, No. 9, p. 301.

[7] Lipetz, Alphonse I., "Transmission of Power on Oil Engine Locomotives," *ASME Transactions*, 1926, Vol. 48, p. 263.

[8] Lemp, Hermann, "Internal Combustion Electric Locomotives," *General Electric Review*, October, 1924, Vol. 27, No. 10, p. 673.

[9] Lemp, Hermann, "Controlling Means for Engine Generator Driven Vehicles," U.S. Patent 1,216,237, February 13, 1917.

[10] "The Concise Encyclopedia of World Railway Locomotives," Hawthorn Books, Inc., New York, p. 111.

[11] "The Engineer," November 14, 1924, p. 554.

[12] Diesel, Rudolph, "The Present Status of the Diesel Engine in Europe and a Few Reminiscences of the Pioneer Work in America," lecture presented to ASME April 30, 1912, published by Busch-Sulzer, Diesel Engine Company, p. 20.

[13] Reck, Franklin M., "On Time, The History of Electro-Motive Division of General Motors Corporation," Electro-Motive Division of General Motors Corporation, 1948.

[14] Kettering, E. W., "History and Development of the 567 Series General Motors Locomotive Engine," ASME paper presented November 29, 1951, Atlantic City.

[15] Addie, A. N., "Design and Development of Turbocharger for the General Motors Corporation Series 567 Engine," *ASME Paper 60-OGP-7*, May, 1960.

[16] Ephraim, Jr., Max and Williams, Jr., H. A., "A Look at a Modern Two-Cycle Diesel Engine," *ASME Paper 68-DGP-10*, May, 1968.

[17] Marta, H. A., Mels, K. D., and Itami, G. S., "Design Features and Performance Characteristics of the High Traction, Three-Axle Truck," *ASME Paper 72-RT-3*, March, 1972.

[18] Ware, James R., "Development of a Turbocharged Two-Stroke Twenty-Cylinder Diesel Engine for Marine Application," *SAE Paper 670949*, presented at Pittsburgh, Pennsylvania, October 30-November 3, 1967.

[19] Ephraim, Jr., Max and Hamilton, A. B., "Locomotive Reliability," *ASME Paper DGP-14*, April, 1973.

[20] "Statistics of Railroads of Class I in the United States, Years 1965 to 1975, 60 Statistical Summary Numbers," Association of American Railroads, Economics and Finance Department, 1977.

5 One Hundred Years of Progress in Railway Mechanical Engineering – Electric Locomotive Development

P. H. BAKER, A. C. BIEBER, J. C. CALDWELL,
R. E. RINEHART, and M. A. SCHLEIGH

EARLY EXPERIMENTS

The development of the electric locomotive in the U.S. can be traced as far back as 1835, when Thomas Davenport of Vermont built a small model to run on a circular track. It was not until 1847, however, that Professor Moses Farmer built a battery locomotive of a size sufficient to carry two people, and in 1851 Dr. Charles Page operated a battery powered locomotive over a five mile stretch of track between Washington, D.C., and Bladensburg, Maryland. Although speeds of 19 mph were achieved, the batteries were destroyed by the rough ride, and further progress in electric locomotives had to await the development of externally powered systems fed by generators.

Thomas Edison and Stephen Field both began developing generator powered electric locomotives in the early 1880's. At his Menlo Park, New Jersey laboratory, Edison laid 1400 ft of track, and operated a small locomotive with a belt drive, Fig. 1-1, which could pull two cars at speeds up to 40 mph. Henry Villard, President of Northern Pacific Railway, showed an interest in this emerging technology, and ordered Edison to build two prototype locomotives which could be used in railroad service, Fig. 1-2. While Edison was building these prototypes, which resembled steam locomotives of the day, Field constructed the first steeple cab locomotive which would prove to be a popular configuration for generations of electric locomotives yet to come. Demonstrated at a small railway in Stockbridge, Massachusetts, this locomotive had a siderod drive and spoked wheels.

In 1883 the NP electrification plan was dropped due to the financial problems of the railroad, and Edison and Field joined forces to form the Electric Railway Company of the United States. The first locomotive built by the new Company was The Judge, Fig. 1-3, which was demonstrated at the Chicago Railway Exhibition where, during the course of the exhibition, it hauled 26,000 passengers around a 1500 ft circular track at speeds up to 12 mph. Its single traction motor was powered by a central third rail pickup, and drove one axle by a combination of gears and belts.

A third important experimenter of the 1880's was Frank J. Sprague, who developed the now familiar axle hung motor/drive arrangement. In 1889 the Electric Railway Company became a part of the Edison General Electric Company. Under the presidency of Henry Villard the Sprague Company was acquired, and with it the technology of tractive controls and drives developed by Sprague.

Another merger with another pioneer in railway traction, the Thomson-Houston Company, followed in 1892, and in 1893 the newly formed General Electric Company produced its first locomotive, Fig. 1-4 and exhibited it at the World's Columbian Exposition in Chicago. A 30 ton unit with a single truck, it could develop 12,000 lb of tractive effort and achieve speeds of 30 mph. It had two gearless traction motors and a quill drive, and collected power from a 500 V dc overhead trolley wire.

FIG. 1-1 EDISON'S FIRST ELECTRIC LOCOMOTIVE

Up to this time, all electric locomotives had been powered by dc traction equipment. In 1895, the Westinghouse Company entered the electric locomotive field by joining forces with the Baldwin Locomotive Works. The two firms produced the experimental ac locomotive of Fig. 1-5, a 32 ft long box cab weighing 46 tons, having two 2-axle trucks, and developing 600 hp. After ten years of experiments with both ac and dc equipment, George Westinghouse had decided that the ac system was superior, and chose to develop ac traction systems for his firm's future locomotive designs.

At this point electric locomotives had proven that they could do the job. The principal body, truck and drive components had been developed, and the stage was set for the first full scale application: the electrification of the B & O tunnel.

GROWTH WITH TUNNELS AND TERMINALS

Early electrification of steam railroads was undertaken largely to solve the problems of steam operation in tunnels, covered terminals, and urban areas where high density train service brought intolerable concentrations of exhaust smoke. Electrification

FIG. 1-2 EDISON'S LOCOMOTIVE OF 1880

FIG. 1-3 "THE JUDGE" BY FIELD & EDISON, 1883

FIG. 1-4 COLUMBIAN EXPOSITION LOCOMOTIVE

FIG. 1-5 BALDWIN/WESTINGHOUSE EXPERIMENTAL LOCOMOTIVE

for these reasons began in 1895 and lasted until 1910, except for three later exceptions. Much of this early development came concurrently with similar designs for self-powered passenger cars for street railway use.

The first of these tunnel electrifications came on on the Baltimore and Ohio Railroad's new mainline through Baltimore, Maryland. The plan was to replace a slow car ferry operation across the harbor with a tunnel under a portion of the city. The electric locomotives pulled east bound trains over the electrified 3.7 mile route. The steam locomotives, with banked fires, provided no tractive effort. Reaching the opposite end, the electric locomotive would cut off and the steam locomotive would resume normal operation.

The initial three 96 ton locomotives were a "steeple cab" type but, unlike later models, these pivoted at the center cab, requiring open cab doors and an apron deck between each cab half. This B-B configuration, Fig. 2-1, used a 360 hp, 600 V dc motor on each axle, and a gearless quill drove each 62 in. wheel by a soft coupling at each end of the armature. Earlier streetcars and locomotives had used trolley poles for current collection, but these bigger locomotives required a higher current density. Power was delivered overhead by a unique off-center double Z-bar section, in which ran a pantograph collector. The collection method was changed within a few years to an over-running third rail shoe. These first three locomotives remained in regular service until 1910.

Meanwhile smoke and cinder problems had reached intolerable levels in New York's Grand Central Terminal with the presence of 700 daily trains. In 1903, the New York Legislature passed legislation prohibiting use of steam locomotives in New York south of the Harlem River after 1908. The New York Central decided to electrify this "forbidden zone", and purchased the 95 ton 1-D-1's shown in Fig. 2-2. These locomotives built by Alco-G.E. had 44 in. drivers and used a new 425 hp design bipolar gearless motor on each axle. In this design, the axle was also the armature shaft, and the stator portion of the motor was mounted on the locomotive's frame. Current collection from the 600 V dc third rail was by an underrunning shoe. Thirty-five locomotives were delivered by 1906 and several are still in service today.

In these early years of the century, two basic types of electrification were developing: low voltage (usually 600 V) dc sytem promoted by G. E., and the high voltage ac system by Westinghouse. The ac-dc controversy was an extension of earlier positions taken by the two firms concerning commercial power generation. In the dc systems, high voltage ac power was converted to dc by rotary converters (motor-generator sets), then fed to a third rail at trackside where the locomotive tapped the source by a sliding shoe. Such were the B & O and N.Y.C. systems. In the other emerging system, high voltage ac power was delivered by overhead catenary and picked up on the locomotive by a pantograph. The voltage was stepped down by transformer on the locomotive and applied to the ac traction motors. The proponents pointed to the greater efficiencies of high voltage ac, but G.E. was about 10 years ahead with dc technology.

In New York, while the ac vs dc question was being debated between the New York Central and the New Haven, another front for the battle of ac vs dc was developing further west. Westinghouse had lost out to G.E. on the N.Y.C. project, but convinced the Grand Trunk Western Railroad to electrify their St. Clair River tunnel (Port Huron to Sarnia). The G.T.W. was sold on ac by the performance of the Baldwin/Westinghouse demonstrator No. 9, built in 1905. The two unit, 135 ton locomotive had six 225 hp ac traction motors and received line voltage at 6600 V ac. The motors were nose suspended, used single reduction gears and were force ventilated from a central source, as are modern traction motors. The G.T.W. subsequently bought six similar units to No. 9 for the 1908 opening of electrified service in the tunnel. Their use continued until 1958 when diesel-electrics took over.

Downstream, under the Detroit River between Detroit and Windsor, G.E. was electrifying the parallel route of N.Y.C. subsidiary Michigan Central with dc equipment. Although in planning at the same time as their Grand Central electrification, some differences arose on the Detroit River tunnel. The locomotives, Fig. 2-3, were 100 ton steeple cab B-B's with four axle hung, force ventilated traction motors, using twin reduction gears at each end of the armature. Like the Grand Central system, Detroit River used 600 V underrunning third rail collection. Initially, six locomotives were delivered in 1909, but more followed in later years. The last two weighed 125 tons and are thought to be the heaviest steeple cabs built to date. These electrics continued in service until 1953 when dieselization made power changing unnecessary.

In these early years another type of railroad electrification was being proposed: the use of three

FIG. 2-1 B&O ELECTRIC LOCOMOTIVE

FIG. 2-2 GRAND CENTRAL TERMINAL LOCOMOTIVE

FIG. 2-3 DETROIT RIVER TUNNEL LOCOMOTIVE

FIG. 2-4 GREAT NORTHERN CASCADE TUNNEL LOCOMOTIVE

FIG. 3-1 NEW HAVEN EP1 LOCOMOTIVE

phase high voltage ac. While extensively used in Italy, it was tried but once in the U.S. in the original Cascade Tunnel of the Great Northern. General Electric supplied the locomotives and electrical power generation equipment. Four B-B locomotives, Fig. 2-4, were delivered in 1909, with one 400 hp 500 V, 25 cycle, three phase, force ventilated traction motor per axle. Since the motors were synchronous, the locomotives would accelerate their trains to about 15 mph and remain at that speed regardless of load. Three phase collection was accomplished by double trolley poles with the third phase delivered through the rail. The two overhead wires were spaced 8 ft apart allowing trainmen to walk the tops of cars between the 6000 V wires. This system remained in use about 20 years until it was supplanted with a more modern single-phase ac system.

MAINLINE ELECTRIFICATION

NEW HAVEN RAILROAD

The New Haven Railroad chose to meet the restrictions of the New York Legislative Act of 1903 by electrifying with an 11,000 V, 25 Hz, single phase system with Westinghouse equipment. However, the New Haven locomotives were designed to operate also from the 660 V dc third rail of the N.Y.C. Grand Central Terminal Zone.

The first New Haven electric locomotive was built in 1905 by Baldwin and Westinghouse as the first of a 35 unit order. The New Haven units, denoted Class EP-1, were a double end, box cab design, 37 ft 6-½ in. overall length, weighing 102 tons, with a B-B wheel arrangement of two swivel trucks. Each locomotive, Fig. 3-1, was equipped with two pantographs for ac current collection, one small dc pantograph for bridging third rail gaps in the terminal, and eight third rail shoes. The shoes and pantographs could be raised or lowered by the engineman to permit transition between power sources without stopping.

Difficulties with nosing occurred as the locomotive began full operation in 1907, so a single guiding axle was added to the outboard end of each truck, and six additional EP-1 units built in 1908 were delivered with a 1-B-B-1 wheel arrangement. All EP-1 units were used in passenger service with 80 mph operation.

The traction motors were series wound commutator type, operable on either dc or ac. Although of gearless design, the armature was not mounted directly on the axle, like the G.E.-built New York Central units, but was on a hollow quill surrounding the axle with a flexible spring drive to the 62 in. diameter driving wheels. The motor frame was mounted to the truck, reducing the unsprung weight. The truck frames were outside the wheels, contrary to steam locomotive practice, providing the maximum possible space for motor length.

To determine the best design for freight service, the New Haven in 1910 evaluated four experimental locomotives by Baldwin-Westinghouse. Three of the units had a 1-B+B-1 wheel arrangement, differing from the EP-1 locomotive by articulation of the trucks. The trucks also carried the draft gear and couplers so that tractive effort was transmitted through the trucks. The fourth design was a rigid frame with 1-D-1 arrangement. Two of the 1-B+B-1's and the 1-D-1 had a new gear and quill design, with the motor completely frame mounted above the axle instead of concentric with it. The locomotive with the rigid frame 1-D-1 arrangement had two motors per axle. The fourth experimental locomotive had two articulated cabs on its 1-B+B-1 wheel arrangement with one 630 hp motor located in each cab. The drive design was similar to that of the Pennsylvania Railroad DD-1. Torque was transmitted from the motor to a jackshaft using two connecting rods, and from the jackshaft to the wheels via siderods.

The wheels on the units with geared quill drive were 63 in. in diameter, while on the side rod drive locomotive 57 in. diameter wheels were used. The locomotives weighed from 124 to 140 tons.

A fifth experimental locomotive, Fig. 3-2 was built by G.E. It also had two articulated cabs, but with two 400 hp motors in each cab and a 2-B+B-2 wheel arrangement. Each motor drove a geared jackshaft with connecting rod drive from the jackshaft crank to a single driving axle with 54 in. driving wheels.

From these experimental locomotives came the 1912 design by Baldwin-Westinghouse of 36 EF-1 freight locomotives, and three similar units geared for passenger service and equipped with train heating boilers. Only the three passenger units had third rail operating equipment. The EF-1 locomotive used the 1-B+B-1 wheel arrangement under a single, double ended box cab with draft forces taken through the trucks. It used the twin motor design with a gear drive to a quill around the axle. The EF-1 weighed 110 tons, had 63 in. drivers, and had 45 mph maximum speed. Each unit carried two pantographs for AC current collection and was equipped for MU operation.

New Haven's next locomotive class was the EP-2, also built by Baldwin-Westinghouse, using twin motors and a geared quill drive like the EF-1. The wheel arrangement was 1-C-1+1-C-1 and cast steel truck frames were introduced. These units were 69 ft overall, providing a "front porch" design for access to the box cab by end rather than side doors. Total weight was 175 tons and maximum speed was 70 mph. Five units in 1919 were followed by twelve more in 1923 and another ten in 1927.

The dc axle hung traction motor was brought to the New Haven in 1926 on five freight locomotives of Class EF-2 built by G.E. The mechanical configuration and 1-B+B-1 wheel arrangement was very much like the EF-1, although weight was in-

FIG. 3-2 NEW HAVEN EXPERIMENTAL LOCOMOTIVE

FIG. 3-3 BUTTE, ANACONDA, AND PACIFIC B+B LOCOMOTIVE

creased to 141 tons. The cab contained a motor-generator set to convert the ac power supplied to dc for the traction motors. This design concept was advanced by G.E. to take advantage of high voltage ac for power distribution efficiency and still obtain the performance benefits of the dc motor for traction.

In 1931 ten passenger units from G.E., in a new Class (EP-3), were built with the geared quill drive with twin motors for each axle. These locomotives had six twin motors on the 2-C+C-2 wheel arrangement using cast steel truck frames. The wheel arrangement had been developed by G.E. for the Cleveland Union Terminal locomotives built in 1929 and was to be continued in the Pennsylvania GGl. Six additional units of the same configuration (except for a streamlined cab) were built in 1938 as EP-4's, and in 1942 ten similar units geared for freight and for AC operation only were built as Class EF-3.

BUTTE ANACONDA & PACIFIC

Direct current electrification took another step with the use of 2400 V catenary supply on the new system installed for ore haulage between mine and smelter in the Butte-Anaconda region of Western Montana. The higher voltage permitted smaller conductors, fewer substations, and lower currents at the collector than earlier low voltage dc systems, and thus overcame some of the arguments for ac electrification systems. With the dc system the performance advantages of the dc commutating pole, series motor could be obtained. The locomotives, Fig. 3-3, were of articulated double truck design in a B+B wheel arrangement weighing 162,000 lb. The cab was a double-ended box type with both side and end doors. Draft gear was mounted on the trucks, which were of fabricated design with semi-elliptic spring and lever equalization. Overall length was 37 ft 4 in.

The traction motors were axle hung with a twin gear drive similar to that used on earlier locomotives for the Detroit River electrification. Driving wheel diameter was 46 in. and maximum speed was 35 mph. The motors were applied at 1200 V by permanent series connections. Control voltage was 600 V dc provided by a motor-generator set in the cab which was coupled to the blower to supply motor ventilation.

Each locomotive carried two pantographs equipped with steel rollers to make the actual trolley wire contact, since it was believed the heavy current for an MU consist was too much for a single sliding contact strip. Initial problems with the rollers losing lubrication and seizing were solved by installation of roller bearings, but in the 1920's a sliding contact strip made of a graphite compound was used, more in keeping with other electrification systems. The BA&P electrification was a complete success and the locomotives operated into the 1960's.

NORFOLK & WESTERN

In 1915 the Norfolk & Western introduced an electrification system to improve the coal haulage performance through their Elkhorn Tunnel in West Virgina, where 2 percent grades limited even the heaviest steam power to a crawl. Westinghouse Electric furnished the system and the locomotives, Fig. 3-4, and chose the 11,000 V, 25 Hz, single phase power supply used in the New Haven. Westinghouse chose three phase induction motors and introduced an on-board rotary phase converter to change the single phase supply to three phase for the traction motors.

Baldwin built the mechanical equipment which was arranged as two semi-permanently coupled box cab units, each with a 1-B+B-1 wheel arrangement. Each B truck carried two traction motors, mounted high on the frame to gear-drive a single jackshaft located between the two sets of drivers. The jackshaft, in turn, was coupled to the drivers with side rods, like a steam locomotive. With this design a low unsprung weight and high center of gravity were obtained to provide reduced lateral rail forces. Single axle, radial type pilot trucks at each end of the unit provided guiding for the extremely curved trackage in the mountain valleys. The rigid wheel base was 11 ft, total wheel base per unit was 43 ft, and overall length per unit was 52 ft. Driving wheel diameter was 63 in. and the draft gear was carried directly on the trucks. These locomotives were joined in 1922 by a second order, Class LC-2, built by Alco with Westinghouse equipment. The wheel arrangement was modified so that the jack-shaft was located at the outer end of each B truck and driven by only one large frame-mounted motor with side rods similar to the LC-1. These units weighed 350 tons, contrasted to 270 tons for the LC-1, and were rated at 4000 hp, compared to 2800 hp for the 1915 units. Both classes of units performed well until electrification was scrapped in 1950.

VIRGINIAN RAILWAY

The LC-2 locomotive design was repeated in most details in 1925 when the Virginian Railway also electrified at 11,000 V, 25 Hz single phase power. The Virginian coupled three units together for heavy coal haulage and used these locomotives for 30 years until replaced by ignitron rectifier EL-C units in 1956.

MILWAUKEE

Main line electrification at 3000 V dc was carried out in the mountainous territory of the Continental Divide during 1915, and service began in 1916 with the first of 30 Class EF locomotives, Fig. 3-5. Twelve similar units geared for passenger service and equipped with train heat boiler were also built. The locomotives were a twin box cab design weighing 288 tons total, with 225 tons on the 52 in. drivers. The wheel arrangement was 2-B+B and B+B-2, with eight axle-hung traction motors and twin spring gear drive. The two unit locomotive was 112 ft long and the height over locked down pantograph was 16 ft 8 in.

The Milwaukee electrification was so successful that a second section over the Cascade Range was

FIG. 3-4 NORFOLK AND WESTERN 1-B+B-1 LOCOMOTIVE

FIG. 3-5 MILWAUKEE EF FREIGHT LOCOMOTIVE

FIG. 3-6 MILWAUKEE BIPOLAR PASSENGER LOCOMOTIVE

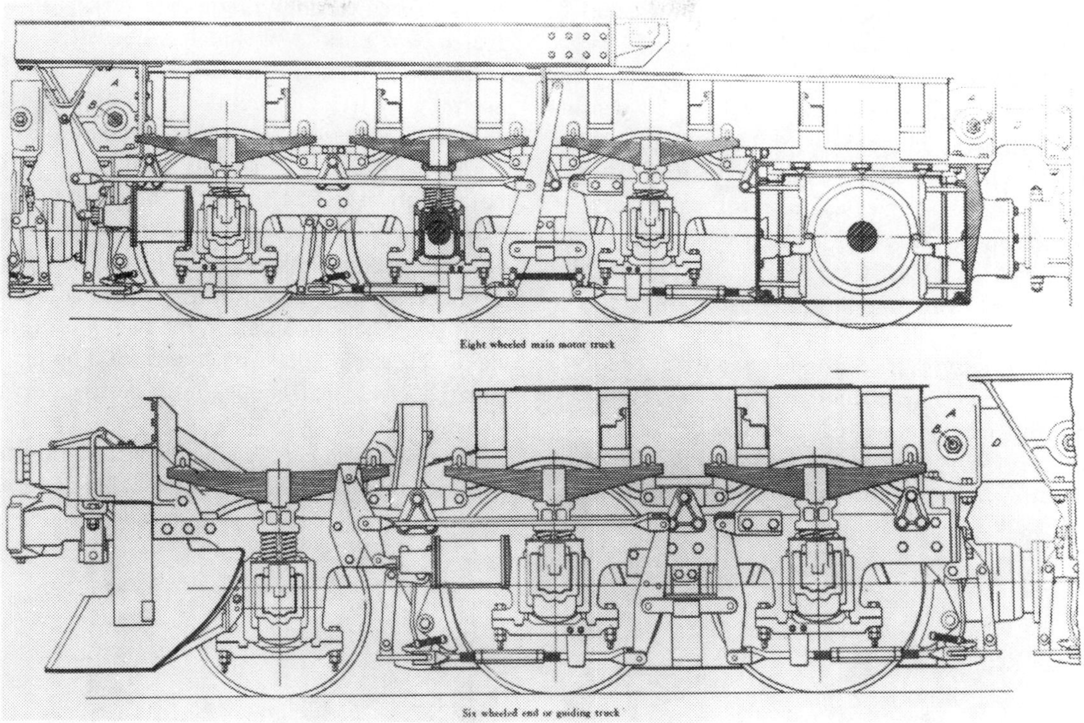

FIG. 3-7 MILWAUKEE BIPOLAR RUNNING GEAR

installed and five unique passenger locomotives, Fig. 3-6, were built in 1918. Classified EP-2, these locomotives had a center steeple cab with rounded contour end hoods for a most distinctive appearance. The carbody was in truth three sections mounted on a 1-B-D+D-B-1 wheel arrangement, Fig. 3-7, which provided 12 traction motors of the gearless, bipolar type first used on the G.E.-built New York Central locomotive in 1906. This wheel arrangement supported 260 tons and made a locomotive length of 79 ft. Rigid wheel base was 13 ft 9 in. and driving wheel diameter was 44 in. The locomotive developed 4000 hp on a one hr basis and 3000 hp continuous, and the maximum speed was 90 mph.

PENNSYLVANIA RAILROAD

Soon after work began on the Pennsylvania Railroad's New York Terminal project in 1903, President A. J. Cassatt appointed a committee to develop an electric locomotive design that could haul passenger trains over the 1.93 percent grades in the Hudson and East River tunnels then under construction. By 1905, the railroad's Juniata Shops had constructed two B-B direct-current experimentals using Westinghouse electrical equipment. Locomotive 10001 was driven by four direct-geared 350-hp motors, one per axle, while 10002 had one truck powered with two 320-hp gearless motors supported from the main journals, and the other truck with two 300-hp motors attached to truck frames.

Initial tests on the Long Island Railroad revealed nosing problems, as well as excessive lateral motion and forces at high speeds. Accordingly, a test site was prepared on the newly-electrified (1906) West Jersey & Seashore (another PRR subsidiary). Eighty special recording ties, employing a hardened steel ball on each tie to register wheel lateral forces in boilerplate, were installed in the track near Franklinville, N.J. The first two experimentals were joined by a third test locomotive, built by Baldwin in 1907 with a 2-B wheel arrangement, as well as a New Haven electric and two PRR steam locomotives. The new 1907 experimental, numbered 10003, was powered by two 375-hp gearless-quill motors.

The track tests proved the poor tracking qualities of the original two experimental designs, while also demonstrating the superiority of 10003 with its leading truck. Therefore, the locomotive chosen to initially power the new Pennsy third-rail electrification was a two-unit, coupled back-to-back, 2-B+B-2: the PRR DD1, Fig. 3-8. From 1909 to 1911 the railroad's Juniata Shops built 33 DD1's with Westinghouse electrical equipment. Mounted above the frame of each unit was a 21-ton, 2000 hp dc series motor, connected to the four 72-inch driving wheels by rods and cranks by way of a jackshaft. Maximum speed was 80 mph, with a continuous rating (at 58 mph) of 1580 hp and 10,200 lb tractive effort for the two-unit locomotive. The DD-1 chalked up an enviable performance and maintenance-cost record, finally being phased out of service beginning in the 1930's after the dc third rail west of Penn Station was replaced by ac catenary.

The Pennsylvania continued to advocate jackshaft-driven electrics through the 1920's, even though large direct-geared types had been built for use elsewhere since 1915. In the PRR's opinion, these other electrics simply lacked the sustained, high-speed capability required for their anticipated New York-to-Washington electrification. Several experimental PRR jackshaft electric locomotives were tried, including a large alternating current 1-C+C-1 with three-phase synchronous motors in 1917 and 24 1-B-B-1's built with ac/dc motors between 1924 and 1928, but each only proved the limitations of the jackshaft design. The technological break-through the Pennsylvania was seeking came in 1927 with the developing of the single-phase ac commutator-type twin motor, capable of delivering 1,000 hp per axle (500 hp per armature), yet small enough to fit between driving wheels.

Electrification plans were now hitting full stride, with three basic locomotive types being designed in

FIG. 3-8 PENNSYLVANIA RAILROAD DD1 LOCOMOTIVE

FIG. 3-9 PENNSYLVANIA RAILROAD CLASS 01 LOCOMOTIVE

FIG. 3-10 PENNSYLVANIA RAILROAD P5A PASSENGER LOCOMOTIVE

collaboration with electrical equipment manufacturers: the Class 01, the P5 and the L6. Class 01, Fig. 3-9, was planned for light passenger work. Eight of the 2-B-2's were built by the railroad's Altoona Shops in 1930 and 1931 to experiment with various motor ratings, gear ratios and manufacturer's equipments. As the first locomotives built to the PRR's master plan, the 01's proved the basic design theory to be excellent from electrical and mechanical standpoints. However, the 2-B-2's were too light for general use, and thus no further units were constructed.

Planned for heavy passenger duties, the first two P5's were assembled at the Altoona Shops in 1931. After some minor changes, 62 additional P5's were built by General Electric, Fig. 3-10 and Baldwin-Westinghouse during the following two years. These 3750 hp electrics headed all passenger schedules with the opening of the New York-to-Wilmington and Paoli electrification in January 1933. However, the P5 exhibited severe lateral forces at high speed and in addition, was not able to meet schedules with heavy trains unless a second unit was added.

The freight hauler of the trio was to be the L6, Fig. 3-11, a 1-D-1. Altoona built the first two in 1932 with single armature motors, direct-geared to the axles. Lima Locomotive Works built 30 more bodies for the 2500 hp freighter the following year, but only one of these ever received the intended Westinghouse electrical equipment and entered service. The L6 proved to be underpowered for hauling heavy freight trains and the box-cab, similar to the P5, was considered unsafe after a P5 hit a truck, killing the crew. Eventually, the 29 motorless L6 bodies were scrapped without running their first revenue mile! However, the three L6's that were completed saw freight service for a few years and then began a long career as passenger terminal switchers.

To resolve the P5's poor tracking qualities, the Pennsylvania again constructed a high-speed testing facility, this time near Claymont, Delaware. Three hundred special impact-recording steel ties were installed in the test track section, and a P5 was equipped with weigh bars and strain gauges. A test equipment car behind the P5 housed multi-element oscillographs and sensitized paper recorders. The data obtained led to changes in the truck equalization.

However, the PRR was not satisfied with just resolving the P5's immediate problem. They next borrowed a New Haven EP3, one of a class of 2740 hp 2-C+C-2's built by General Electric in 1931. Regeared to a top speed of 120 mph for testing, the EP3 recorded substantial reductions in lateral forces over even the corrected P5. These trials gave equal credence to two differing theories about basic locomotive designs. One opinion gave EP3's low axle weights the credit for good riding characteristics; the second theory stated that certain wheel arrangements (regardless of axle weights) fostered good tracking qualities.

Two sample locomotives were ordered with design emphasis being placed upon different wheel arrangements and axle loadings. Already perfected, twin-armature traction motors were improved to give the same power in half the size. Baldwin-Westinghouse built the R1, Fig. 3-12 a 5000 hp 2-D-2 rigid frame locomotive, which was essentially an enlarged P5 with one more driving axle but lower axle loadings. General Electric constructed the prototype 4620 hp 2-C+C-2 GG1, Fig. 3-13 with higher axle weights than the EP3, but still lower than the 2-D-2. Both had streamlined superstructures, and were driven by geared quills. The R1's traction motors had the same rating as those used in the P5's, while the GG1 employed more motors with lower ratings.

Both locomotives proved to have adequate power, but close analysis of the lateral force data disclosed that the GG1 possessed a slight edge over the R1. The GG1's articulated chassis was responsible for at least a portion of that difference and as a result, the GG1 was nominated for fleet production as the standard heavy electric class of the railroad.

General Electric built fourteen more in 1935,

FIG. 3-11 PENNSYLVANIA RAILROAD L6 LOCOMOTIVE

FIG. 3-12 PENNSYLVANIA RAILROAD R1 LOCOMOTIVE

FIG. 3-13 PENNSYLVANIA RAILROAD GG1 LOCOMOTIVE

while between 1935 and 1943 the Altoona Shops constructed another 124 with G.E. and Westinghouse electrical equipment. Without question, the GG1's became the most famous and successful fleet of electric locomotives in the country. Thirty-five years after the last GG1 left the erecting shop, 76 percent of the class still see service, including the original GG1 of 1934.

POST WORLD WAR II ELECTRIFICATION

In 1947 General Electric built two 5000 hp units for the Great Northern, Fig. 4-1, rated at 180,000 lb starting tractive effort, and weighing 720,000 lb with a B-D+D-B wheel arrangement. These MG locomotives could be run in multiple and individually out-performed two 4-8-4 steam locomotives. As with several other electric railroad operations, run-through trains powered by diesels displaced these locomotives because they eliminated the need to change motive power at either end of the electrified zones.

The Virginian Railroad also took delivery of electric locomotives in 1947. These units, designated EL-2B's, were two semi-permanently coupled cab units rated at 6800 hp. They had a maximum speed of 50 mph and were equipped with regenerative braking equivalent to 7800 hp. Many diesel locomotive features were included, such as side equalized swing hanger trucks, and axle hung traction motors.

The Pennsylvania Railroad, with one of the major electrified zones in the country, began thinking about new power in the late 40's and early 50's. In 1951 both Westinghouse and G.E. produced prototypes for evaluation on the Pennsylvania Railroad. General Electric produced three 2-unit, single phase ac electric locomotives ratėd at 5000 hp (for each 2-unit consist) designated E2b, Fig. 4-2. These locomotives had ac commutator motors, a concept used on previous PRR locomotives. These locomotives introduced the stressed side sheet concept, where a 5 percent increase in side sheet thickness to allow it to be used as a load carrying member allowed a 40 percent saving in platform weight.

Westinghouse also produced two 2-unit locomotives. The E3b had a B-B-B wheel arrangement while the E3c had a more conventional C-C arrangement. These locomotives, rated at 6000 hp (for two units), were the first to introduce mercury arc ignitron rectifiers in this country.

The rectifier locomotive became a production

FIG. 4-1 GREAT NORTHERN B-D+D-B LOCOMOTIVE

FIG. 4-2 PENNSYLVANIA RAILROAD E2B LOCOMOTIVE

FIG. 4-3 NEW HAVEN RECTIFIER LOCOMOTIVES

FIG. 4-4 NEW HAVEN RECTIFIER LOCOMOTIVE TRUCKS

FIG. 4-5 VIRGINIAN EL-C LOCOMOTIVE

reality with a run of ten locomotives built by G.E. for the New Haven Railroad in 1955, Fig. 4-3. They were dual voltage, operating either on the 11 kV 25 cycle catenary northeast of New York City, or on 600 dc third rail while entering Grand Central Terminal. Figure 4-4 shows the unique truck used with this locomotive: a tri-mount, side equalized, swing hanger, with the equalizers in the side frames, and designed for a 90 mph top speed.

In 1956 the Virginian Railroad added to their electric locomotive fleet by purchasing 12 EL-C locomotives, Fig. 4-5. These locomotives were the first to employ the road switcher configuration which was then popular for diesels. They had a tri-mount side equalized truck and could MU up to four in multiple with each other and with the earlier EL-2B's. Unlike the regenerative braking of the EL-2B's, the EL-C's had dynamic braking (dissipating generated energy in resistors).

The largest production run of electric locomotives in the post World War II period was the construction of 66 4400 hp locomotives for the Pennsylvania Railroad. These E44 locomotives, Fig. 4-6, were very similar to those produced in 1956 for the Virginian. Developments in solid state rectification led to replacement of the ignitron rectifier and its water cooling system by an air cooled silicon diode recitifers on the 36th unit. They were successful and so the last five units also had silicon diode rectifiers.

Another electric locomotive milestone occurred in 1968 with the construction of two E50C locomotives for American Electric Power, Fig. 4-7. These thyristor control locomotives were designed for 25 kV, 60 Hz single phase operation, the first commercial frequency venture in the United States. In addition, the locomotives are fully automated and operate on a 15 mile railroad in Ohio, transporting coal between a mine and the power generating station. A logical extension of electrification was the use of 60 Hz power at a higher voltage. The Black Mesa and Lake Powell Railroad was opened in the early seventies with a catenary voltage of 50 kV. Since transmission losses were minimized, only one substation was needed for the 78 mile automated system. In 1973, General

FIG. 4-6 PENNSYLVANIA RAILROAD E44C LOCOMOTIVE

FIG. 4-7 MUSKINGUM RAILROAD E50C LOCOMOTIVE

FIG. 4-8 BLACK MESA AND LAKE POWELL E50C LOCOMOTIVE

FIG. 4-9 AMTRAK E60CP LOCOMOTIVE

FIG. 4-10 GM6C DEMONSTRATOR LOCOMOTIVE

FIG. 4-11 GM10B DEMONSTRATOR

Electric supplied three single end box cab units, shown in Fig. 4-8, and delivered another three in 1976.

Since the famous GG1's were approaching retirement age, in 1974 Amtrak ordered 26 E60C locomotives, Fig. 4-9. Rated at 6000 equivalent diesel hp, these locomotives are capable of running on the several present and future power supplies used on the Northeast Corridor, including 11 kV, 25 Hz; 12.5 kV, 60 Hz; and 25 kV, 60 Hz. A first for a production run domestic locomotive was the use of a rotary air compressor rather than the usual reciprocating type. Stressed side sheet cab construction was used, with the locomotive riding on floating bolster three axle trucks. Lateral as well as yaw dampers were added to the truck for operating up to 120 mph. Axle hung dc traction motors with antifriction bearing axle suspensions were employed on this locomotive.

Recently, General Motors has produced two prototype electric locomotives: the GM6C (Fig. 4-10) and the GM10B (Fig. 4-11), both featuring thyristor control by ASEA of Sweden. The mechanical design of the GM6C is similar to existing diesel electric locomotives but the GM10B uses a B-B-B wheel arrangement. The fabricated truck of the GM10 locomotive has frame mounted traction motors to reduce unsprung weight. Both locomotives are now in revenue demonstration freight service on Conrail.

SUMMARY

Most of the electrified railroad lines have now disappeared as diesel electric locomotives overcame the tunnel and grade problems and offered overall operating economies. However, the basic design developments – running gear, traction motors, MU control, – which are features of today's diesel fleet, were a product of electrification. Without the traction motor, which by succeeding stages of improvement on electric locomotives became compact enough to fit between the wheels in high horsepower axle hung design, diesels would not have so readily supplanted the steam locomotive. The recent electric locomotives built have borrowed back those technologies further developed by the diesel-electric over its 40 years of main line service. The future for electrification in the United States is clearly not certain, but it is certain that the locomotives of today owe fundamental design principles to the years of invention and development on electric locomotives.

ACKNOWLEDGEMENT

We wish to thank the History and Heritage Committee, Erie Section, ASME, for their cooperation in helping us put this article together.

BIBLIOGRAPHY

1. Middleton, W. D., *When the Steam Railroads Electrified,* Kalmbach Publishing Co., Milwaukee, Wisconsin, 1974.
2. Haupt, F. J. G., *The Early History of the Electric Locomotive,* publisher and date of publication unknown.
3. Baldwin Locomotive Works and Westinghouse Electric and Manufacturing Company, *Electric Locomotives,* J. B. Lippincott, Philadelphia, 1896.
4. Davis, J. H., "Review of Operating Data on B&O RR Electrification," *General Electric Review,* January 26, 1925.
5. Knoll, Charles M., *The Water Level Route,* Rochester Chapter of the National Railway Historical Society, Rochester, New York.
6. Middleton, William D., "Classics Under Catenary (and beside third rail)," *Trains,* July 1970, pp. 22-25.
7. Droege, John A., *Passenger Terminals and Trains,* New York, McGraw Hill, 1916.
8. Pinkepank, Jerry A., "A Tale of Two Tunnels – 2," *Trains,* October 1964, pp. 43-47.
9. "The Electrical Equipment of the Detroit River Tunnel," General Electric Company, May 1911, Bulletin No. 4834.
10. "The Electrification of the Cascade Tunnel of the Great Northern Company," General Electric Company, June 1910, Bulletin No. 4755.
11. Byron, Carl R., "Hoosac Tunnel – The Mohawk Trail by Rail," *B&M Bulletin,* Fall 1973, pp. 9, 10.
12. Middleton, William D., "CN (of All Roads) Under Catenary," *Trains,* March 1975, pp. 20, 21.
13. Staufer, Alvin F., Pennypacker, Bert, and Flattley, Martin, *Pennsy Power,* Alvin F. Staufer, Medina, Ohio, 1962, pp. 246-298.
14. Staufer, Alvin F. and Pennypacker, Bert, *Pennsy Power II,* Alvin F. Staufer, Medina, Ohio, 1962, pp. 118-154.
15. *The Pennsylvania Railroad Electrification (New York – Washington),* General Electric Company, Schenectady, 1937.
16. Gowans, F. D., Widell, B. A., Bredenberg, A., "A New Electric Locomotive for the Pennsylvania Railroad," *AIEE Technical Paper 52-23,* Dec. 1951.
17. "Flat Tops, Round Noses, and Square Heads on the Vgn," *General Electric Review,* May 1958, pp. 34-36.
18. Ephraim, Jr., M. and Quinn, H. E., "Electrification and New Electric Locomotive Designs," *ASME Paper 76-WA/RT-7,* 1976.

19. Gowans, F. D., "Rectifier Locomotives for the New York, New Haven and Hartford RR.", *AIEE Transaction Paper 55-206,* 1955.

20. "The Electrics are Doing All Right," *General Electric Review,* Sept. 1955, pp. 56-60.

6 The Development of Remote Multiple Unit Locomotive Control

C. W. PARKER

INTRODUCTION

The economy of railroad transportation of bulk commodities requires their movement in trains as long and heavy as can be handled safely and expeditiously.

As the strength of conventional couplers and drawgear elements effectively limits the tonnage behind head-end locomotives, the time-honored method of increasing train tonnage has been to introduce one or more locomotives within the body of the train. Prior to the 1960's, such in-train locomotives were manned by an engine crew and operated independently, depending originally on whistle signals but more recently on voice radio for communication between the enginemen to synchronize their handling of the locomotives.

Helper or assisting locomotives, as these in-train locomotives are termed, continue in use in many countries including Canada, the U.S. and the U.S.S.R. where certain freight trains grossing 10,000 tons or more are handled over undulating territory with the assistance of a manned helper locomotive. However, many railroads, principally in Canada and the United States, have adopted the radio-controlled remote locomotive to increase train tonnage while at the same time improving brake system effectiveness and train handling.

HISTORICAL DEVELOPMENT OF RADIO CONTROL OF REMOTE LOCOMOTIVES

As early as 1957, Southern Pacific Railroad engineers discussed with Electro-Motive Division and Union Switch & Signal Engineers the possibility of applying radio control to an assisting locomotive in the body of a train. The result was a scheme employing coding devices as used in centralized traffic control to obtain eight steps of throttle control in forward and reverse. Dynamic brake was similarly controlled, but no proposal was made to radio commands to the air brake controls in the assisting locomotive.

Before this project attained implementation, the first actual demonstration of radio control of a remote locomotive was carried out on the Great Northern Railway in 1959, utilizing radio to relay operations of the lead unit throttle to a remote unit although a rider was employed on the remote unit to govern the independent brake. The system was operated successfully in freight trains working through the Rocky and Cascade Mountains.

Southern Pacific then instituted radio remote control on helper locomotives operating through Tehechapi Pass, and extended the control system to the independent brake on the radio-controlled helper consist. The demonstration in June 1961 was successful but was terminated due to representations by organized labor. Later in the same year, a similar demonstration was made on the Santa Fe, which was concluded for the same reasons, but furthered the state of the art by proposing the inclusion of independent motoring and brake control for the remote locomotive consist.

About this time, certain steel and iron ore producers queried the Westinghouse Air Brake Company on the possibility of operating ore trains grossing 35,000 to 40,000 tone from Minnesota to Pittsburgh with as many as four separate radio-controlled in-train locomotives to minimize tension and compression forces in the draft rigging throughout the train. Exploratory engineering development conducted by Westinghouse Air Brake and Union Switch & Signal Divisions concluded that the proposal was impractical from a braking standpoint, but that the synchronized control of one in-train brake valve was indeed feasible.

The Southern Railway, having followed the earlier demonstrations, then commissioned a demonstration set of radio control equipment through EMD and Westinghouse Air Brake, which resulted in the first operation of a through train employing radio control in July 1962 between Bulls Gap, TN and Asheville, NC. Remote control was applied to the throttle and independent brake but not to the automatic brake. Handling 200- to 220-car trains of 18,000 tons instead of the conventional 12,000 to 13,000 tons, the road tests were successful, but the operating environment proved too severe for some of the electrical equipment and the trials were discontinued in 1963.

The basic feasibility of remote radio control of in-train locomotives having been established by this time, the Southern Railway ordered a set of prototype equipment from North Electric Company of Galion, Ohio, specifying the use of solid state electronic apparatus where possible in place of the relays used in earlier equipment. Using Westinghouse pushbutton-type brake valves, brake control centres, and interface units, and a solid-state keying system developed by Southern Railway communications engineers for the radio transmitters, North Electric produced a prototype that was installed on diesel locomotive No. 2600 and control car R-3 and made its first run on train No.

156 between Birmingham and Atlanta on March 12, 1964. The failure of soldered joints marred operation until corrective measures were introduced, but from July 1964 onward the system operated successfully. In December, 1964, North Electric Company sold its controls manufacturing to Radiation, Inc. of Melbourne, Florida, and shortly afterward the Southern Railway ordered 16 sets of equipment, named Locotrol, from Radiation, Inc., which included solid-state Motran radios from Motorola. This order, the first for production equipment for radio control of remote locomotives, was followed before the end of 1965 by an order for 50 more sets. On January 1, 1966 the Southern Railway operated its first radio train down Saluda Mountain, beginning an era of large-scale operation of radio-controlled remote locomotives.

Wabco, meanwhile, having engineered and produced all the braking components for the Locotrol equipment ordered by the Southern Railway, completed its own version of remote control equipment, the RMU 580, incorporating solid state logic and coding. Three sets were sold to the Louisville and Nashville Railroad. Demonstrator sets were operated successfully on the New York Central, Southern Pacific, Pennsylvania Railroad, and CP Rail. None of these sets remains in service, but two sets which were sold to the Vale Rio Doce (CVRD) ore-hauling railroad in Brazil are understood to be still operating. Locotrol equipment, however, went on to secure the major portion of the North American market with large orders from the Burlington Northern, CP Rail, British Columbia Railway, Kansas City Southern, Santa Fe, Union Pacific, and the Quebec North Shore & Labrador, and by 1976 had delivered or were negotiating for a total of 408 lead unit and 386 remote sets of equipment. The Wabco RMU production was discontinued in 1970, leaving Radiation Inc.'s successor, Harris Controls, using principally New York Air Brake's braking components, in sole possession of the radio remote control equipment market.

During the testing phase some interesting feats were recorded. In 1967 the Pennsylvania Railroad operated a 341-car coal train grossing 25,500 tons using three diesel units on the head end assisted by six radio-controlled units positioned two-thirds of the way back in the train. Before the year was over, the Norfolk & Western marshalled 500 loaded coal cars totaling 48,000 tons into a train four miles long. Using three 3600-hp diesel units on the head end and three more 300 cars back under Locotrol control, the train was operated over a distance of 159 miles from Iager, W. Va. to Portsmouth, Ohio without undue difficulties. In 1969 CP Rail, accustomed to marshalling five remote units together in its unit coal trains over its mountainous route through British Columbia, increased the remote consist to six 3000-hp units and, with four units on the head end and a manned pusher consist of three more units trailing, operated successfully over a 90-mile stretch including ascending and descending 2.2 percent grades, under both RMU and Locotrol radio control.

Train movements of the latter variety invariably give rise to the suggestion that not one, but two or more separate locomotive groups be spaced back in a long train under radio remote control. Indeed, such an operation was demonstrated in 1966 when the Southern Railway placed in service a Model 203 Locotrol equipment handling two remote consists in one train. The train handling proved too cumbersome, however, and the trials were discontinued after a few runs.

Overseas the use of Locotrol equipment has extended to Australia, where seven sets have been delivered to the Queensland Government Railway and one set to the Mount Newman Mining Company Railway; to the CVRD in Brazil (two sets) and the National Railways of Algeria (one set); and negotiations are proceeding for the sale of Locotrol equipment to railways in Iran and Liberia.

DESCRIPTION OF EQUIPMENT

The equipment added to the leading unit of the head-end locomotive consist and to any one of the units in the remote consist (or installed in a separate control car connected to the remote consist) includes the following:

Head end equipment	Remote equipment
Remote control console	Power supply
Air brake pushbutton console	Brake control center
Power supply	
Control supply	Logic cabinet no. 1
Brake control center	Logic cabinet no. 2
Logic cabinet no. 1	Relay interface cabinet
Logic cabinet no. 2	Two radio transceivers
Relay interface cabinet	
Two radio transceivers	

The operation of this equipment is depicted diagramatically in Fig. 2. Radio signals between the two locomotive consists perform the same

FIG. 1 REMOTE RADIO-CONTROLLED UNITS WITH CONTROL CAR IN CP RAIL COAL TRAIN CROSSING STONEY CREEK BRIDGE

functions as multiple-unit hoses and jumper cables between coupled locomotive units. Movements of the lead unit throttle or of the brake pushbuttons are translated into electrical responses that are sensed in the logic cabinets and processed into coded signals that are transmitted by radio to the remote locomotive.

At the remote consist the radio signals are received by a transceiver, processed by logic equipment, and by means of the interface relays are applied to the power control or air brake control circuits to cause the remote locomotive to operate in synchronism with the lead consist or in an independent mode if such is selected by the engineman.

All the transceivers in use today operate on one of two standard frequency bands, either 160 or 450 MHz with a band width of 3 kc. So that one train does not conflict with another, each set of master and remote transceivers is mated through an address code. One of 2046 different printed circuit address selector cards is installed at the master station. At the remote station the corresponding code is selected by means of four thumbwheel switches in the logic cabinet. This arrangement allows as many as 2046 separate systems to operate, if necessary, in the same territory without mutual interference.

OPERATION OF EQUIPMENT

In the operating cab of the lead locomotive all the normal controls including the throttle, reversing lever, sanding control, whistle, and electrical switches are retained and are operative normally. The conventional automatic and independent air brake controls remain in place but are locked in running position. Their function is supplanted by a brake control panel which employs pushbuttons to apply and release the locomotive and train brakes through intervening electrical controls.

In Fig. 3 is illustrated a representative arrangement of the controls at the engineman's station. Except for locking pins applied to the air brake valve handles, the conventional control equipment is unchanged. On top of the control stand are mounted the air brake pushbutton panel and the remote control console.

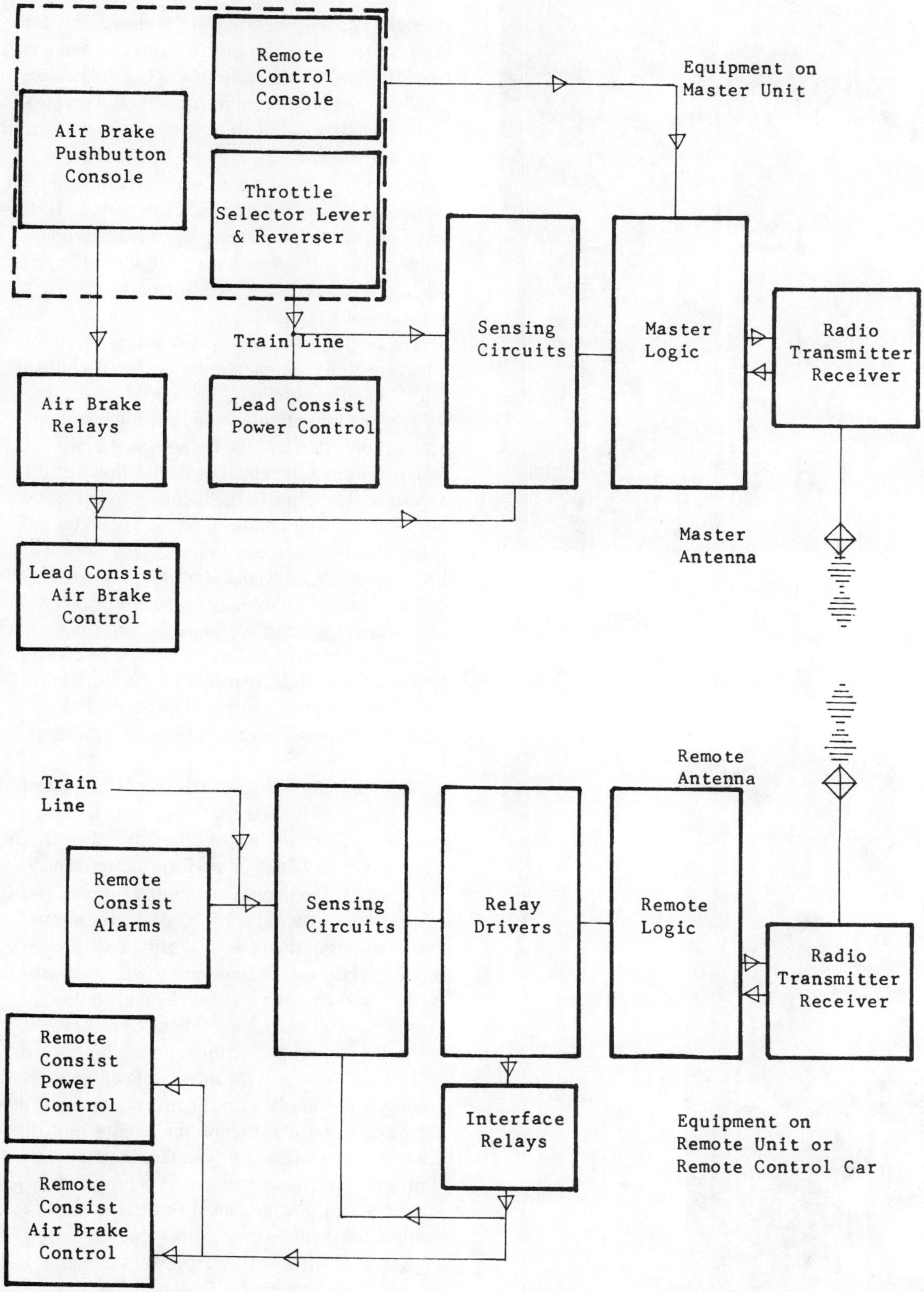

FIG. 2 DIAGRAM OF LOCOTROL EQUIPMENT

FIG. 3 CONTROLS AT THE ENGINEMAN'S STATION

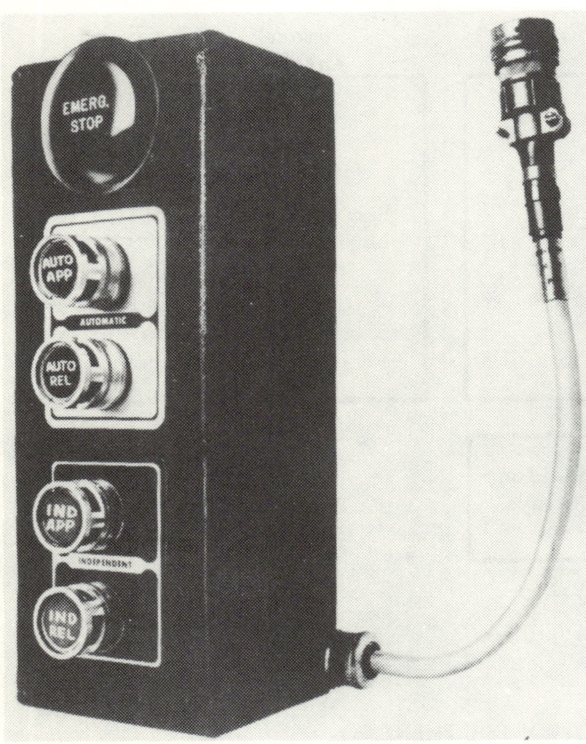

FIG. 4 AIR BRAKE CONSOLE

As illustrated in Fig. 4, the top button of the air brake pushbutton group is the emergency brake button. Depressing this button causes an emergency brake application throughout the train regardless of whether the remote equipment is turned on or off.

Immediately below the emergency brake button is the automatic brake application pushbutton. Depressing this button for 2 sec causes a minimum reduction. A longer depressed time causes a further service reduction, the extent of which is proportional to the interval that the button is depressed. When this button is released, the brake valve is automatically lapped.

The next lower button is the automatic brake release button; a momentary push on this button causes an automatic brake release.

The two lowest buttons are the independent brake application and the independent brake release. Pushing the upper or application button applies at least 5 psi to the locomotive brakes on both the lead and remote consists. The extent of the application increases as long as the button is held depressed, up to the limiting setting of brake cylinder pressure. Pushing the lower button releases the locomotive brakes on both the lead and remote consist. If, however, the lead consist brakes were applied by means of the independent brake valve, a push on the independent brake release button will release only the remote consist brakes.

Including the Wabco RMU console, a total of five versions of remote control console have been installed on locomotives of different railroads. The basic functions which all perform are to display the status of the remote locomotive consist, to cut in or cut out its air brake feed valve, to indicate malfunctions in the remote consist, and through a mode selector switch to govern the throttle settings of the remote engines. The differences between remote control consoles are dictated by the operating preferences of the railroads concerned.

It is not necessary for the remote consist to react synchronously with the master consist. When the mode selector switch on the remote control console at the engineman's station is set at "MU" position, synchronous operation is provided. When it is advanced to a numbered throttle setting, the remote units will operate at that throttle setting regardless of the throttle setting on the master consist. When it is set to IDLE position, the remote engines remain in idle and do not follow the throttle changes made at the master consist, but will obey all air brake commands. When the

FIG. 5 REMOTE CONTROL CONSOLE AT THE ENGINEMAN'S STATION (ONE VARIANT)

mode selector switch is in ISOLATE position, the remote engines remain in idle, but there is no response to air braking commands except an emergency brake application.

In normal operation on most railroads, the mode selector switch is kept in MU position, only assuming an independent mode when physical characteristics of the territory require that the remote consist be operated at a throttle position different from the master consist. By contrast, one railroad does not permit independent operation of the remote consist, preferring MU operation at all times. Another railroad does not use MU operation but requires the throttle setting of the remote engines to be governed at all times by adjustments of the mode selector switch.

Radio continuity is, of course, the essential link in remote locomotive control. When all the equipment is functioning properly, the engineman will have a small number "1" light up intermittently on the remote control console as the radio on the master consist transmits data to the remote consist every 21 sec and immediately receives an acknowledgment as a continuity check.

If a radio malfunction occurs, the second pair of transceivers comes into operation and the number "2" shows on the control console. In the event the radio link is entirely disrupted, as in the case of a radio breakdown, a defective printed circuit, or a physical obstruction such as a long tunnel, and if the remote unit does not receive a message from the master consist for 45 sec, the air brake feed valve at the remote station would "drop out" or close and the remote engines would step down to idle unless some overriding control were actuated to keep the remote engines operating.

Radio breakdowns are not infrequent in the operation of remote control systems, usually due to frequency shift. Temporary impairment of communication can occur in territory characterized by rock bluffs and severe curvature, or in operation through tunnels. Whatever the reason for the loss of continuity, protection against automatic shutdown of the remote consist is desirable in most cases. Early installations of Locotrol had a manual override feature which the engineman could operate to ensure continued operation of the remote consist for approximately 30 min if he pressed the override button before loss of continuity occurred. This feature was acceptable under conditions when the engineman could anticipate loss of continuity, but there were many occasions when continuity was lost unexpectedly.

Remote systems now have an automatic override feature which does not require the engineman to anticipate loss of radio continuity, but will keep the remote engines operating for about 30 min in whatever operating mode – power or braking – they were employing when continuity was lost. During the no-continuity interval, however, the override feature does not provide any means of altering the operating mode of the remote consist. If the engineman desires to alter the remote consist operating mode during a no-continuity interval, he has no means of communication with the remote consist except through the train brake system. Accordingly, the remote apparatus is designed so that during operation in override, any significant flow of air through the brake pipe will cause a shutdown of the remote consist. The engines will be reduced to idle and the feed valve will drop out. This action can be initiated by making a train brake application or, if one is already in effect, by releasing it or by making a heavier application. This is not an ideal solution to the no-continuity problem, but it is the best available under existing circumstances.

OPERATIONAL ADVANTAGES OF REMOTE MULTIPLE UNIT CONTROL

The possibility of increasing train tonnages is not the only advantage due to the use of remote in-train locomotives. The potential for operating the two locomotive consists in different modes offers opportunities for vastly improved train handling over undulating territory. Finally, radio control of brake system operation at the remote consist provides faster charging of the train brake system, improving braking response, and shorter stopping capacity.

As the remote locomotive implements a brake application or release at virtually the same time as this action is initiated on the lead locomotive, the overall time required to apply or release the train brakes is considerably shorter than in a conventional train. Depending on the location of the remote locomotive – and two-thirds back is the optimum location for braking purposes – the time required to initiate a brake application can be reduced as much as 50 percent while the release time can be reduced by as much as 65 percent. In the case of long trains with considerable brake pipe leakage, the use of a remote locomotive improves brake cylinder pressures at the rear under heavy brake applications.

A considerable saving can be made in the time required to charge the brake system initially. Here the reduction can be as much as 70 percent.

The operation of remote in-train locomotives is attractive to railroads operating in very cold climates where the lengths of trains in winter are drastically restricted because of the increased brake pipe leakage induced by cold weather which causes the brake pipe gradient in conventional trains to exceed the 15 psi (100 kPa) permissible maximum.

LOCATION OF THE REMOTE LOCOMOTIVE

Since the two-thirds location in the train provides optimum braking support, most remote locomotives from the earliest days of radio operation have been inserted at the two-thirds point. For operation on severe mountain grades, the usual method was to place the remote consist so as to pull its portion of the total tonnage equal to its proportion of the total tractive rating.

The latter procedure, however, generates a slack condition immediately ahead of the remote consist which can be potentially disruptive of drawgear components. Few trains proceed with all cars moving at all times at the same speed; when differences occur, slack action develops. In the case of the radio-controlled train marshalled as "two trains connected," a "node" or point of zero drawbar force normally exists at the front of the remote consist but moves forward or rearward depending on changes of grade, curvature, throttle position or rail surface condition. If the head end of the train encounters a brief adverse gradient it will slow down. For a short interval the remote consist will be moving faster than the head end and will close up the slack, causing the node to move toward the head end. Once the head end is past the point of restriction it will speed up, especially since the remote consist is slowing down under its increased burden. When the head end is moving faster than the rear, it will start to "pick up" the cars that the remote locomotive was pushing, accelerating each in turn to the speed of the lead consist. Backward moves the node until it reaches the front coupler of the remote consist. If at that instant there is a significant difference in speed between the two parts of the train, the resistance of the heavy remote locomotive to being suddenly accelerated will develop a peak coupler force of considerable magnitude. Many knuckle or coupler failures have occurred under this condition. The fact that all drawgear parts of the train behind the front coupler of the remote consist are in tension at this instant contributes to the likelihood of drawgear failure.

There are two solutions to this problem. Either relocate the remote locomotive forward so that the entire train should always be in tension, or move it rearward to place the node in normal operation at a safe distance ahead of the remote consist.

Fig. 6 illustrates a marshalling arrangement in which the remote consist is inserted ahead of the point at which the drawbar pull of the lead consist would be exhausted. In the traction mode, therefore, the train is kept in tension throughout its length and there is normally no "node" to surge back and forth, although with rough handling a temporary slack condition conceivably could develop fore or aft of the remote consist. This

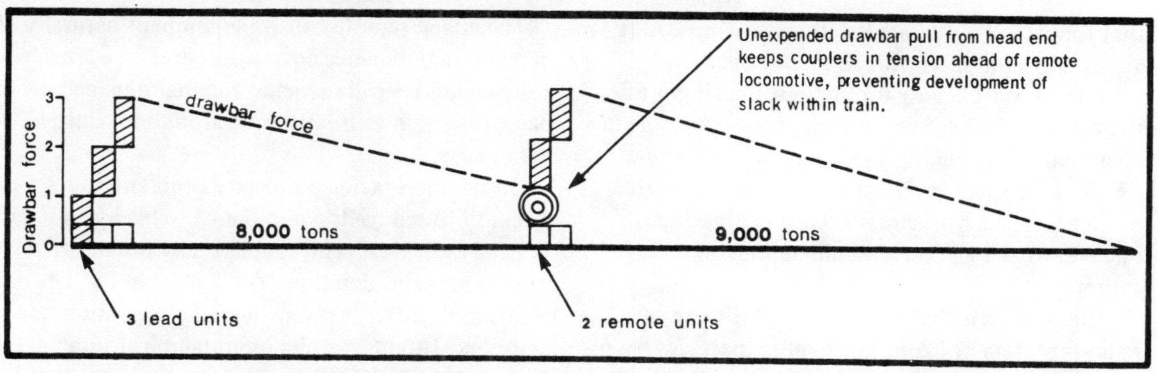

FIG. 6 WITH THE REMOTE UNIT PLACED AHEAD OF THE POINT WHERE A NODE WOULD DEVELOP, THE ENTIRE TRAIN IS NORMALLY IN TENSION. A MINIMUM SLACK CONDITION PREVAILS.

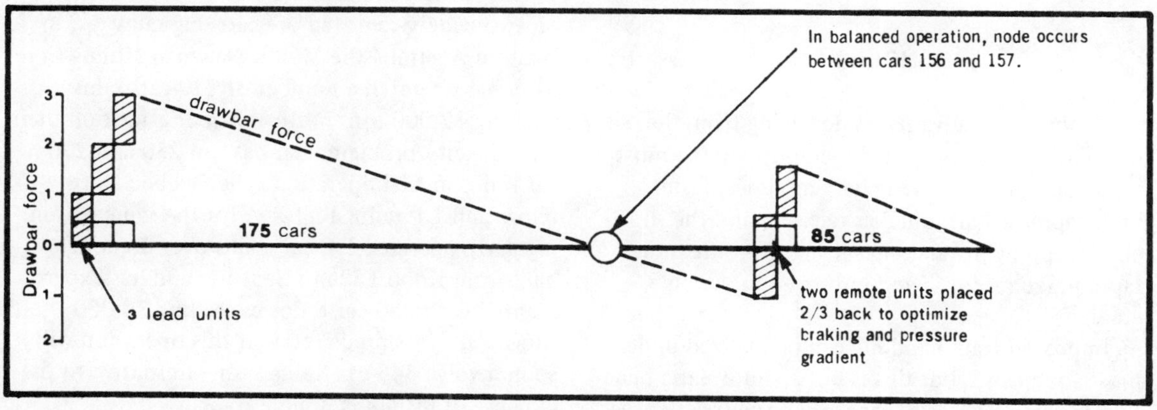

FIG. 7 ORE TRAIN OF 260 CARS HAS REMOTE CONSIST PLACED AT ABOUT THE TWO-THIRDS POINT FOR OPTIMUM BRAKING. MOTIVE POWER IS DIVIDED 60/40 TO PLACE THE NODE WELL AHEAD OF THE REMOTE CONSIST.

arrangement is recommended for trains of moderate length.

For extremely long trains, especially in winter, having a need to place the remote consist at the two-thirds location for optimum brake pipe support, the no-node condition can be attained only if the tractive capacity ratio of the lead consist to the remote consist exceeds 2:1. For a very heavy train such as that illustrated in Fig. 7, requiring five or more 6-axle units, the foregoing requirement regarding power ratio may be difficult to establish, especially if the railroad does not want more than three or four units on the head end. Under these conditions it is accepted that a node will occur ahead of the remote consist. To minimize its knuckle-breaking potential, the total number of units is split so that the remote consist contains more than one-third of the total power. This forces the normal position of the node to be at some distance ahead of the remote consist.

OPERATION THROUGH TUNNELS

Short tunnels, particularly on relatively level territory, present no problems to the operation of remote locomotives as long as the train keeps moving and the override feature is operative. If the train stops with one of the locomotive groups within a tunnel, and more than 100 ft from the tunnel mouth, effective radio contact may not be retained. At the conclusion of the automatic override interval (usually 30 min) or if a train brake application is made, the remote locomotive will shut down and its feed valve will close and cannot be cut in again by radio as long as continuity is lost. Under these conditions it is necessary to place a man on the remote locomotive to handle the controls manually or to release the brakes so that the train can be moved far enough to regain continuity.

As radio contact can usually be maintained within the first 100 ft of tunnel length, it follows that tunnels of moderate length exceeding 200 ft are usually equipped with an antenna suspended from the tunnel ceiling throughout the length of the tunnel. Much longer tunnels, say over 1 mile in length, and particularly those located on a grade where train brake applications may be made, are often equipped with a radio repeater station at one end to assist in maintaining radio communication.

The Southern Railway developed the use of radio repeaters in 1966. Logic radio messages from the consist in the tunnel were conveyed via the fixed antenna to a radio station outside the tunnel and "repeated" to the other consist, and vice versa.

CONCLUSION

Of the many advantages devolving from the use of remote radio-controlled locomotives, the most significant are improved charging times, faster brake applications, quicker releases, and the shortened stopping distances resulting from the more rapid brake response and improved brake pipe gradient.

Improved train handling can be realized under most conditions, but this is not an automatic benefit. On certain kinds of territory only experienced and skillful handling will avoid worse train slack conditions than the average conventional train incurs. On the whole, however, given intelligent handling, a train employing radio-controlled in-train locomotive units should obtain better control of slack than a conventional train under similar circumstances.

The added remote control equipment naturally increases maintenance costs. Moreover, the apparatus demands special servicing and maintenance facilities as well as frequent attention by skilled technicians.

On rail lines facing a capacity problem, the necessity to reduce the number of trains by increasing individual train tonnage may outweigh every other consideration. The Canadian Pacific, for example, after introducing radio operation on its unit coal trains over its mountainous route through British Columbia, converted all of its bulk commodity trains westward from Calgary to the Pacific Coast to radio operation to relieve a traffic glut on this single-tracked line.

The longest radio trains, however, are operated on two widely-removed ore carrying railways. In Western Australia the Mount Newman Mining Company has operated a number of 230-car trains grossing 32,000 tons in this form of attack on their line capacity problem, and has run 250 and 270-car trains in trial operation. The Quebec North Shore and Labrador Railway, for the same reason, regularly operates 230-car trains over the long haul from Knob Lake to Sept Iles and has experimented with 280-car trains exceeding 39,000 short tons. It is predicted that this operation will soon involve 345-car trains as an alternative to the expense of installing double track.

In general, when used strictly to gain the operating advantages for which it is best fitted, and when applied to operations where these advantages can be exploited to maximum effect, remote radio-controlled locomotives can provide excellent benefits and produce an operation unexcelled by any other means.

7 Canadian Rolling Stock Developments

K. J. ARREY and W. G. BLEVINS

Development of railways in Canada generally proceeded apace with those in the U.S. in the Windsor-Toronto-Montreal area from the 1850's onward. Canadian railroads were active participants in the Master Car Builders Association from its formation in the late 1860's. With construction and completion of Canadian Transcontinental lines in 1880-1915, interchange of rolling stock with the U.S. became even more important. Geographical conditions are similar and through-put requirements matched those of the central and northern U.S. transcontinentals. However, winter operation physical conditions generally have longer periods of severe weather than normally experienced south of the border.

A number of interesting developments in locomotive and car design were introduced to North America or first tried out on modest to large scale in Canada.

The Rotary Snow Plow was invented in the early 1870's by a Toronto dentist. The principle is the same as now employed world wide. A rotating set of adjustable cutter blades pulverizes packed snow and imparts velocity to it in order to throw a stream of snow many feet clear of the right of way. The rotary fan blades were driven by a multi-cylinder steam engine, generally with a medium size locomotive type boiler providing steam supply. The typical rotary plow is pushed by one or more road locomotives. Modern day versions have substituted a couple of diesel electric locomotive traction motors for the steam engine and electric power supply is taken from a typical locomotive style 2000-3000 hp diesel driven generator set. The D.E. rotary plow is also pushed by an appropriate consist of diesel electric road locomotives.

Rotary plows are not as essential as they were with steam locomotive power since it is fairly easy to equip road diesel locomotives with "V" wedge plows on the end sills which can usually handle 1-3 ft of loose snow. However, rotaries are essential in areas with long cuts where snow and dirt tends to fill and pack hard or in mountain areas where very heavy snow fall and snow slides are experienced.

The first road locomotive application of the super heater in North America was made by the Canadian Pacific to 4-6-0 (ten-wheeler) locomotive No. 548 in 1901. At this time North American Locomotive Builders were going heavily to compounding (2-stage expansion) in order to improve locomotive efficiency. Common were two-cylinder Cross Compounds and Tandem compounds (4 cylinders) with large and small pistons mounted on the same piston rod. The gain in thermal efficiency was obtained with some sacrifice in mass of pistons and rods, counter balancing problems, and much more complex valve gear mechanisms. Thus the superheater offered major increase in efficiency by allowing major increases in heat content per pound of steam, without the mechanical mass, complexity and dynamic balance difficulties.

The Canadian Pacific originated the design of

the all weather or enclosed cab for common steam locomotives with the cab at the firebox end of the boiler. This first appeared on 2-8-2 (Mikado) locomotives of the 5000 class in 1912. The Canadian Government Railways followed suit in 1914. This type of cab design became the hallmark of Canadian steam locomotives built after that time. It was also used on many U.S. locomotives for northern roads.

Streamlined 4-8-4 high performance passenger locomotives designs, such as CNR's 6400 series, were created in the early 1930's, using scientific research methods, rather than concentrating on aesthetic considerations at the expense of functionality.

Dual service passenger and freight heavy duty high performance 2-10-4 locomotives were produced in quantity for the CPR Pacific Region mountain grade serivce in the late 1920's-1940 period. These were notable for their high steaming capacity and could move a 500 ton train along at 30 mph up a 1 percent grade.

Canadian National had designed and operated the first main-line diesel (oil) electric road locomotive in North America in 1928. This was a two unit box cab 2600 hp locomotive and was equipped with Beardmore-Westinghouse diesel (oil) engines and Canadian Westinghouse generators and traction motors. Both CP and CN became effectively dieselized by the mid 1950's.

Both major Canadian roads have seriously considered electrification for their high density lines. However, other than the Canadian National's 20 mile long Deux Montagnes, suburban route and Central Station electrification, no other significant lines have been put under the wire. The main reason has been the broad flexibility of multiple unit diesel-electric motive power. When combined with radio remote multiple unit control upwards of 30,000-35,000 tractive horsepower can be practically employed per train and many trains can follow each other on minimum signal system headway. In the Pacific region very heavy additional coal, grain and chemical traffic has been absorbed without broad need to double track or electrify.

Regarding car equipment, the Canadian Railways have led in introducing large numbers of unique high capacity cars.

First interest is directed toward the round body covered hopper car design. This was created by a team of Canadian Railway and Carbuilders Engineers in 1959 to optimize construction and obtain highly cost effective designs of aluminum and low alloy steel cars in true high capacity configuration. To creatively solve the light roadbed and bridge problem on branch lines with a high capacity car, several thousand aluminum body covered hoppers have been built which can carry 88 tons of load within a 110-ton maximum weight on four axles. Over 10,000 102-ton capacity steel covered hoppers have been built for grain service within 131.5 ton gross rail load.

To minimize the rock and roll dynamic problems on typical staggered 39 ft long jointed rail in North America, these and other designs have used basically longer truck centers than are common in the U.S. This allows reduction in height of a given lading even in open top gondola type cars. Illustrated are two types of unit coal cars having 46 ft truck centers. The CNR design employs a center sill and still gives only an 84 in. center of gravity with 103 tons of coal or sulphur. The CPR design is without a center sill, has a depressed bottom or "bathtub" between the trucks and can attain an 80 in. center of gravity with 105 tons of coal. All of these covered hoppers and coal-sulphur cars are equipped with empty-load brakes; the fully loaded braking ability is such that solid loaded or unit trains of these cars handle as well as lighter general mixed interchange consists.

Perhaps the heaviest average train weights in the world (outside Western Australia) are regularly handled by two major Canadian Iron Ore railroads: The Quebec North Shore and Labrador and the Quebec Carter Mining Company. These roads average about 16,000 tons (15,000 mt.) of lading and over 20,000 total tons of cars per loaded train. This type of service started 25 years ago on the QNS & L. A portion of this service includes powdered ore concentrate which is carried in specially equipped covered hopper cars owned by the Wabush Mining Company. These three unit cars have highly automated loading and unloading equipment allowing a complete train to be loaded or unloaded while in motion. The Wabush Service has been in operation over 14 years.

In the passenger car area, Canadians led in the development of several unique light weight high capacity designs. A fleet of light weight all aluminum main line passenger cars have been operation in "TEMPO" service since 1968. The United Aircraft Design High Speed Turbo Trains with single axle trucks between articulated car bodies have been operating since the early 1970's between Toronto and Montreal. The UAC trains have car body suspension above the center of gravity for improved curve negotiability and comfort. A new design of powered banking style tilting body passenger car is being produced for both Canadian VIA and U.S. Amtrak National Passenger Service. This is called the LRC type train for Light-Rapid-Comfortable.

For suburban services, a new design full-width true double-decker car is now operating in locomotive hauled commuter trains in the Toronto Region. This car design employs a combination disc and wheel tread type air-hydraulic friction braking system to provide full performance under a wide range of conditions.

FIG. 1 CPR #548 FIRST ROAD SERVICE INSTALLATION OF SCHMIDT TYPE SUPERHEATER IN 1901

FIG. 2 CPR #5018 FIRST TRUE ALL WEATHER CAB DESIGN IN NORTH AMERICA 1912

FIG. 3 CNR 6401 4-8-4 STREAMLINED MODERN HIGH POWERED STEAM LOCOMOTIVE DESIGN DEVELOPED BY CNR IN 1931-1933 AND BUILT IN 1936

FIG. 4 CNR-GTW ST. CLAIR TUNNEL PORT HURON-SARNIA 3300 V ALTERNATING CURRENT LOCOMOTIVES FROM 1907

FIG. 5 MONTREAL ELECTRIFICATION 1914 2400 V DC

FIG. 6 OIL (DIESEL) ELECTRIC RAIL CAR 1925. THIS CAR MADE A NONSTOP 2937 MILE TRANSCONTINENTAL RUN IN 67 HR

OIL-ELECTRIC 4-8-2 + 2-8-4

Canadian National's first 9000 — and North America's first diesel-electric locomotive designed for road service — was assembled at the Kingston, Ontario, plant of the Canadian Locomotive Company in 1928. William Beardmore and Company of Glasgow, Scotland, supplied the two 12-cylinder 1330 HP power plants for the twin unit *Oil-Electric* locomotive, while major electrical components came from the Canadian Westinghouse Company. Each of the two sections was forty-seven feet in length, and the locomotive weighed 325 tons.

The units — both numbered 9000 at first — were tested on CNR passenger and freight runs between Montreal and Toronto. Later the second unit was re-designated as 9001, and both engines were used independently on various lines in Central Canada. During the mid 1930s, number 9001 was used to haul the Toronto-Hamilton daily commuter trains.

In October 1939, both 9000 and 9001 were placed in storage, slated for the scrapyard, but the war years intervened, and 9000 was re-engined and camouflaged to provide power for an armoured train. During 1944, the unit was returned to civilian CNR service, again rebuilt, and assigned to local passenger trains between Edmundston and Quebec City. It was retired and dismantled in May 1946.

SPECIFICATIONS

Class	*Numbers*	*Builder*	*Date*
V-1-a	9000-9001	C.L.C.	1928

Oil-electrics — Beardmore engines
Two units . . . . 1330 h.p. each.
built by Canadian Locomotive Co. 1928.

Locofolio 12

FIG. 7 OIL (DIESEL) ELECTRIC MAIN LINE ROAD LOCOMOTIVE 1928

FIG. 8 GAS TURBINE HIGH SPEED TURBOTRAIN CIRCA 1970 BY UAC

FIG. 9 BILEVEL FULL WIDTH UPPER DECK SUBURBAN CARS, TORONTO AREA

FIG. 10 FIRST ROUND BODY COVERED HOPPERS IN NORTH AMERICA, 1959. ALUMINUM CONSTRUCTION 35,100 LB EMPTY 210,000 LB GROSS RAIL LOAD. DESIGN BY CANADIAN RAILROADS AND SUPPLIERS. 87.5 T CAPACITY.

FIG. 11 MODERN 90 TON CAPACITY ROUND BODY 4100 CU FT COVERED HOPPER OF ALUMINUM CONSTRUCTION.

FIG. 12 CANADIAN WHEAT BOARD 101 TON CAPACITY ROUND BODY COVERED HOPPER OF STEEL CONSTRUCTION FOR SERVICE ON CANADIAN NATIONAL.

FIG. 13 CANADIAN WHEAT BOARD 101 TON CAPACITY 4550 CU FT COVERED HOPPER OF STEEL CONSTRUCTION FOR SERVICE ON CANADIAN PACIFIC.

FIG. 14 CANADIAN NATIONAL 102 TON CAPACITY UNIT COAL TRAIN CAR WITH CENTER SILL CONSTRUCTION HAVING 84 IN. CENTER OF GRAVITY. FOR ROLL OVER DUMPING.

FIG. 15 CANADIAN PACIFIC 105 TON CAPACITY BATHTUB TYPE CONSTRUCTION UNIT COAL TRAIN CAR OF CENTER SILL LESS DESIGN HAVING LOW (81 IN.) CENTER OF GRAVITY. FOR ROLL OVER DUMPING.

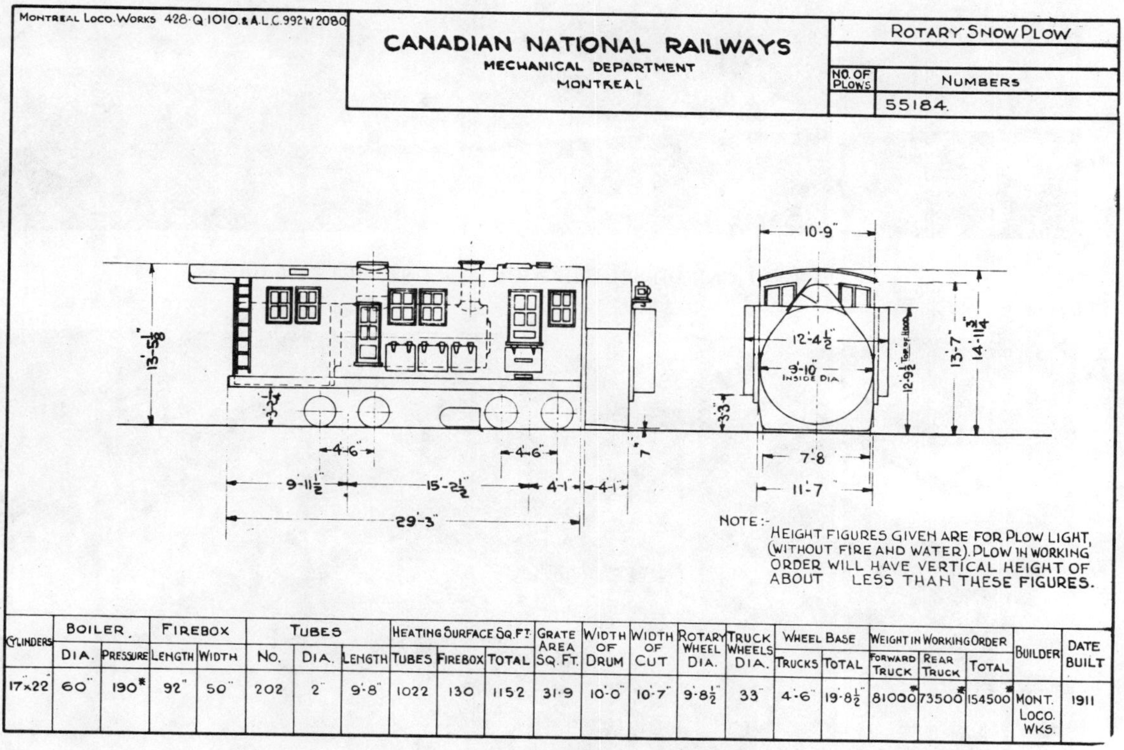

CYLINDERS	BOILER		FIREBOX		TUBES			HEATING SURFACE SQ. FT.			GRATE AREA SQ. FT.	WIDTH OF DRUM	WIDTH OF CUT	ROTARY WHEEL DIA.	TRUCK WHEELS DIA.	WHEEL BASE		WEIGHT IN WORKING ORDER			BUILDER	DATE BUILT
	DIA.	PRESSURE	LENGTH	WIDTH	NO.	DIA.	LENGTH	TUBES	FIREBOX	TOTAL						TRUCKS	TOTAL	FORWARD TRUCK	REAR TRUCK	TOTAL		
17"×22"	60"	190*	92"	50"	202	2"	9'-8"	1022	130	1152	31·9	10'-0"	10'-7"	9'-8 1/2"	33"	4'-6"	19'-8 1/2"	81000*	73500*	154500*	MONT. LOCO. WKS.	1911

FIG. 16 ROTARY SNOW PLOW

FIG. 17 CNR DIESEL-ELECTRIC 9000 (1928 BLT 2600 HP)

FIG. 18 CPR 2-10-4 STEAM LOCOMOTIVE 5900

FIG. 19 CPR ROTARY SNOW PLOW 300811

8 The Evolution of Rail Freight Vehicles in the United States

W. P. MANOS

In tracing the evolution of rail freight cars, it appears to have started in the United States in 1826. The Granite Railway Company of Massachusetts, which was three miles long, hauled granite blocks to the Bunker Hill Monument. Within the next ten years several railroads came into existence. By 1829 the Delaware and Hudson Railroad initiated an anthracite coal haul in New York. The cars used for this service were eight-wheeled wooden open top hopper cars. The wooden structure was a Howe type and the truck side frames and bolsters were also made of wood. An updated version which appeared in the 1878 Carbuilder's dictionary is shown in Fig. 1. The length was 22 ft, width 10 ft, maximum height 7 ft 3 in. It was a center sill-less design with the two bottom dump gates. This car had a capacity of 16-25 gross tons and a tare weight of 18,480 lb.

Another hopper car whose date of origin is unknown appears to predate the one shown in Fig. 1. It is shown in Fig. 2. This four-wheeled car was also used to haul anthracite coal. It had a capacity of 13,000 lb and a tare weight of 7600 lb. Its construction was also a Howe type wooden truss with wood sheathing. It had an axle spacing of 5 ft, height of 7 ft 7 in. and a cubic capacity of approximatel6 210 ft^3.

By 1888 the capacity of coal hopper cars had

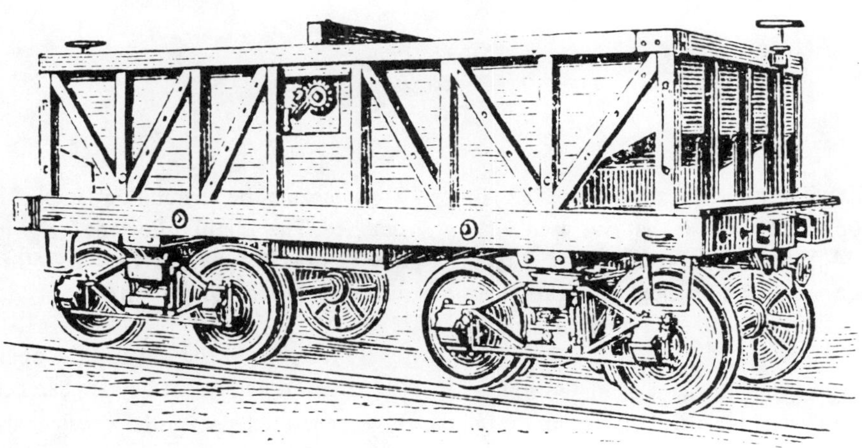

HOPPER-BOTTOM COAL CAR, *Eight-wheeled.*

FIG. 1

FOUR-WHEELED HOPPER-BOTTOM COAL CAR.

FIG. 2

COKE CAR.

FIG. 3A

reached 40,000-50,000 lb. Shown in Fig. 3 (A, B, and C) is a hopper bottom car that was used for coke. It was 24 ft long over the end plates, 7 ft 6-1/2 in. inside width and 8 ft high. Its construction was basically a flat bed with sides anchored in stake pockets.

The capacity of freight cars was related to the journal bearing size. This set the maximum overall car weight (i.e. 5 X 9 in. journals had a maximum loaded weight of 143,000 lb. If the empty car weighed 43,000 lb, capacity was 80,000 lb or 40 tons.).

Over the years *Normal Car Carrying Capacity in Tons* became the terminology in North American practice. This is still followed: 5 X 9 in.–40 tons; 6 X 11 in.–70 tons; 5½ X 10 in.–50 tons; 6½ X 12 in.–100 tons; and 7 X 14 in.–125 tons.

By the turn of the century the open top hopper car achieved its present day appearance (Fig. 4). It was a riveted steel girder type construction having a 50 ton capacity, a light weight of 43,000 lb, and a cubic capacity of 2035 ft^3. This car was built by Pressed Steel Car Company and appeared in the 1909 Carbuilder.

Another car of the same vintage built by the

P.R.R 53 001
AIR BRAKE
60,000 LB. CAPACITY COAL CAR.
LIGHT WEIGHT 26,700 LBS.
BUILT IN 1886

FIG. 3B

FIG. 3C

FIG. 4

FIG. 5

same car builder but of a side dump configuration is shown in Fig. 5. It, too, is a steel riveted construction; however, it is a combination Howe truss and girder design. Its capacity was 50 tons. Between 1912 and 1925 the coal car grew in size from 50 tons to 70 tons to 100 tons in capacity. Fig. 6 shows a 70 ton triple hopper car having a lightweight of 52,700 lb and a 2500-ft^3 capacity. This car was built by A.C.F., Bethlehem Steel, Pressed Steel, and Standard Steel Car Companies. Fig. 7 shows a 100 ton hopper car with a slightly different gate configuration built in 1917 by the Norfolk and Western Railway. It had a lightweight of 60,000 lb and a 2828 ft^3 capacity. This car had the first application of 6-wheel trucks of a flexible cast steel side frame configuration. Note the wrought steel wheels, rather uncommon in freight service at the time.

The first all welded 70 ton hopper car was built by Pullman Standard in 1931. This triple hopper car shown in Fig. 8 was an inside post construction, 2863 ft^3 capacity and had a lightweight of 45,900 lb. Actual load capacity was 82 tons within the 210,000 lb gross rail load allowed with "70 ton" capacity trucks at the time.

Present day coal cars are now primarily 100 ton

FIG. 6

FIG. 7

FIG. 8

FIG. 9

FIG. 10

capacity, riveted or welded, three hopper or four ("Quad") hoppers, manual or automatic door operation with cubic capacity from 3500 to 4000 plus ft^3. An automatic motion unloading or dump-on-the-run coal car is shown in Fig. 9. This car built by Ortner Freight Car Company is 61,000 lb empty and 263,000 lb gross rail load. Extreme height is 13 ft and capacity is 3626 ft^3. A similar "automatic unloading" car was built by Pullman Standard for the B&LE to haul either coal or ore. This car (Fig. 10) built in 1969 had a cube of 3300 ft^3 and 100 ton capacity trucks with 6½ X 12 in. roller bearings. Later versions are longer with a cubic capacity of 3850 ft^3.

To date, open top hopper cars have been built to capacities in excess of 7000 ft^3 for wood chips, Fig. 11 and as low as 1000 ft^3 for iron ore, Fig. 12. These cars were built by Pullman Standard. Note that the ore car has load bearing side sills which carry all the draft and buff forces of a 150-200 car train.

FIG. 11

FIG. 12

FLAT CARS

Undoubtedly the first freight car built must have been a simple flat bed on four wheels upon which goods were stacked and covered with tarpaulins. Information on these early cars in United States is difficult to find; however, a pre-1878 flat car shown in Fig. 13 was an eight-wheeled wooden vehicle approximately 27 ft long and having a capacity for 15 tons. Note that this car has stake pockets for either stakes or stake sides to form a gondola. The car deflections and strength were controlled by steel-tensioned tie rods adjusted by turn buckles. Such a construction is termed "queen post truss" and it was commonly used in other wooden car construction to control deflection and strength of underframes.

At the turn of the century the flat car became a riveted tapered steel girder design with stake pockets and wooden floors. Fig. 14 shows Pressed Steel Car Company's 40 ton capacity flat car which was 30 ft long and weighed 32,000 lb. By 1912 the flat car grew to 70 ft 7in. weighing 90,000 lb and having a capacity for 100 tons. The 8 ft 6 in. wide car shown in Fig. 15 was built by McGuire-Cummings Mfg. Company.

By 1915 the flat car was already being used to haul containerized freight. Fig. 16 shows a 44 ft

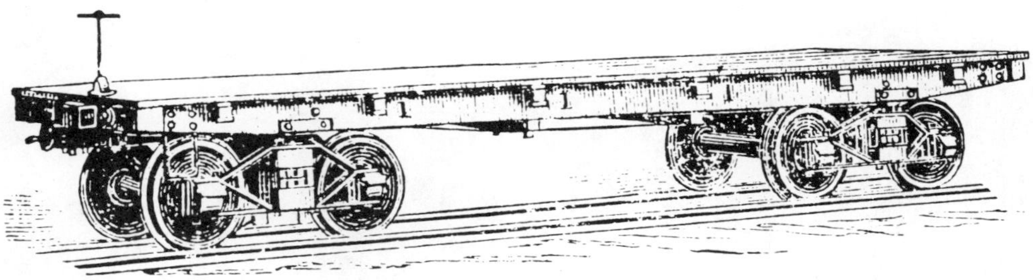

FLAT CAR.
Stake-pocket.

FIG. 13

FIG. 14

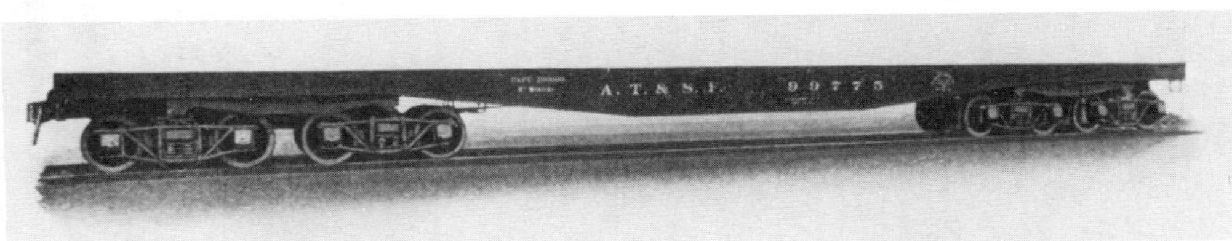

FIG. 15

FIG. 16

long 50 ton capacity flat car carrying five containers. The lightweight of this car is 64,000 lb and it was built for River and Rail Transportation Company by A.C.F. During that time the Merchants Dispatch Transportation Company built its own container car for BX service. Their car carried nine containers for handling express lading, and it is shown in Fig. 17. This car was basically a bulkhead flat car that gave the appearance of a cut down passenger car. Note that it used equalized inside swing hanger cast steel passenger type trucks and had passenger car brake control valve with clasp brake rigging.

The flat car continued to become specialized as early as 1915. Shown in Fig. 18 is an all steel 110-ton capacity well-hole flat car. Its lightweight was 91,900 lb, length of platform 55 ft 7 in., width 9 ft 10 in. The well opening was 26 ft 2 in. by 6 ft 1 in., and it was built by the Lehigh Valley Railroad. Six-wheel passenger type trucks were used.

In 1934, Merchants Dispatch Transportation

FIG. 17

FIG. 18

FIG. 19

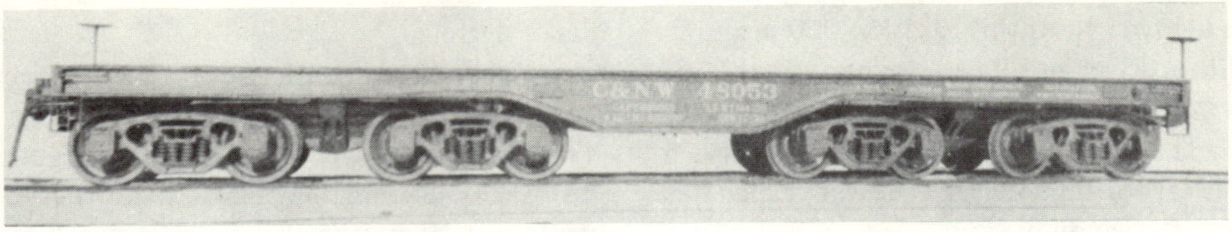

FIG. 20

FIG. 21

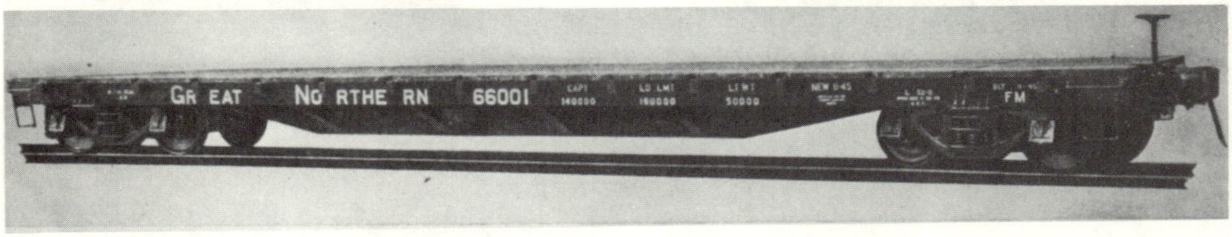

FIG. 22

FIG. 23

FIG. 24

FIG. 25

Company developed a special purpose flat car for carrying two milk tank trailers (Fig. 19). This concept appears to be one of the first freight hauling highway trailers in piggyback operations, although some electric interurban railways had piggyback on a small scale in the early 1920s. The car was equipped with passenger trucks and had a 30 ton capacity with a 70,000 lb lightweight. The length of this car was 50 ft and it used typical equalized swing hanger baggage car style passenger trucks of that period.

A unique construction for high capacity flat cars for the steel industry was designed and built by Commonwealth Steel in 1933. The car shown in Fig. 20 has a one-piece cast steel underframe. Weighing 104,700 lb, this car had a 200 ton capacity, a length of 44 ft and a deck height of 4 ft 2 in. From conventional cast steel spring plank type

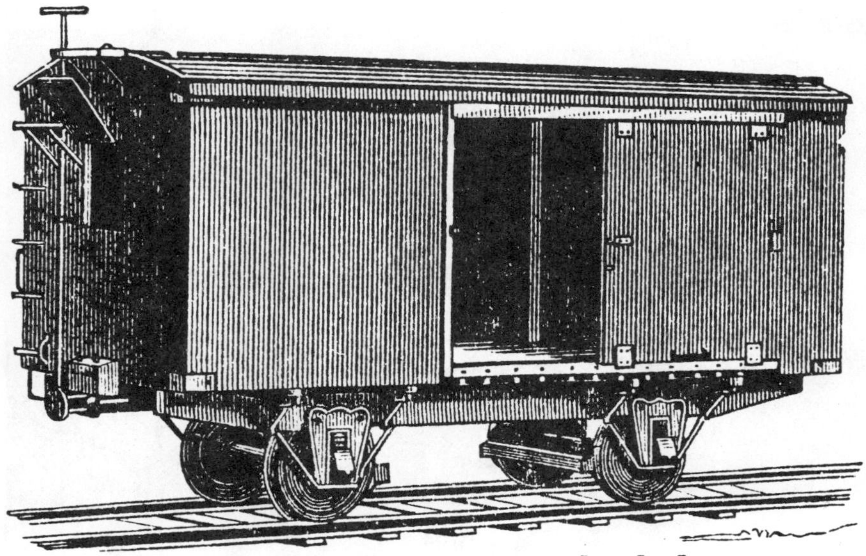

Box Car, *Four-wheeled.*
(Little used; the same type is built with inside sheathing only, leaving framing exposed.)

FIG. 26

freight car trucks were used. Each set of trucks was connected with a cast steel span bolster.

Another special purpose cast steel underframe car built by Commonwealth Steel Company in 1942 was the depressed center transformer car shown in Fig. 21. This car had a 125 ton capacity, 126,000 lb lightweight, 57 ft 9 in. length and a 2 ft high 22 ft long depressed center.

The 1940's also saw some changes in the fabrication of standard flat cars. In 1945 A.C.F. built the 52 ft 70 ton general purpose flat car shown in Fig. 22. This car was all welded and had a light weight of 50,000 lb. It was equipped with a wooden deck and a tapered beam center sill with cross-members supporting side sills equipped stake pockets. Flat cars similar to this one are still being built today in various lengths up to 70 ft with or without bulkheads for pulpwood, gypsum board, lumber, and machinery.

During the 1950's the flat car grew to 85 ft and then to 89 ft, and the era of auto rack cars and piggyback or TOFC got underway. The most recent construction of this car type is shown in Fig. 23. This car built by Pullman Standard for Trailer Train is designated "low deck-flush deck" car. It uses 3-piece freight car trucks with 28 in. diameter wheels. This car is 89 ft 3 in. long and can be equipped for auto rack service (Fig. 24), Verti-pak service (Fig. 25) or TOFC. As an auto rack car it can carry 12 large automobiles or 15 compacts. In TOFC service, it can carry a 45 ft and a 40 ft trailer. Presently there are in excess of 65,000 flat cars of these two types in service today. They consist of high deck cars (42 in. high decks) using freight car trucks with 33 in. diameter wheels, having a 70 ton capacity and low deck cars (34-1/2 in. high decks) with 28 in. wheels having a 55 ton capacity.

BOX CARS

In tracing the evolution of box cars in the U.S., available information prior to 1878 is scant and exact dates of equipment are difficult to find. Probably, one of the first box cars on U.S. roads is that shown in Fig. 26. By 1878 this two-axle four-wheeled-wooden box car was little used. It is approximately 19 ft long, and it had a wood underframe and truss sides with inside sheathing.

A more typical box car of style dating back to 1850's is that shown in Fig. 27. This 28 ft car was an externally sheathed wood Howe type truss side with a queen post wood underframe construction that utilized turnbuckles to control the underframe deflections. It had a 15-ton capacity with a lightweight of 20,900 lb and it was equipped with grain doors to facilitate the shipment of grain. It was 8 ft 7-3/4 in. wide, and had an extreme height of 11 ft 10-5/8 in. The double four-wheel truck

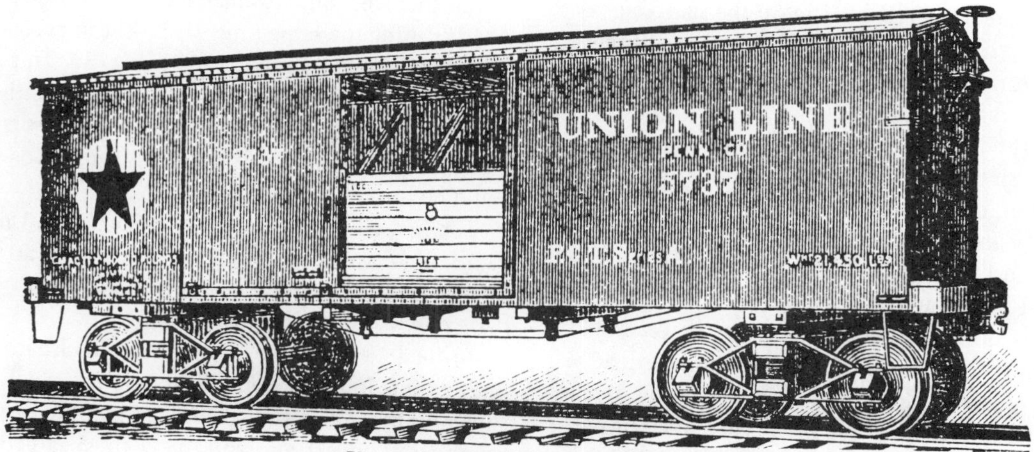

Box Car, *showing* Grain Door.

FIG. 27

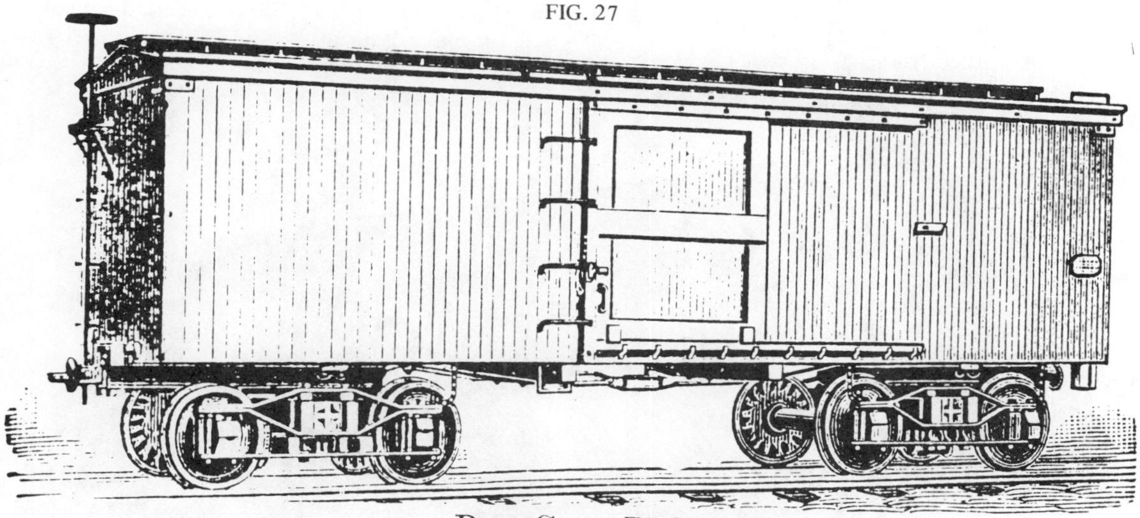

Box Car, *Eight-wheeled.*

FIG. 28

FIG. 29

freight car was better adapted to the sharp curves staggered rail joints and general track conditions of North American railroads than the long wheel base 2-axle cars.

By 1888 the eight-wheeled wooden box car grew in size from 28 ft to a new standard of 34 ft 11-1/4 in. over the running boards having a 25 ton capacity and a lightweight of 24,800 lb. This car is shown in Fig. 28. Cars of this type were built by Haskell-Barker and others.

Early in the 1900's steel began to be used in box car construction. By 1905 the box car load capacity jumped to 50 ton capacity and the steel underframe was developed. The car shown in Fig. 29 was built by A.C.F. and it weighed 44,300 lb, was 36 ft long and it had a steel center sill as well as pressed-steel side sills. The rest of the body was still a wood-sheathed truss.

Another steel underframe concept developed by A.C.F. during the same time was a 40 ton capacity box car for automobile traffic, Fig. 30. This 39,100 lb 40 ft 6 in. long car had a wood-sheathed steel Howe type truss side and hinged end doors. With special stowage-loading devices four autos could be loaded in this type of car.

The first all-steel box car was built by the Union Pacific Railroad (Fig. 31). This riveted 40 ft 50 ton car of 1908 design was a steel sheathed Howe type truss side with a riveted steel underframe.

Around 1915 the Northern Pacific Railroad built a 60 ft long wooden 40 ton box car which is shown in Fig. 32. Although this car had the queen post tie rod underframe and the wood-sheathed truss side it did feature two-door openings on either side. This car weighing 43,500 lb was used for collecting L.C.L. freight from team tracks.

FIG. 30

FIG. 31

FIG. 32

FIG. 33

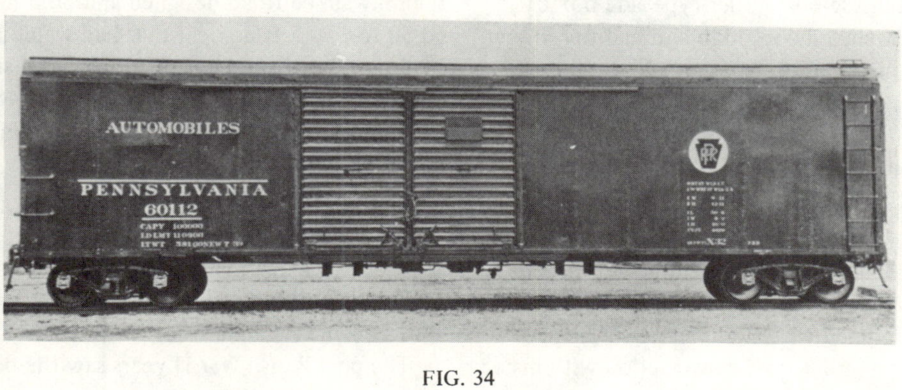

FIG. 34

FIG. 35

FIG. 36

Shortly before 1919 A.C.F. introduced an all-steel riveted inside-post girder-type side box car (Fig. 33). It weighed 43,100 lb and had the following dimensions:

Inside length – 36 ft.
Inside Width – 8 ft 6 in.
Inside height – 8 ft 4 in.

The Pennsylvania Railroad in 1933 designed and built an all-riveted steel 50 ton 50.5 ft inside post box car having a 12 ft door opening with double doors on each side (Fig. 34). Five hundred similar cars were equipped with the then new AB type brake. It was a girder type construction with inside posts and wood-lined. Just prior to World War II the Milwaukee Road designed and built an all welded 50 ton 50.5 ft box car that had welded construction, horizontal corrugations and inside posts, (Fig. 35). This car had a light weight of 48,200 lb, a height of 10 ft 9 in., a width of 9 ft 6 in. and a cubic capacity of 5157 ft^3.

In 1948 Pullman Standard built one of its first all welded 40 ft, 50 ton all purpose box cars (Fig. 36). It had a lightweight of 47,100 lb and a cubic capacity of 3898 ft^3. This car designated the PS-1 became a standard for many years.

The post World War II years saw the box car

FIG. 37

FIG. 38

grow from 40 ft to 50 to 60 to 86 ft and from standard cushioning devices to 30 in. travel hydraulic devices with sliding center sills. In the late 1950's Pullman Standard produced the first long travel hydraulically cushioned underframe box car. A 70 ton 50.5 ft box car having a cushion underframe, double doors with a 16 ft opening and DF-2 rails is shown in Fig. 37. This car had a light weight of 67,400 lb.

Later in 1963 the cushioned underframe box car increased in capacity to 100 tons and increased in length to 60 ft 9 in. A typical car of this type having 20 in. of hydraulic cushioning (shown in Fig. 38) was produced by Pullman Standard. Note that this car also has double plug doors, outside stakes and is single sheathed.

Following the continued demand for a higher cube box car, the automotive industry had a need for an even higher cube box car for transporting low density automotive parts. A committee was established to standardize on a design which resulted in establishing an 86 ft 6 in., 10,000 ft^3 cushioned box car in either 70 or 100 ton capacity. A typical car of this type was built by many car builders. Pullman Standard's version is shown in Fig. 39.

REFRIGERATOR CARS

Refrigerator cars were already in existence prior to 1878, and they too saw many changes. Solid trains of water ice refrigerated cars operated from Sacremento to Chicago and St. Louis in the early 1870s. A typical pre-1878 refrigerator car is shown in Fig. 40. This car was 28 ft long, insulated, and double sheathed. It had a 15 ton capacity with cooling accomplished by circulating air passing over ice stored in bunkers.

Note that this car utilized the same type of queen post underframe construction as seen on its contemporary box cars.

By 1895 the refrigerator car grew to 36 ft, weighing 41,500 lb and having a 20 ton capacity. This design shown in Fig. 41 utilized an automatically ventilated ice bunker system termed the

Hanrahan System of Refrigeration. Other than the double set of doors the car construction was similar to that shown in Fig. 40.

In the early 1900's the "reefer" grew to 40 ft and the steel underframe was developed. A car of this type built by A.C.F. is shown in Fig. 42. It had a lightweight of 44,500 lb and a 30-ton capacity.

A riveted steel frame Howe truss reefer was built by the N&W Railway in 1915 (Fig. 43). This car was capable of carrying 30 tons of lading and 5 tons of ice. It had a lightweight of 55,800 lb and had end ice bunkers for better distribution of cold air.

In 1936 A.C.F. developed a heavily insulated refrigerator car (Fig. 44) that used solid carbon dioxide in container bins for refrigeration. This riveted steel car was 66,100 lb and had a capacity of 40-tons. Present day refrigerator cars range from 50 ft 70 ton to 60 ft 100 ton capacities and are insulated with polyurethane foam for better heat transfer properties. Although some liquid nitrogen cooling systems have been used, (Fig. 45), the most commonly used today are the diesel driven mechanical refrigeration systems which came into

FIG. 39

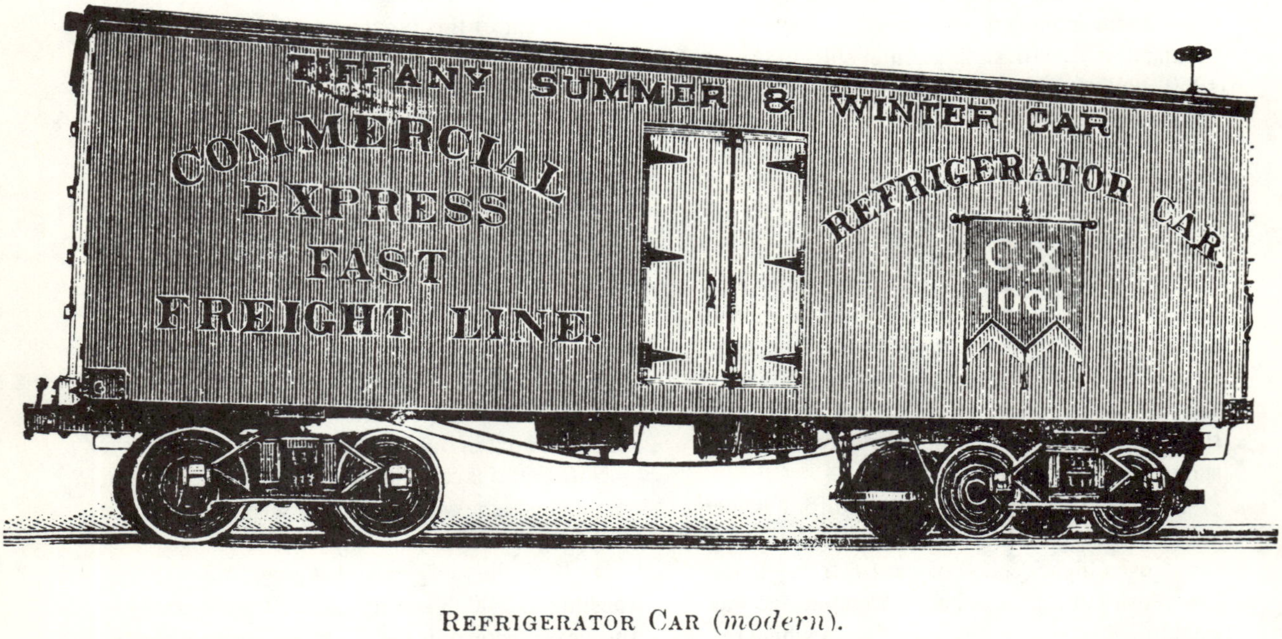

REFRIGERATOR CAR (*modern*).

FIG. 40

existence after World War II. Virtually all temperature controlled cars use this system. Water ice refrigerator cars are now practically extinct.

GONDOLAS

The early gondola appears to have started from a side-stake flat-car. Various configurations of this car type exist which range from hopper bottoms such as the pre-1878 24 ft 19,800 lb 25-ton capacity wooden car, (Fig. 46) to a side-dump 1905 50 ton 36 ft car weighing 38,900 lb, (Fig. 47), and even more elaborate geometries such as the 1905 40 ton two-way side-dump car shown in Fig. 48. The more common type is the low side-drop end gondola (Fig. 49) or that shown in Fig. 50. The latter two are also from the 1905 era. This car type also grew from a 24 ft 25 ton wooden car to lengths in excess of 60 ft with capacities of 125 tons. By 1917 the flat bottom gondola was 50 ft riveted steel weighing 73,900 lb and having a capacity of 120 tons (Fig. 51).

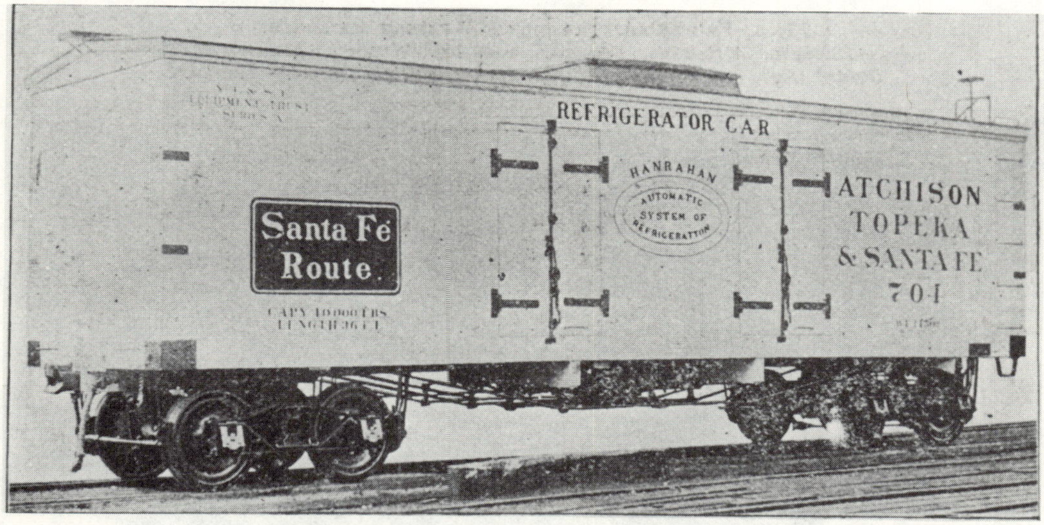

FIG. 41

FIG. 42

FIG. 43

FIG. 44

FIG. 45

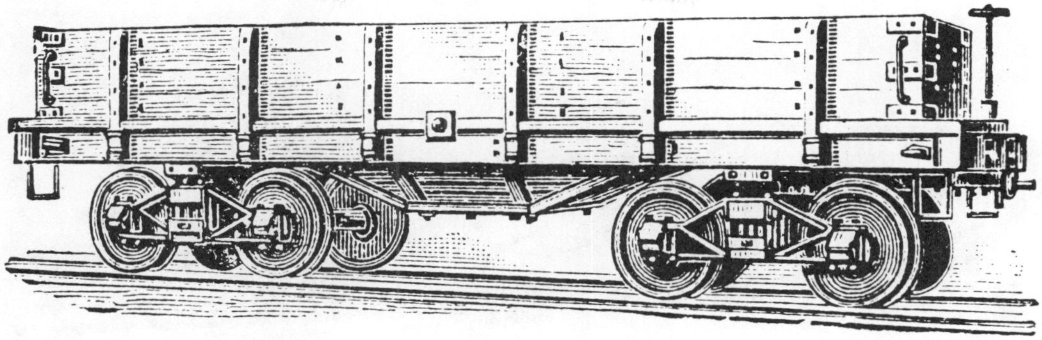

HOPPER-BOTTOM GONDOLA CAR.

FIG. 46

FIG. 47

FIG. 48

FIG. 49

FIG. 50

FIG. 51

FIG. 52

FIG. 53

FIG. 54

FIG. 55

FIG. 56

Gondolas have been used for hauling bulk commodities such as coal, ore, or scrap metal, or used to haul coiled steel, flat structural shapes and other commodities. For bulk materials the cars are either bottom-dump or equipped with special couplers for rotary dump service. Typical examples are shown in Figs. 52 and 53, which are Pullman standard's 1950's 100 ton rotary-dump gondola for ore and for coal. Thrall's 1960's 100 ton 4000 ft^3 rotary gondola for coal, (Fig. 54) and 100 ton coil steel car with lift-off covers with long travel hydraulic cushioning built for the Pennsylvania Railroad (Fig. 55) are also typical examples.

COVERED HOPPERS

The covered hopper car is a 20th century rail car. The first one appeared in the literature about 1912 (Fig. 56). This car built by A.C.F. was a riveted all-steel 50 ton car weighing 43,500 lb, 36 ft inside length and was used to haul brewery products. This car had a cubic capacity of approximately 2500 ft^3. Since that time the covered hopper car grew to capacities up to 125 tons and cubic sizes close to 6000 ft^3. In addition to size increase, special devices have been developed to facilitate loading and unloading of the various commodities. In

FIG. 57

FIG. 58

1935 the N&W built a 50 ton 1460 ft^3 cement car (Fig. 57) that had square hatches for loading. Four side hoppers with sliding gates were provided for discharging cement. It had a light weight of 42,300 lb. Fig. 58 shows a 40 ton, 3000 ft^3 car built by GATX for the transport of carbon black in 1938. Circular hatches for loading and special outlet gates were provided to facilitate the materials handling. Another special type of covered hopper was also developed by GATX in 1958 (Fig. 59) to transport difficult to unload bulk materials. This car was equipped with air pads to promote flow. It was a 50 ton car 2600 ft^3 weighing 55,500 lb. An enlarged version of this concept was built in 1964 which was a 100 ton 4180 ft^3 car (Fig. 60).

In the 1960's Marine Industries of Canada built the tank type tear drop shaped aluminum covered hopper car to basic designs by Alcan (Fig. 61). This car weighed 35,300 lb and had a cubic capacity of 3000 ft^3, and could carry 77 tons of bulk materials. This car had a stub sill center and load-bearing side sill structure. Canadian railroads have constructed over 20,000 cars of this type since the early 1960's. A year later A.C.F. introduced a tank type covered hopper that facilitated flow with air pressure. This 125 ton capacity pressure differential car (Fig. 62) weighed 70,200 lb empty and 315,000 lb fully loaded.

In the middle and late 1960's, Pullman Standard increased the covered hopper size from a 70 ton 4000 ft^3 to 100 ton 4427 (Fig. 63) to 4750 to 5840 ft^3. The 5840 ft^3 car shown in Fig. 64 is currently used to transport plastic pellets. The latter car utilizes special outlet gates that facilitate vacuum unloading. Concurrently, A.C.F. produced a family of covered hoppers from 2000 ft^3 to 5700 ft^3. Their configuration is a curved side center sill-less, center discharge type. A typical car is their 4650 ft^3 plastic pellets car shown in Fig. 65 which was built in 1964. Their current size plastics

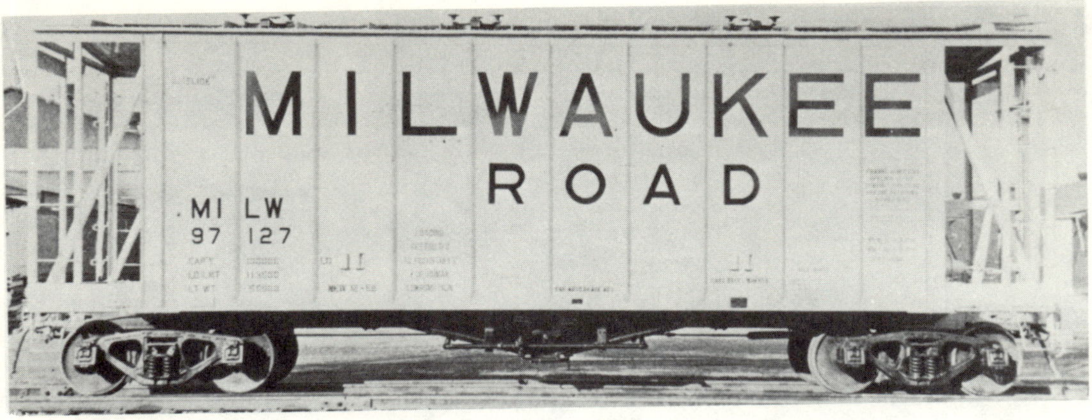

FIG. 59

FIG. 60

FIG. 61

FIG. 62

FIG. 63

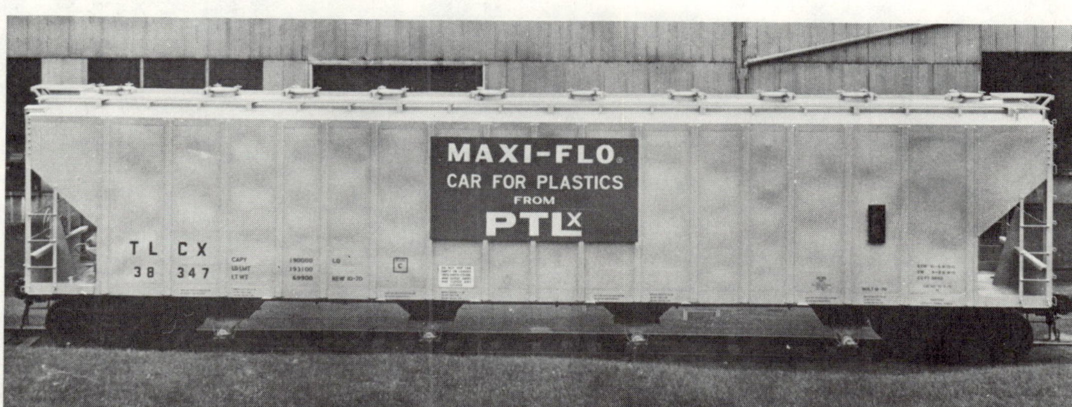

FIG. 64

FIG. 65

FIG. 66

car is 5701 ft^3, shown in Fig. 66.

TANK CARS

The tank car configuration was basically established from its beginning. As can be seen in Fig. 67, it consisted of a steel-riveted tank strapped to a wooden underframe with wooden crossmembers and a king post struss to strengthen it. This car, which was in existence prior to the Civil War, had a light weight of 28.000 lb and a capacity for 3700 gal of oil. By 1905 the tank car based on the Van Dyke patent grew to 10,000 gal capacity, and was a center sill-less design of riveted steel (Fig. 68).

The Standard Tank Car Company in 1926 designed and built a 20,000 gal 100 ton tank car that had a through center sill, three compartments and weighed 88,000 lb (Fig. 69). The tank length was 43 ft.

One of the first welded tank cars was built by A.C.F. toward the end of World War II (Fig. 70). It was a 40 ton 6000 gal six-compartment car that weighed 55,700 lb. Each compartment was insulated and it was used to transport wine.

Although the tank car continued to be improved each year in the loading and unloading devices, no major changes took place until the 1960's, when the capacities more than doubled. General American Transportation's entry, shown in Fig. 71, had a light weight of 110,700 lb and a 30,300 gal capacity. This car was approximately 89 ft long and was used to carry liquified petroleum gas. Concurrently Union Tank Company built a 50,000 gal

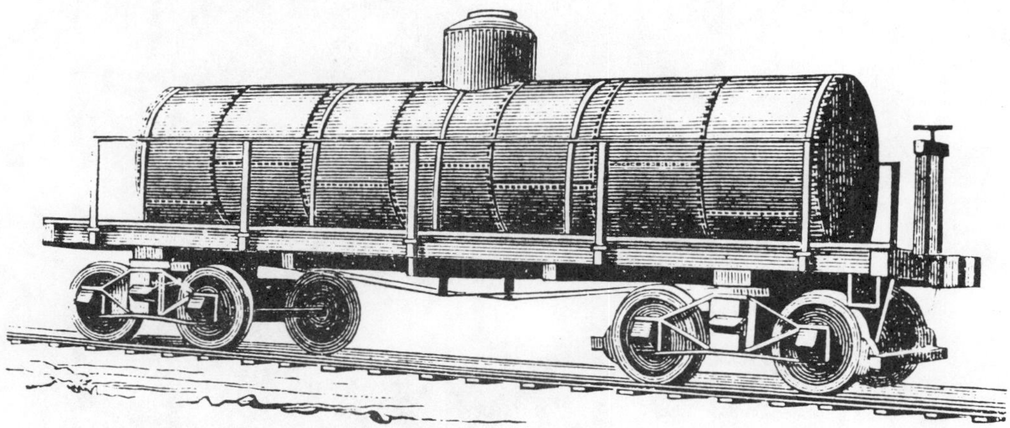

TANK CAR OR OIL CAR.

Usual capacity 3,700 to 3,850 gallons, weighing 7.3 lbs. per gallon.

FIG. 67

FIG. 68

FIG. 69

FIG. 70

FIG. 71

FIG. 72

tank car for the same service. This car, shown in Fig. 72, had double trucks and a span bolster at each end and was 89 ft long. Note that both cars are center sill-less and do not have an underframe. The tank itself transmits all longitudinal forces. Many of these four-truck tank cars are in service with load capacities ranging from 150-200 tons.

OTHER TYPES OF CARS

Many other types of cars have been built for special services, and some have gone into extinction. A partial list of these cars would include cars for hauling hot metal, steel billets, poultry, pickles, vinegar, military devices, milk, live stock (hogs, cattle, horses), atomic waste, bridge components, and many other commodities that required specific configurations. Since these car types are numerous and the quantities are limited, they are excluded from this presentation.

ACKNOWLEDGEMENT

The author wishes to express his appreciation to the American Association of Railroads, Simmons-Boardman Publishing Company, RAILWAY AGE and to Pullman Standard for supplying the photographic services. The author also wishes to thank Mr. Ernie Derjan of Pullman Standard R&D for his excellent photographic services, Mrs. Susan Colias and Miss Cheryl Pratt for their typing and proof reading and Mrs. Margaret Manos for gathering and selecting the historical source material.

9A Railroad Passenger Cars
A Century of Mechanical Engineering

WILLIAM VAN DER SLUYS

By the time of the formation of the ASME, the railroad passenger car had already evolved into a form which is still recognizable as directly related to present day cars. The car body was basically a rectangular tube, with a row of windows on each side, supported usually on a pair of 4 or 6-wheel swivel trucks. Although there had been a number of sleeping cars built and operated in the 1840-1860 period, George Pullman had boldly built his Pioneer A in 1864, with close to a 10 ft width – a foot wider than existing clearance limits permitted. He thus approached present day standard widths when the Chicago and Alton widened clearance along its right-of-way between Chicago and Springfield, possibly for Lincoln's funeral train in early 1865.

Accommodations were crude by 20th century standards. George M. Pullman did not invent the sleeping car, but he did combine efforts and improve or develop the idea into a national operation. Braking was almost entirely manual. Couplings between cars were completely nonstandardized between railroads, but were usually link-and-pin type –most hazardous to the yard switchman. Heating was by stoves in each car. Lighting was by candles or oil (kerosene) lamps.

Trucks, on the other hand, were remarkably similar to configurations still in most common use in the United States. Axles, while small with 3½ in. × 6 in. or 3¾ in. × 7 in. outboard journals, fitted precisely into the series of plain bearing axles which still persist. Wheels were cast iron, but tread contours were similar to present day practice. Equalizer bars (wrought iron) were usually used to support a rigid frame on primary springs, which were either steel coils or India rubber blocks. Since the broad adoption of English truck design ideas in the early 1860's, a spring plank was carried on swing hangers for lateral suspension, with a bolster carried on elliptic springs and, in turn, supporting the body on a center plate and side bearings. This configuration persisted almost untouched except for for materials and details until the 1930's and 1940's.

At this time, the railroad industry had already become and would persist for at least 50 years as the most fertile field for mechanical inventions in the country. By 1904 there had been over 8000 patents granted relating to the development of automatic couplers alone. Standardization was in its infancy. Even screw threads varied widely in pitch and thread shape with each manufacturer. The Master Car Builders' Association was formed in 1867 to establish interchange rules of acceptance and repair which would permit cars to travel anywhere within the North American continent. The MCB was the direct predecessor of the Mechanical Division of the Association of American Railroads.

The development of the railroad passenger car can be broken into periods which roughly correspond to major changes in the basic construction of the car itself. While the introduction of improved specialties and amenities has not coincided precisely with these changes in body construction, such a division does provide a convenient means for cataloging the innovations in a systematic manner.

The period of 1865 to 1880 can be characterized as the period of transition from cars which were all small, uncomfortable wooden people boxes which cost $2000 to $4000 complete to the development of luxury cars costing $30,000 to $40,000, which would induce the public to make long distance trips in comfort. In fact, in 1877, Frank Leslie organized for Leslie's Magazine a five-month tour of the West from New York by luxury Palace car. Dining at 20 mph by the light of oil lamps was low cost and superb. The latest amenities were provided in the washrooms, with a hand-pump to draw water from under the car (or overhead tanks were used which were filled through the roof). Each car carried its own coal or wood-burning stove, although the luxury cars were equipped with circulating hot air or hot water systems from the Spear or Baker heaters. A peppering of cinders or a layer of soot over everyone was standard procedure, but patented roof or clerestory ventilators were being used which attempted to filter the fresh air entering the cars. However, without pressurization, such attempts were at best moderately successful.

Walking between cars with the train in motion was extremely hazardous, since the platforms were narrow and wide open. There were no buffers or step plates between end sills of the cars. Couplers were generally link-and-pin types with simple spring draft gears. Various styles of semiautomatic "hook" couplers, such as the Miller, were used on some roads. Large free slack existed between cars.

The car body structure itself was a relatively simple construction of a typical wooden bridge and included wood side sills, posts, belt rails, carlines, etc., with wrought iron diagonal tension rods in the side framing. Support and control of vertical reflection was by wrought iron truss rods under the floor. The bridge truss principle was widely used in other railroad car components such as brake beams, crane beams, wooden truck bolsters, etc.

Brakes were primarily manual, although some cars were equipped with elementary nonautomatic vacuum introduced from England in the early 1860's or straight air brakes introduced by Westinghouse in the 1860's and early 1870's. There was still very little standardization between roads, with even the gauge of the track not yet fully standardized. One Canadian road operated with three rails from Detroit to Buffalo, so as to handle the new through passenger cars on the wide gauge while the local freights still used narrow gauge.

The period from 1880 to 1905 was the height of the all-wood car structure, even though some wood (or wood and steel) cars continued to be built up to 1920. This period saw the development of the construction whereby the car side, rather than the underframe, was used to support the car weight. Elaborate wood trusses or arch constructions were developed for use between the side sill and the belt rail, although truss rods continued to be used under the cars.

While the plain automatic brake had been invented by 1872, it was not applied in quantity until the late 1870's. The quick-action automatic air brake became standardized following MCB approval in 1889. By the early 1890's, the automatic locking knuckle type coupler with a friction draft gear finally settled down to the point where knuckle contours could be standard and cars could be interchanged between different roads. In fact, any other type became illegal with passage of the U.S. Safety Appliance Acts in 1893, with complete phase-out by 1900. The Safety Appliance Acts outlawed hand brakes for train speed control and required an automatic quick action air brake on *all* cars in interstate (interchange) commerce. It was during this period that the enclosed vestibule was invented by Mr. Sessions in 1887, for the first time permitting passengers to walk safely through a moving train and thereby encouraging the use of dining, lounge, and other special purpose cars. At first, this vestibule was only between step wells, but by 1893, the trap door was added and the full width vestibule was in use. From the first, this construction used the metal face plate with the canvas multifold diaphragm and the vertical curtain as the safe means of moving between ends of adjacent cars.

Car lighting grew early in this period from oil lamps to Pintsch gas lights. The gas lights were so pleasing that this type of lighting persisted for at least 40 years, using petroleum gas, acetylene gas stored in tanks with acetone and asbestos, or acetylene gas generated on the car by use of carbide powder.

The early electric lights were battery operated only; in fact, the first use was on an American car on an English railway using French "accumulator cells". The first "Twentieth Century" had electric lights with power trainlined from a steam-driven generator on the locomotive. Electric lights did not become common until the axle-driven generator with automatic voltage control had been developed.

The years 1900-1920 encompassed the transition from wooden cars to steel cars, although half the cars in use in 1925 still had some wood, with another 15 percent of composite steel and wood construction. By the early 1900's, the increased size of the cars, along with the elaborate embellishments within, had led to cars of excessive weight (80-100 tons) for the running gear available.

The first use of steel was actually an attempt to reduce car weight while improving safety. Accepted technical opinion held that the energy absorption properties of wood were sufficiently superior to stiff steel structures as to render the wooden cars safer in a collision than steel. Unfortunately, with stoves and ruptured oil or gas tanks involved in a wreck, the wooden cars usually burned, with disastrous effects on passengers. However, when the Interborough Rapid Transit tried to buy steel cars for the New York City subway in 1902, the company had severe difficulty in finding a carbuilder to take on the task.

Early use of steel consisted of exterior sheathing over wood structure or of relatively long thin steel sections wrapped around and stabilized locally by solid wood members. This added greatly to the vertical support strength of a car side when properly used at side sills, belt rails, and side plates. Cast or riveted steel construction was used quite early for the bolster and draft sill area at the end of the car, thereby giving a rugged attachment for couplers and draft gears. With the fully-vestibuled cars becoming standard, buffing gears were added in these car end sections which, by applying heavy pressure between adjacent car ends, reduced the uncomfortable effect of slack in the couplers and added energy absorbing friction to reduce motion between cars. This had a stabilizing effect on the cars and greatly improved the riding quality, especially laterally.

In 1904, the first technical paper was published on the design of an all-steel car structure. While the first all-steel Pullman train was on the rails in 1910, steel underframe cars with wood superstructure were built to the end of this period. Canvas-covered wood roofs were still applied to some extent through the mid-twenties. The composite types had so many variations that there actually were cars with wood structure covered by steel sheets, as well as steel structures covered by tongue-and-groove wood side sheathing.

Telescoping of cars in train wrecks had been a severe problem from the beginning of railroading. The use of steel end sections on underframes soon led to the use of end posts mounted on these car ends to reduce the likelihood of one car overriding another. In 1912, the U. S. Post Office, which had for many years specified the construction of wood cars, issued a steel car specification calling for 400,000 lb static compression strength at half yield, 65 in.^{3} section modulus in end posts, and a minimum strength requirement for side posts, among many other features. Originally, these figures applied only to post office cars, were extended to all cars used in a train including a post office car, later became AAR Recommended Practice, and are now included in DOT requirements.

Steam heating of passenger cars was developed during this period. At the turn of the century, conventional heating involved one or two stoves in each car, usually with hot water circulation through pipes along the floor at each car side and under the seats. Early attempts to use steam involved a cumbersome two trainline pipe system. By 1905, a regulator had been invented which kept high pressure locomotive steam under the car, with atmospheric pressure steam within the car. This is the Vapor system. Many variations of multiple pipes, finned tubing, manual controls, and thermostatic controls were later developed for improved passenger comfort.

1920-1930

This was a decade wherein the so-called "standard heavyweight car" took over almost completely on the nation's railroads. All cars were built of low-strength, low-carbon steel, using riveted construction with a heavy belt rail below the windows producing a plate girder from side sill to belt rail. Truss rods disappeared, and fishbelly center sills were customary. While the clear story roof was still standard—aimed at providing ventilation—the roof was still occasionally made of canvas-covered wood, on the theory that this construction survived better under the sulphuric acid treatment that was created by the interaction of coal ashes and rain water. Even aluminum roofs were used on Illinois Central commuter cars. Operable window sash was still usually wood, as were some other details of car interiors. Floors were trowelled cement, linoleum covered or bare in coaches, carpeted in luxury cars. This was a period of consolidation of all the developments of the preceding century, with passenger trains operating services in and to practically every community in the country, whether the service be classed as main line, commuter, interurban, street car, elevated, or subway. Catalogs of railroad equipment supplies show hundreds of suppliers and an almost infinite variety of special hardware, lighting fixture or plumbing items readily available. This period also saw a great proliferation of gasoline engine self-propelled single railroad cars, primarily for branch line service.

1932-1942

This is the decade of air-conditioning and streamlined lightweight trains. The B & O, the Santa Fe and the Pullman Company began to research the possibility of air conditioning railroad cars in 1930. The next year, some air-conditioned diners were put in service. With the cessation of car building due to the Depression, Pullman and the railroads concentrated on converting existing cars to air conditioning, thereby delaying for at least a decade the general public's switch for long distance travel from trains to automobiles. The early air conditioning systems were in some cases rather crude (fans blowing air over large cakes of ice). The awkward ammonia brine systems with water flow condensers, such as had been developed for stationary building installations, required elaborate duct and thermostat systems, and, above all, usually required a greatly expanded electrical system. An axle-gear driven 35 kW generator eventually came to be normal vs the 5 kW belt-driven generator required for lights only. The added weight, especially with a water-cooled condenser, was appreciable, and there was now a recurrence of the 1900 problem–the cars were getting too heavy, with 90-100 ton cars not uncommon.

The outstanding development of the thirties (in addition to airconditioning) was the use of high-strength, lightweight materials in streamlined high-speed trains.

The lightweight car program was carried out as a three-pronged attack during the early thirties. The first new streamliner was riveted aluminum, using a smaller than normal cross-section and articulation between cars. Simultaneously, the Budd Company, which had developed the "Shotweld" process for spotwelding stainless steel, had been developing a stainless train since 1931, in an effort to combine light weight construction with maximum corrosion elimination. Similar efforts resulted in a welded high-strength, low-alloy steel construction, using arc welding for structure and spot welding for sheathing. A number of these special crosssection, integrated trains with limited interchangeability of equipment were built, but by 1938, the roads had all reverted to full-size dimensions and an 85 ft long car with 10 ft width and 13½ ft height became the standard car for U. S. railroads. Both the stainless steel and the Cor Ten welded structure began as a welded truss with sheathing applied as nonload carrying. The later versions used the whole side and part of the roof as a plate girder, with pier panels at the window level designed for shear transfer between the lower girder sheet area and the upper letter board area. There was still a weight reduction of approximately 25 percent of the shell weight (7500 lb) for the stainless car and the HSLA steel car was not far behind. The stainless structure itself seems to be impervious to the passage of time, whereas the steel structure has only survived for the normal life of the car where corrosion protection has been carefully applied.

Coincident with the structural developments, changes were also made in trucks and brakes for improved high-speed operation. A new appreciation of the need for a low natural frequency in the vertical suspension led to much softer suspensions, although the basic form of the truck did not change. A vertical natural frequency of approximately 1 Hz became standard. Lateral suspensions became even softer, with swing hangers having an effective length of 20 in. or more not uncommon, with the hangers and secondary coil springs outside the truck frame for greater stability. A large percentage of the dry friction which had always been tolerated was eliminated, and hydraulic damping took its place. Unfortunately, the damping was grossly overdone in many instances, until it was realized that with the excellent track then common, minimum damping yielded the best ride. Also, the brake cylinders were moved from the car body to the trucks, with a consequent reduction in weight and friction plus improvement in braking efficiency and effectiveness. Roller bearings had been tried as early as the 1880's, but their use now became universal in the late 1930's and early 1940's.

The high-speed operation of many name trains during the thirties (moving at well over 100 mph) required the development of better brakes. The cast iron brake shoe rubbing on the steel wheel tread had been used for many years. However, it had an undesirable variation in friction coefficient with speed, this being very high under about 25 mph. The HSC or High Speed Control brake with speed governor control permitted much higher brake shoe pressures at high speed, with automatic reduction in pressures as the speed decreased. Straight air brake control by electric trainlines operating application and release magnet valves also gave much faster response to the engineman's signal. This permitted train deceleration rates close to practical wheel rail adhesion. Automatic air brake override still maintained the fail-safe condition. Wheel slide detection was also included, since the operation close to the limit of adhesion would otherwise result in excessive numbers of slid flat wheels.

While the conventional tread brake was thus being modified to meet the new demands, the Budd Company was engaged in a competitive development to accomplish the same purpose—the first off-tread main line railway passenger brake. Specifically designed for braking trains from very high operating speeds to obtain smoother, shorter stops with composition friction materials the disc brake made possible higher braking economy and effectiveness than the cast iron clasp brake with speed governor control. This permitted a faster schedule due to its high rate of retardation. The North American disc is usually bolted to the inner or outer rib of the wheel. Other applications have the disc mounted onto an axle hub. The disc includes a bell-like pressed steel center or bell around which is cast a special iron disc which has two parallel braking surfaces connected by many ribs. As the disc rotates, these ribs act as a centrifugal fan for drawing cooling air past the backs of the two braking surfaces. The shoes that press on each side of the disc are faced with a composition material similar to the friction material used in very heavy duty truck braking. By changing the braking from cast iron shoe tread brake to disc, wheel life is extended, thermal cracked wheels are eliminated, sparking and the resulting fire hazard from cast iron tread shoe breaking are eliminated and overall operating and maintenance costs are greatly reduced.

This decade also produced the tight lock coupler. One of the most uncomfortable moments in a sleeping car is that instant when, due to coupler slack, the locomotive is moving considerably faster or slower than the cars, especially the last car. The high acceleration or deceleration which can result is not conducive to a good night's sleep. The type-H coupler with controlled minimal clearance at the knuckle, and universal pin-joints at the shank, alleviates this problem. With the adoption of the "H" (or the "controlled slack") coupler, heavy buffing gears were abandoned, replaced by springing just sufficient to maintain face plate contact at all times.

1945-1955

Very few trains incorporating these new features had been built when World War II brought a halt to all passenger car building. The war's end brought on a stampede to completely re-equip all the main line trains in the U. S. and Canada. The simultaneous sale of the Pullman Company to the railroads brought an end to standardization of the sleeping cars, as each railroad insisted on its own desires and whims.

By this time, the AAR had achieved a great measure of standardization of individual car parts, especially as to material and interchangeability between railroads. When the operation of Pullmans had been a monopoly, a restricted number of floor plans and room arrangements had satisfied everyone Now there came a time when the average number of cars built to a given set of drawings was reduced to seven! Actually, the only really new development in main line cars during this period was the dome car, where the passenger interested in sightseeing could ride well above the ordinary car on adjacent tracks, thereby getting an excellent view in all directions. While the types of accommodations were not new (except roomettes developed in 1938), there was a tremendous amount of engineering ingenuity exercised to getting all the equipment – plumbing, lighting, ventilating, sleeping – tucked out of the passenger's way when not in use, yet readily available when required, done with the goal of squeezing as many passengers into a car as possible, without letting one passenger feel cramped. While new heights of comfort and luxury were thus being achieved without the ornate embellishments of previous eras, the railroad passenger business was falling apart. Attempts were made to repeat the miracle of the thirties by the introduction of ultra-lightweight, very high-speed trains for operation between cities 300 to 400 miles apart – a long distance for auto travel, but a very short distance for plane travel. Technically successful revolutionary trains were introduced in the mid 1950's, but to no avail. Except for the electric multiple unit Metroliner cartrains, almost no new main line intercity equipment was to be built for over 15 years.

While the lightweight train program did not restore a profitable main-line passenger business, it did instigate a large number of developments which have proved useful in the rapid transit and commuter market, and now with Amtrak.

Double-ended lightweight trains were frequently operated in back up moves inadvertently at high speed with the pushing locomotive only. This led directly to the development of push-pull commuter service, which provides the operator with the versatility of electric MU operation but with diesel locomotive power, operated in one direction from the locomotive as usual and in the other from a cab in the lead passenger car. Welded aluminum and more efficient spot-welded stainless steel structures were developed. Paint was eliminated from the interiors of cars, and eventually even from the outside, except for decorative touches. Suspension systems have had the most radical developments in over a century. New wheel arrangements have been experimented with, and a unit brake mechanism for each wheel was developed, using composition shoes on the wheel tread as well as with disc brakes. High voltage, three phase ac power generated on the locomotive and efficiently trainlined became available for economic electric heat, as well as air conditioning.

The efforts to build lighter weight trucks, while improving riding quality, have been concerned with improving the efficiency of the load path from the car body to the wheels. Since the car body load is carried primarily on the ends of the body bolster, secondary springs supporting this bolster at an extreme outboard position minimizes body weight while maximizing stability. The Budd Pioneer III truck and the General Steel GL-70 trucks use this feature. Transferring the load around the swivelling area of the truck through the side bearings minimizes truck weight, as on the Budd and BART trucks. Further simplification and weight reduction can be achieved by designing the truck elements to provide equalization, thereby eliminating primary springs except for the rubber elements at the journal boxes which provide the necessary flexibility, as in the Budd and Rockwell designs. While the Rockwell truck has yet to prove itself in mainline use, it has been used extensively in transit for up to 80 mph operation. The Budd truck has passed the 4000 in service level. All of these trucks can now be provided with air springs. Air springs are hardly needed for the relatively small load variations in sleepers. However, they are very useful in coaches, and an absolute necessity in transit with a passenger load up to 75 percent of body weight, if a modern soft suspension is to be provided.

No radical changes in concepts of construction are apparent for the foreseeable future, at least as long as long trains of the vehicles operate on wheels. Should the U. S. track system be restored to its former high standards, plus welded rail, there will be no trouble providing equipment (or using present newly built equipment) for operation in excess of 150 mph. Present efforts will be devoted toward improving the economics of passenger train travel by moving more people comfortably in shorter trains.

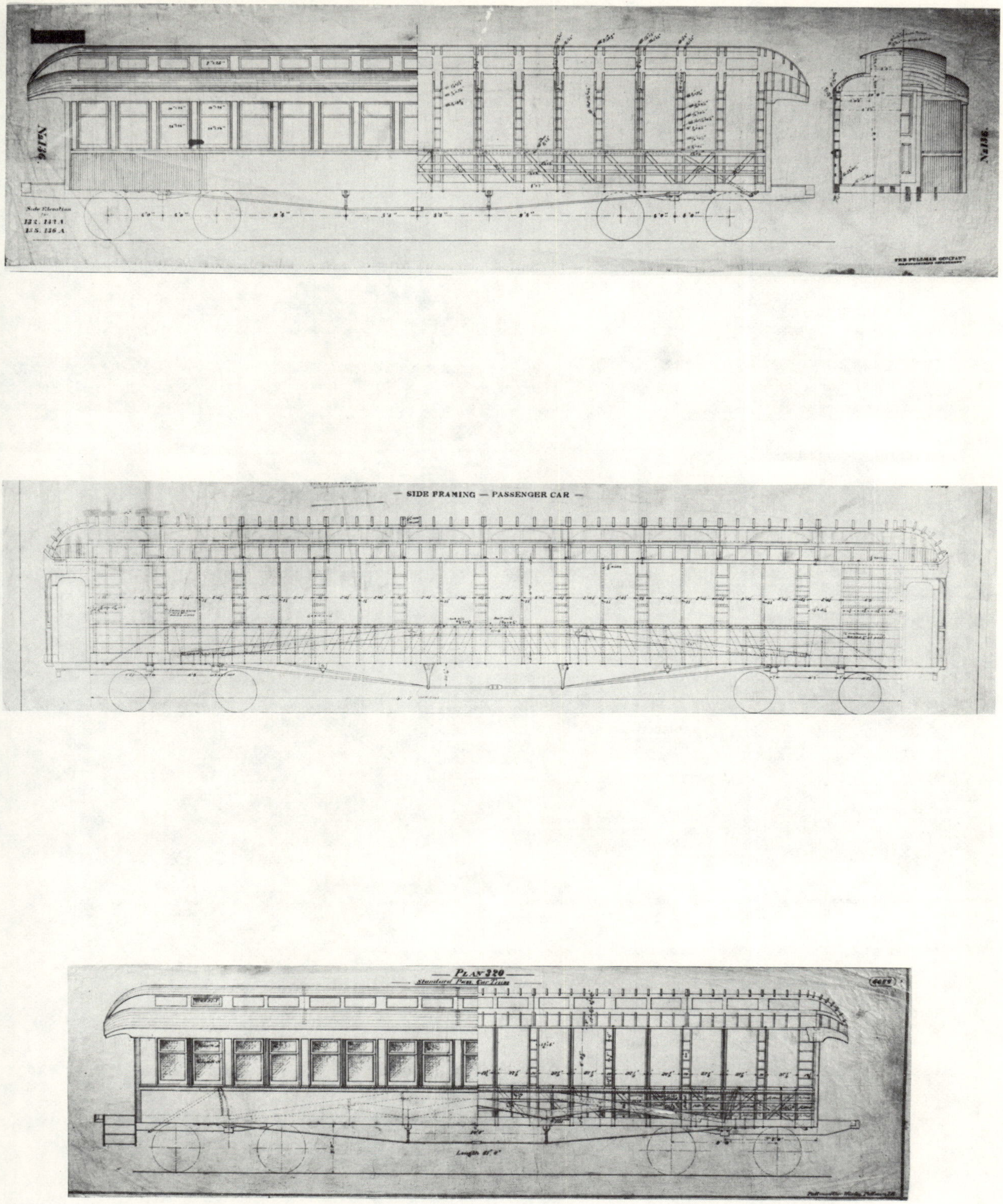

FIG. 1 WOODEN TRUSS DEVELOPMENT

FIG. 2 NARROW VESTIBULE OF 1887

FIG. 3 THE ROCCO PERIOD OF FLORID DESIGN IN SLEEPERS

FIG. 4 THE FIRST TRAIN OF ALL STEEL PULLMAN CARS

FIG. 5 EARLY STEEL CAR STRUCTURE

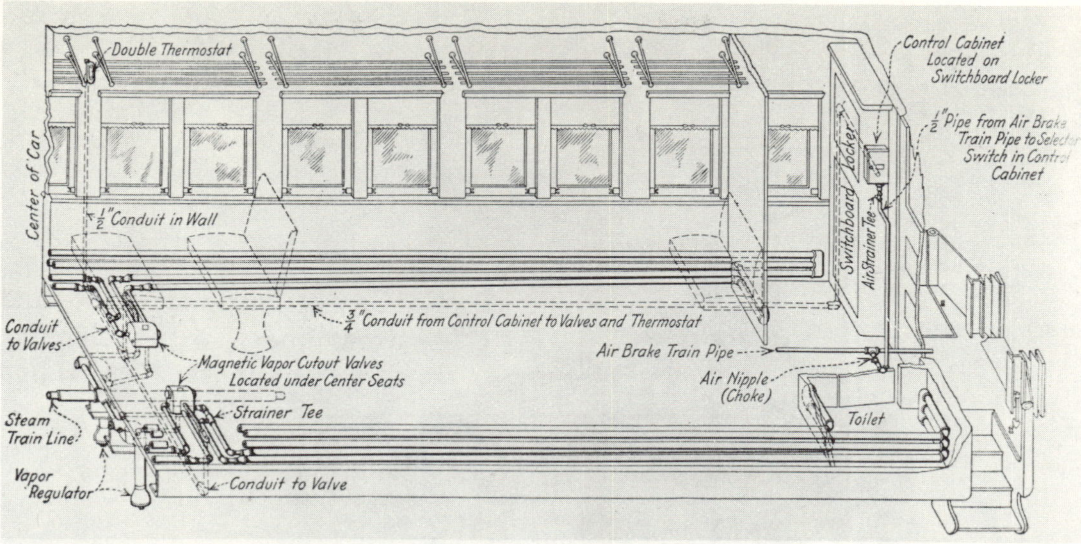

FIG. 6 EARLY STEAM HEAT DEVELOPMENT.
HIGH PRESSURE STEAM UNDER CAR. ATMOSPHERIC PRESSURE STEAM IN CAR.

FIG. 7 TRUSS STRUCTURE OF THE 30's

FIG. 8 ALUMINUM TRAIN OF THE 30's

FIG. 9 STAINLESS STEEL TRAIN OF THE 30's

FIG. 10 MODERN AMTRACK HIGH LEVEL (TWO FLOOR LEVELS) CARS FOR LONG DISTANCE INTER-CITY SERVICE (DAY COACHES, DINERS, SLEEPER, LOUNGE (RECREATION) TYPES)

FIG. 11 TRUCKS (BOGIE) FOR HIGH LEVEL CAR WITH DISC BRAKES AND ANTIWHEEL SLIP CONTROL

9B A Short History of the Railroad Passenger Car

JOHN H. WHITE, JR.

The majority of railroad passenger cars have always been locomotive hauled, nonself-propelled units. They have grown remarkably in size, comfort, cost, strength and complexity in their century and a half history. A car of 1830 measured under 20 ft, weighed three tons and could seat as many as 30 passengers, if the roof top benches were occupied. It had only four wheels, a wooden body, side curtains, but no window glass and it was entirely devoid of toilet, heating, or lighting facilities. Nature provided a free ventilating system. A modern coach by contrast measures 85 ft long, seats 50 and weighs about 55 tons. Its stainless steel body is carried by eight heat-treated steel wheels. Passengers enjoy a splendid view through wide, double safety glass windows. Reclining seats, air conditioning, electric heating and lighting and flush toilets are now common place luxuries. But these advances have come at a price. They represent generations of research and development by car builders, railway engineers and the related railway supply industry—a debt that can never accurately be calculated. Each advance in safety and comfort also represents an increase in initial investment and maintenance costs. For example, the 1830 car cost about $800 while a modern coach sells for around $400,000. Passenger cars have grown ever more complicated and heavy, for they now embody thousands of feet of wiring, conduit and piping. Wire closets, air conditioners and other auxiliary apparatus have added their burden. Car designers have made a determined effort to trim weight; however, the passenger car remains the heaviest land vehicle per passenger, unless one wishes to include army tanks in the contest.

The first passenger cars appeared in this country in 1830. They were produced mainly by carriage builders and accordingly tended to be exact copies of the mellon-shaped stage coaches of that day. Some popular accounts claim that second hand bodies were mounted on railway running gears, but in all documented instances it is clear that the entire car was new. The stage coach style of car, while familiar, did not prove suitable for railway purposes. The curved body was difficult and expensive to fabricate and it could not accommodate glazed windows. The leather strap suspension produced a nauseating rock and roll motion. Other pioneer U.S. lines turned to the British compartment style of car as an improvement over the stage coach variety. These small four wheel coaches—modeled after city omnibus designs—had three compartments with side doors (Fig. 1). The lower body panels were curved, but in general the cars were rather boxy. Glass windows and steel springs were used. Even so, they were very small with limited seating and the four-wheel undercarriage had the unfortunate galloping ride characteristic of all single truck rail vehicles.

The introduction of the eight-wheel or double truck car in 1831 hearlded a marked departure by American mechanics from the established British model. Two independent four-wheel trucks or wheel sets carried each end of the car (Fig. 2). Because they were free to swivel, the car could negotiate sharp curves and at the same time it was possible to materially lengthen the body. Eight-wheel cars could seat nearly as many passengers as three four-wheel cars—this represented a saving in dead weight because of the elimination of extra bulkhead end partitions, couplers and draft gears. The interior of the double truck car was a long, open single compartment. Entrance doors were placed at either end and the center aisle was lined with double seats. End platforms offered egress to the passenger compartment and a means of passage between the cars for the train crew. This basic arrangement prevails to the present time.

Ross Winans claimed exclusive credit for this important innovation and most encyclopedias and popular histories accord him all honors for the eight-wheel car. However, there is ample evidence that he did not work alone and may indeed have played only a small part in evolving the plan; however, it is also clear that the scheme was developed by employees of the Baltimore and Ohio Railroad. The first eight-wheel car, the *Columbus,* was ready for service in the summer of 1831. It was followed by several other experimental prototypes that led to a perfected, production model by 1835. The merits of the design were soon recognized and within five years it was adopted as the national standard. European railways were slow to recognize the superiority of the double truck design and clung to four and six-wheel cars for many years afterwards.

During the first decade of the American type coach, rather little change took place. The railroad industry was relatively dormant during this period because of the Depression following the panic of 1837. Most railroads were short and most trains operated during daylight hours. Hence large, luxurious cars were not generally needed. In addition, most lines had little money for track or rolling stock improvements. Accordingly, cars were kept as small, plain and light as possible. There were instances of luxury cars during this period, but the typical coaches of the 1835-1850 period were short and narrow with an arched roof. The inside ceiling height was often no greater than six feet. To further reduce weight, no center sills were employed. The toilet was a dry closet, a single candle lamp was placed at each end of the car, and a wood burning stove replaced one of the center seats during the winter. Short wheel base trucks with spoked, cast iron wheels offered an acceptable ride at the slow schedules (25 mph) of the period. Pale yellow was a favored color for passenger equipment from the 1830s through the 1860s. Ornamental scrolls and artistically rendered scenes and stripings enlivened their appearance.

By 1855 many factors were at work which resulted in larger and more commodious passenger equipment. The economy, population, and rail network had greatly expanded. Long distance trips were now possible. A monied class was developing that demanded better overnight accommodations than those offered by a stiff backed coach seat. Many trunk lines began to face competition from parallel railroads. Rival lines hoped to woo passengers by offering better trains and faster schedules. Day coaches began to grow in size and appointments. Ceilings were now eight feet high, aisle and seating space was enlarged, window blinds, foot rests, ice water and other small luxuries were added to the passenger's comfort. Wooden trusses built into the car's low side panels permitted the construction of 50 and 60 ft long bodies. More roads began to use six rather than four-wheel trucks. After 1860, ventilation was vastly improved by the adoption of the clerestory roof.* The raised deck probably did more than any one thing to alter the appearance of the passenger car and it remained a distinctive feature of car architecture until the light weight era began in the 1930s.

The general introduction of luxury cars was a more important consequence of expanding rail travel than upgrading the ordinary coach. Care should be taken, however, not to suppose that sleeping, parlor or even dining cars were new ideas in the 1850s. And even greater care should be taken not to ascribe these improvements exclusively to George M. Pullman (1831-1897). Many well appointed sleeping and parlor cars entered service as early as 1838 in this country. A buffet car was tried in 1835 and several parlor or stateroom cars were operated during the 1840s. While well furnished, these case were uncommon and generally ahead of the market demand that would make them economically feasible. T. T. Woodruff was the first to reintroduce the sleeping car idea. His first car was ready in 1857 and within five years, as the largest sleeping car operator in the U.S., he organized the Central Transportation Company. Many other businessmen were attracted to this promising field, among them was George Pullman. He began in a small way with two makeshift cars

*The clerestory was first used by Richard Imlay, a Philadelphia car builder, in 1835, but like so many other new ideas, it was not generally adopted until many years later.

in 1859. Within a few years, Pullman was convinced that a fortune could be made in the sleeping car trade. His talents as a mechanic were actually rather modest, but many historians and popular writers have assumed he came to dominate the business because of his remarkable inventions. This is surely one of the greatest and most persistent misconceptions in railroad history. Pullman gained control of the basic sleeping car patents by leasing the Central Transportation Company in 1870. Amongst the C. T. C. assets were a number of patents for berths and other specialized interior furniture. Yet few of the ideas covered in these patents were new because most had already been used by Richard Imlay who began building sleeping cars in about 1838. However, Pullman is generally given credit for such basic ideas as folding upper berths. Pullman's genius was as an entrepreneur, not an inventor. He overwhelmed his competitors because of his financial and managerial acumen and in those areas he was without equal. Regrettably, he did not live to see the completion of his monopoly in 1899 when the Pullman Company absorbed his last major rival, the Wagner Palace Car Company.

The Palace car began an era of stylish and elaborately finished rolling stock that started in the middle 1860s. Beyond the decorative cabinet woods, bronze hardware, nickel plated lamps, Wilton carpets and other luxurious ornamental trappings, some important engineering changes were introduced. These improvement involved basic car construction, safety devices, heating and lighting. And much that was done first for the Palace cars eventually was taken over by the common day coach. Hot water heating was introduced on a commercial, rather than an experimental basis, by William C. Baker in 1865. Kerosene lighting was widely used by that date and steel tired wheels—R.N. Allen's paper centered wheel was particularly popular—were just beginning to be accepted in America. The air brake and Miller's platform and coupler improved safety—both were widely used on passenger trains by 1875, although it was many years before air brakes or any form of automatic coupler would be widely used in freight service. In 1873 Eli Janney obtained a patent for a swing jaw coupler that would eventually be the nation's standard.*

While these improvements were underway, wooden cars were growing in size and weight. By 1890 most cars measured sixty feet long, seated sixty and weighed 35 to 40 tons (Figs. 5 and 7). A few wooden cars were built as long as 80 ft. Spliced timbers, iron reinforcing plates and an abundance of tie and truss rods were necessary for these long, complex wooden structures. Ends were also reinforced with steel plates to prevent telescoping. By the middle 1880s gas lights began to replace oil lamps in the better classes of cars; some lines were already experimenting with electric lighting, although it did not see much general use until about 1910.

Dining cars had been tested in the late 1860s and buffets even earlier but they were not in general use until 1885. The high cost of operation slowed their acceptance but as one line put them on, their competition felt obliged to do likewise. Much the same can be said about the vestibule. It was a very old idea, but it did not become popular until the late 1880s when reintroduced by the Pullman Company. It permitted safe passage between cars and thus opened a path between the sleeper and diner or the buffet-observation at the end of the train.

The wooden passenger car was fully mature when the 20th century opened and yet it had many critics who claimed it was a menace to the traveling public and that it was long since obsolete. Agitation for fireproof cars began early in the railroad era. It was partially quieted by the gradual adoption of electric lighting and steam heating; however, most cars were still inhabited by the open flame for many first class cars had stove heated, hot water systems and gas lights. Most coaches continued to use common coal stoves and oil lamps. Iron car advocates succeeded in building a full size test car in 1859 and several production models were produced soon thereafter. More patents and sample cars followed, but the railroad industry remained indifferent to these efforts. The projected opening of a midtown Manhattan terminal finally prompted a major railroad to seriously consider the need for a fireproof car in 1904. Officials of the Pennsylvania Railroad decided that the operation of wooden cars through the long approach tunnels was too hazardous—even a small fire could result in a major disaster. Fire resistant cars meant by necessity abandoning wooden construction.

Economics to a far lesser extent hastened the end of the wooden passenger car. The prime

*Readers should beware of statements that Janney's first patent of 1868 covered the basic design of his coupler. It in no way resembled the swing jaw coupler and was never used.

timbers, long, straight, free of large knots or shakes, were growing scarce. American forests continued to provide an abundance of lumber, but the prime trees were fast disappearing, particularly the large white oaks favored for floor frames. As the supply dwindled, the price rose. At the same time (1900), steel and steel fabricating machinery had become more common and lower in price. Concern for passenger safety was surely the chief motivation for introducing steel cars, but the timber supply played a minor role as well.

By 1907 a practical design had evolved from test cars built for that purpose. Large scale production began and soon other major railroads began to purchase steel cars. By 1913 no more wooden cars were produced for domestic service. Although hundreds of new steel cars were produced each year, the wooden car was slow to disappear. In 1920, sixty percent of the passenger car fleet remained wooden. Steel underframes helped to correct this situation and gradually the remaining wooden cars were down-graded to secondary trains. Even so, some remained in commuter service as late as the 1950s.

Superficially the steel cars looked very much like the late model wooden cars. Floor plans, overall size, window arrangements and roof lines were nearly identical (Fig. 8). But weight made a dramatic increase, and while the needs of safety were well served by the conversion, the passenger car's tonnage problem was only aggravated. The cars were built from common steel, meaning that heavy plates and massive underframe sills were needed to support an 80-ft car. To meet or exceed the Railway Mail Service's minimum buff limit of 400,000 lb (imposed in 1911), car builders resorted to cast steel end frames of large I and Z beam collision barriers. Concrete slabs provided a cheap, fireproof floor, but again at the expense of the dead weight ratio. Six-wheel, cast steel trucks with massive wrought steel wheels added to the burden. But the heavyweight, or standard era steel cars were as durable as they were ponderous and many of these Pullman-green, rivet encursted giants provided 50 and more years of service.

During the 1920s, attempts were made to reduce passenger car weight by substituting aluminun wherever possible, but this effort was largely limited to commuter or self-propelled cars. Automotive and aircraft engineering suggested that light metals, high strength steel alloys and new methods of fabrication could bring about remarkable improvements in vehicle weights. Car builders were finally pushed into trying these techniques by the streamlining craze that began in the early 1930s. The benefits of aerodynamic design proved largely imaginary for rail service, but the experimental motor trains like the *City of Salina* showed that passenger cars could be made from something other than ¼-in. steel plate. Light weight aluminum, stainless and Cor-ten steel cars began to appear after 1936. Carefully designed trucks helped trim more unwanted pounds; however, the simultaneous introduction of air conditioning and its attendant power supply made streamliner weights nearly equal to those of an old fashioned heavy weight car. Few full size, ac, light weight cars weighed much under 60 tons.

Ice-cooled car apparatus had been tried in this country during the 1880s and the Santa Fe persisted with such experiments in the first decade of this century. However, no serious attempts were undertaken until the late 1920s. Practical electromechanical units were devised that began a new era in passenger train travel. Steam locomotives prevailed at this time and there was no more disastrous combination than a hot summer day, an open car window and a belching coal burner roaring away at the head of the train. Air conditioning produced a cool, quiet and cinder-free ride for the weary traveler. Large scale installations began in 1930 and within six years 5800 cars were air-conditioned. Most other main line equipment, excepting head end and commuter cars, were similarly equipped by 1950. Air conditioning resulted in a radical change in passenger car appearance. The clerestory was no longer needed and car builders returned to the ancient arch roof design.

Other advances begun in the 1930s were tightlock couplers, roller journal bearings and disc brakes. Here again, predecessors for all of these devices can be found, but in the main they had their commercial beginnings during this decade except perhaps for roller bearings, which were in regular service on a few roads by the late 1920s. Their main contribution was a reduction in hot boxes and easier and less frequent maintenance of journals. Tight lock couplers were an outgrowth of heavy traction technology that offered greater security in attachment of one car to the other. The Budd Company began to market disc brakes in 1938-1939. They reduced wheel wear and helped train stopping to some degree, but they were not generally adopted until about 1955.

The post World War II period promised much in the way of passenger car improvements, but in reality little of real substance was produced. Three thousand new cars were ordered in 1945-1946, but the designs were rubber stamps of prewar equipment. The Dome car and the Santa Fe's hi-levels were welcome innovations, yet they represented a small portion of the national passenger car fleet. In addition, only a few dome cars saw service in the eastern states, and the expected retirement of the heavy weight fleet never took place. By 1950, only 15 percent of our passenger cars were light weights. The ratio improved somewhat in later years, but this was due more to a reduction in the size of the overall fleet rather than in the addition of more streamliners. This lack of progress is again explained by economic rather than technical reasons. The general collapse of the rail passenger market after 1955 essentially ended new car orders and encouraged the retention of obsolete equipment. As more travelers took to the road or the airways, more trains were withdrawn and ever more surplus passenger equipment became available. American Car and Foundry quit the passenger car field entirely in 1961 and Pullman and Budd held on with an erratic diet of transit car orders.

By 1971 the situation was so hopeless that a government sponsored corporation, Amtrak, took over the operation of the remaining long distance passenger trains within the United States. The Pullman Company's sleeping car operation, controlled by an association of operating railroads since 1947, expired in 1968. By this time, America's passenger fleet was no longer the envy of all nations. It was little more than a collection of veteran to vintage rolling stock. Many of the better Santa Fe cars had already been purchased by the privately operated Auto Train Corporation. Only the Metroliners (1969) offered the modern image that Amtrak felt it needed to succeed. In 1973 Federal funds were provided to buy some new cars. The design was a nonpowered version of the Metroliners equipped with Pioneer III trucks—an excellent light weight, inside bearing design dating from 1956. The stainless steel cars were arranged as coaches and coach-cafes. In 1975 and 1976 more luxurious hi-level cars were ordered from Pullman to fill Amtrak's desperate need for new sleepers and diners. While both groups will be welcomed by rail travelers, neither represents a departure from design or fabrication techniques established nearly forty years earlier. Perhaps this is for the best, since the new generation of transit and gas turbine cars produced by our forward thinking aerospace industry have proved to be something less than a remarkable success, despite their technical innovations and futuristic styling.

A NOTE ON SOURCES

No footnotes have been appended to this paper, but readers will find ample documentation together with more details on passenger car construction in *The American Railroad Passenger Car* by John H. White, published in 1978 by Johns Hopkins University Press.

FIG. 1. FOUR-WHEEL WOODEN CARS, WITH SIDE DOORS PATTERNED ON BRITISH DESIGNS, WERE AT FIRST FAVORED BY AMERICAN RAILROADS. THE CARS SHOWN HERE WERE PLACED ON THE BOSTON AND LOWELL RR. IN 1835.

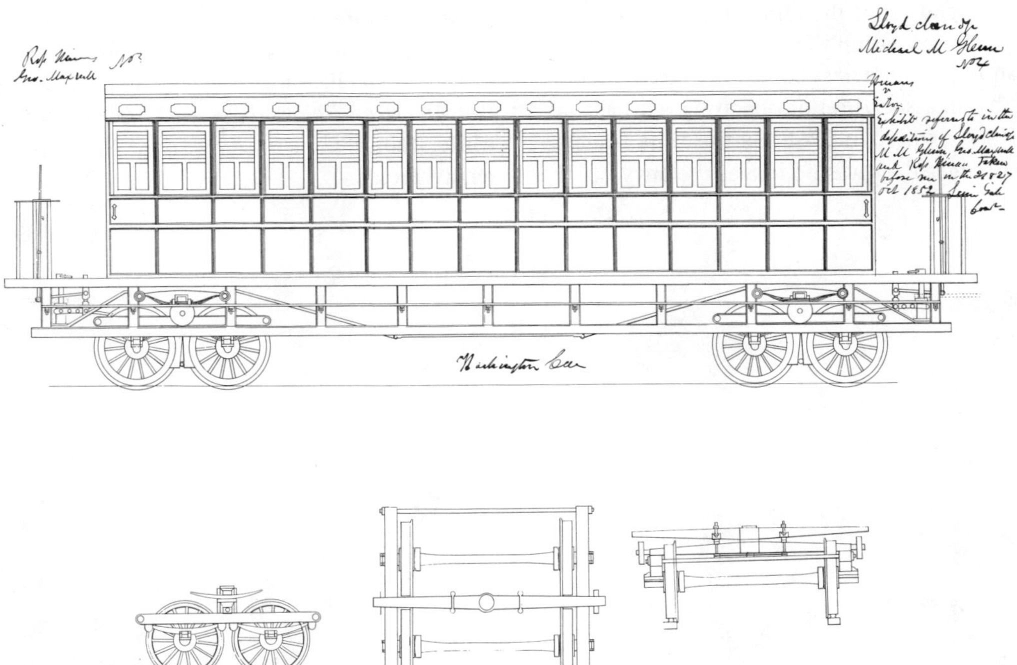

FIG. 2. EIGHT-WHEEL OR DOUBLE TRUCK CARS WITH END PLATFORM ENTRANCES AND CENTER AISLE SEATING, THE BASIC AMERICAN DAY COACH PLAN, WERE DEVELOPED BY THE BALTIMORE AND OHIO RR BETWEEN 1831 AND 1835. SHOWN HERE IS A CAR OF 1835.

FIG. 3. THE INTERIOR OF A TYPICAL AMERICAN COACH OF THE 1850s IS SHOWN IN THIS ENGRAVING FROM THE *ILLUSTRATED LONDON NEWS,* APRIL 10, 1852.

FIG. 4. AFTER 1860 THE ELEVATED OR CLERESTORY ROOF BECAME POPULAR WITH AMERICAN CAR BUILDERS. THE CLERESTORY IMPROVED VENTILATION. THE CAR SHOWN HERE WAS BUILT IN ABOUT 1867.

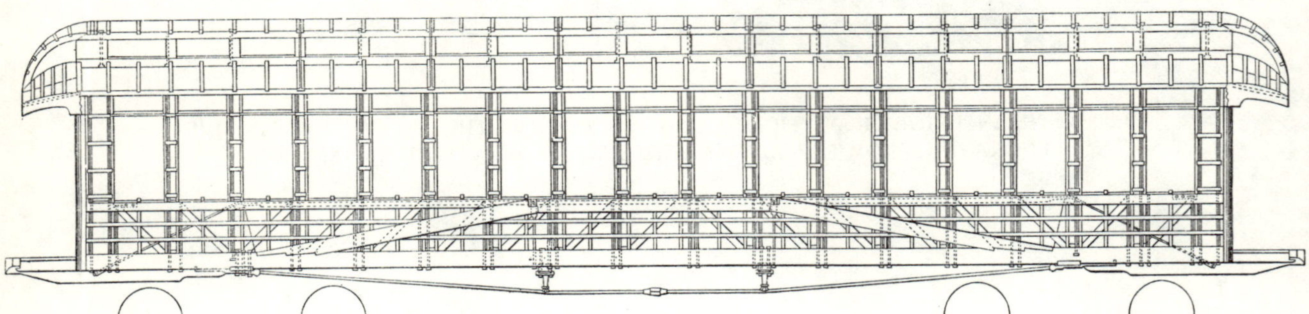

FIG. 5. WOODEN CARS WERE SUPPORTED BY ELABORATE SIDE FRAME TRUSSES, WHICH WERE REINFORCED BY GLUED, MORTISE JOINTS AND IRON TIE RODS. TRUSS RODS BENEATH THE FLOOR SILLS ADDED ADDITIONAL STIFFNESS.

FIG. 6. SLEEPING CARS WERE OPERATED ON AMERICAN RAILROADS LONG BEFORE GEORGE PULLMAN ENTERED THE BUSINESS IN 1859. THE INTERIOR OF A WOODRUFF SLEEPER IS SHOWN HERE. WOODRUFF WAS THE LARGEST U.S. SLEEPING CAR COMPANY BEFORE ITS LEASE TO PULLMAN IN 1870.

FIG. 7. LATER IN THE WOODEN CAR ERA, CARS GREW IN SIZE AND WEIGHT. VESTIBULES, HOT WATER HEATING AND GAS LIGHTING BECAME COMMON FEATURES ON THE BETTER CARS. THIS EXAMPLE WAS BUILT IN 1895 BY JACKSON AND SHARP OF WILMINGTON, DELAWARE.

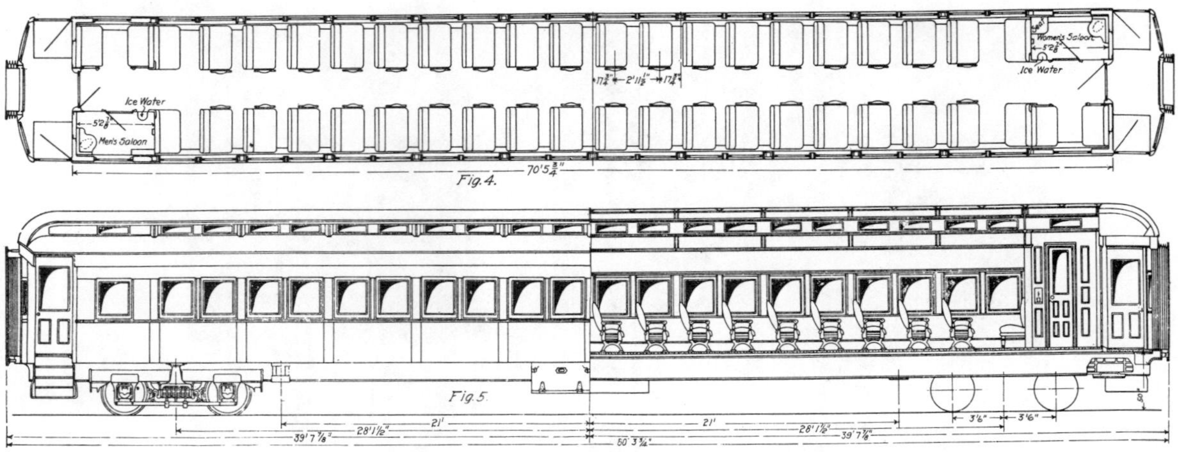

FIG. 8. QUANTITY PRODUCTION OF STEEL PASSENGER CARS BEGAN IN 1907, WHEN THE PENNSYLVANIA RAILROAD STARTED AN AMBITIOUS PROGRAM (NOT COMPLETED UNTIL 1928) TO CONVERT ITS FLEET FROM WOOD TO STEEL. OTHER RAILROADS PROCEEDED AT AN EVEN SLOWER PACE.

FIG. 9. MASSIVE AND SOLID, MOST HEAVY WEIGHT ERA STEEL CARS OPERATED FOR UPWARDS OF HALF A CENTURY. THIS B & O OBSERVATION CAR WAS PRODUCED BY PULLMAN IN 1929.

FIG. 10. COACH INTERIORS IN THE HEAVYWEIGHT ERA (1910-1930) WERE PLAIN AND UTILITARIAN. SEAT BACKS COULD BE REVERSED BUT DID NOT RECLINE.

FIG. 11. IN 1933 THE NEW FAD OF STREAMLINING SWEPT AWAY THE TRADITIONAL IMAGE OF THE RAILROAD PASSENGER CAR. BUT THE BASIC ELEMENTS OF ART DECO WERE SOON COMPROMISED AND WERE ONLY SUPERFICIALLY USED IN LATER LIGHT WEIGHT CARS.

FIG. 12. THE USE OF RUST RESISTANT ALLOY STEELS AND WELDED BODY FABRICATION DID MUCH TO SMOOTH AND LIGHTEN THE MODERN AMERICAN PASSENGER CAR. THE 1721 WAS BUILT BY PULLMAN IN 1941.

FIG. 13. THE MOST DURABLE AND EXPENSIVE CARS WERE BUILT FROM STAINLESS STEEL. THE BUDD COMPANY FABRICATED THE ENTIRE STRUCTURE INCLUDING THE UNDERFRAME FROM STAINLESS. OTHER BUILDERS PRODUCED ALLOY STEEL AND ALUMINUM CARS.

FIG. 14. ONE OF THE FEW IMPORTANT POST WAR PASSENGER CAR INNOVATIONS WAS THE DOME CAR. PULLMAN BUILT THIS CAR IN 1947 FOR THE TRAIN OF TOMORROW.

10 The History and Development of Railway Passenger Car Trucks in the United States

F. T. BURGESS and K. L. JACKSON

To cover as complex a subject as truck development in a brief article presents a real challenge. So before outlining what the paper covers, we should perhaps begin by stating what it does not. It will make no reference to locomotive trucks, nor will those special, exotic passenger trucks of limited usage be included. No attempt will be made to outline features of foreign trucks. The authors will concentrate on the development of the main line passenger truck and certain important developments in commuter and rapid transit applications in the U.S. In general, our research tells us that the patented art provides a good chronology of the important milestones in truck development, and we have drawn generously from that art for most of the illustrations.

In considering modern truck designs, anyone with a passing knowledge of conventional running gear probably could set down a list of the important features in a passenger truck. Such a list might include a *pivot of some sort*, a *vertical suspension, means for keeping the truck in tram, means for transferring the longitudinal braking and traction forces* through the truck, a system for *equalizing wheel loads,* and a *lateral suspension*. The designers of yesteryear, however, did not have the advantage of hindsight and so the above items were invented and applied one at a time. During the period from 1830 to 1860, all of these elements were brought together, resulting in a truck concept frequently referred to as the M.C.B. truck (from the Master Car Builders' Association). From 1860 to the present, a period of refinement took place in response to the changing weights, speeds, and operating conditions. This included the introduction of inboard bearing trucks for rapid transit cars in the 1950's, along with new developments in braking and propulsion systems which made this form of truck feasible. But the same principles which enabled the M.C.B. truck of the 1860's to do its job are still found in the successful trucks of the 1970's.

Now let us go back to the beginning. Prior to 1831 the few railway passenger cars in use in the United States were largely of the stagecoach-body type mounted on four wheels and, if sprung at all, used the thoroughbrace suspension common on road carriages of that period. Longer cars were subsequently built, frequently consisting of three compartments, each similar to an individual stagecoach body, but also mounted on four wheels. In many cases, the single truck consisted of a rectangular wooden frame with the axles or journal boxes clamped to its side beams. Serious accidents frequently occurred when one of the two axles broke.

Considering curvature conditions on the few American railroads in existence at that time, any attempt to extend the rigid wheelbase of the four-wheel cars to a great enough length to provide a longitudinally stable support would have caused interference with curving movements of the car, resulting in wheel flange and rail wear. The idea of a long car mounted on a pair of four-wheel trucks arranged to swivel about their centers with respect to the body was proposed as early as 1825, but the

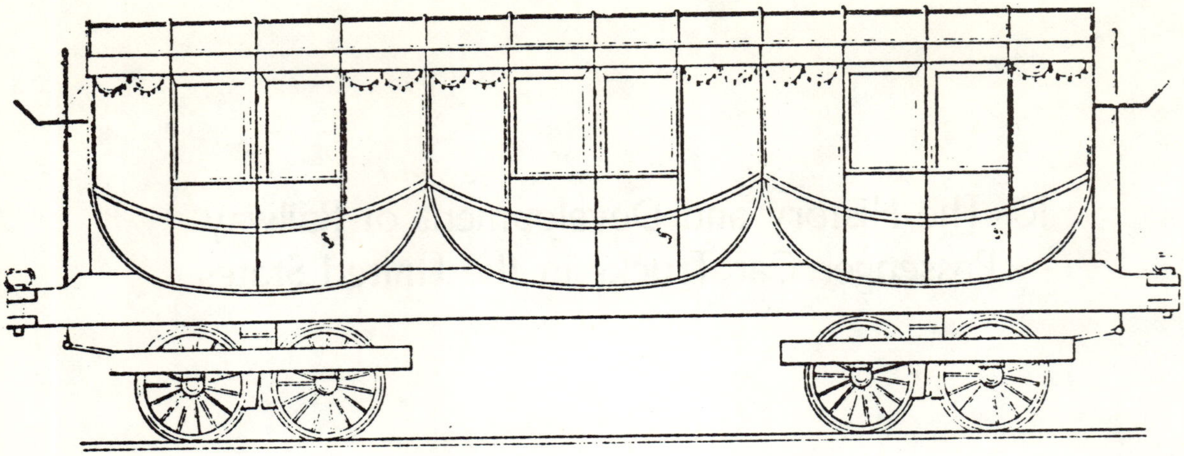

FIG. 1

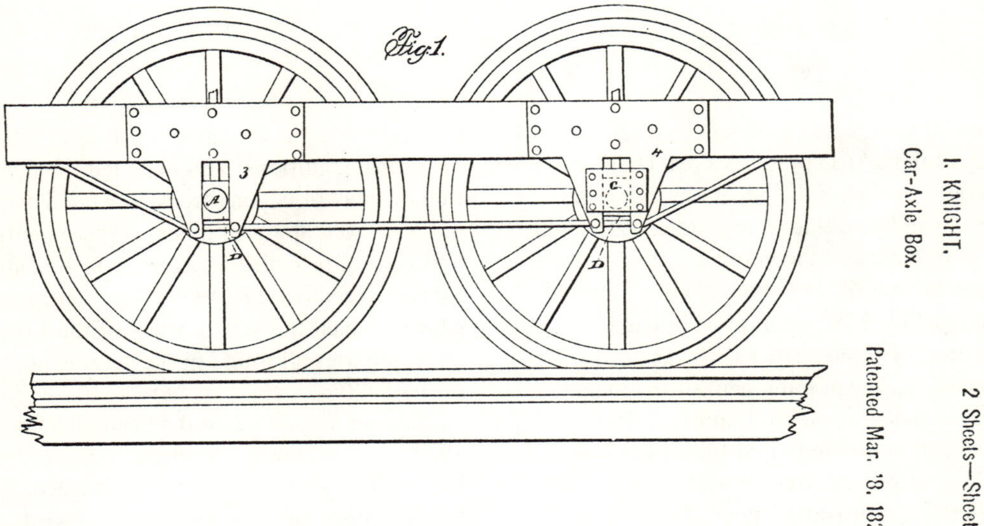

FIG. 2

first passenger car to be constructed in this manner appears to have been the "Columbus" of the Baltimore & Ohio in 1831. An early application of swivel trucks to a multi-compartment car built by the B&O in 1832 is illustrated in Fig. 1. Apparently no resilient suspension was included in these trucks.

On October 1, 1834 a patent was granted to Ross Winans on the concept of a relatively long car body mounted on a pair of swiveling four-wheel trucks. The truck as described consisted of two wheeled axles, semi-elliptic springs supported at their opposite ends on the journal boxes and forming in effect resilient side frames, and a truck bolster supported at its ends on the midpoints of the springs with a center pin swivelly connecting the truck bolster to a body bolster resting on it. An interesting feature is that if the spring at one side deflected more than that on the other, the axles would not remain parallel. The need for a resilient suspension, both for the comfort of the passengers and to avoid damage to the car body and track, was recognized at an early date. In the Winans patent, this was provided solely by the springs supporting the bolster from the journal boxes.

As early as 1829, means for positively maintaining the axles parallel with each other, while permitting relative vertical movement of the axles to accommodate track irregularities, were well-known and were shown in a patent granted to Isaac Knight (Fig. 2) disclosing a pair of pedestal jaws for guiding the journal boxes, thus fixing the position of the axles in the horizontal plane such that the axles would be parallel and the wheels in longitudinal alignment with each other. In the Knight patent a stem extends upwardly from each of the boxes

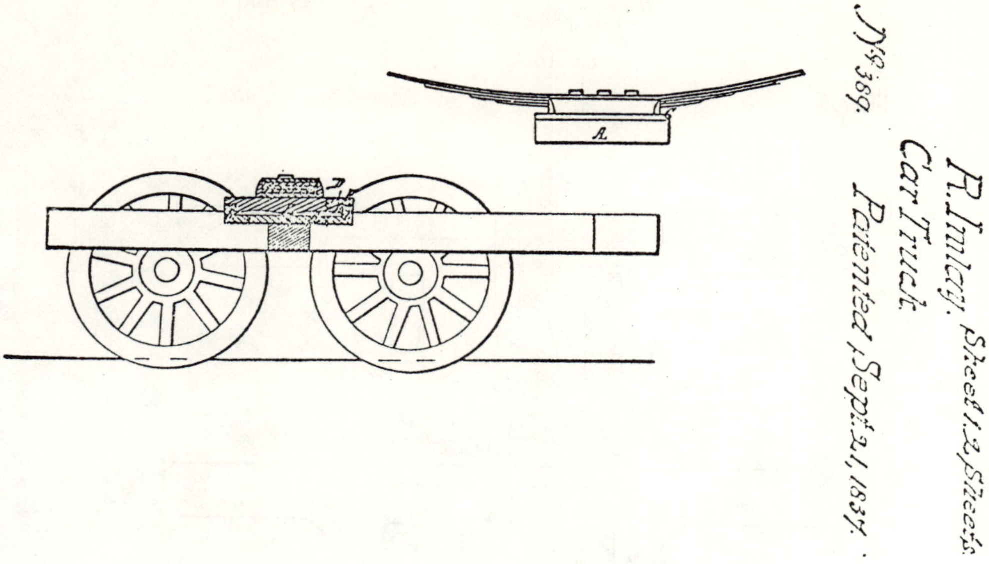

FIG. 3

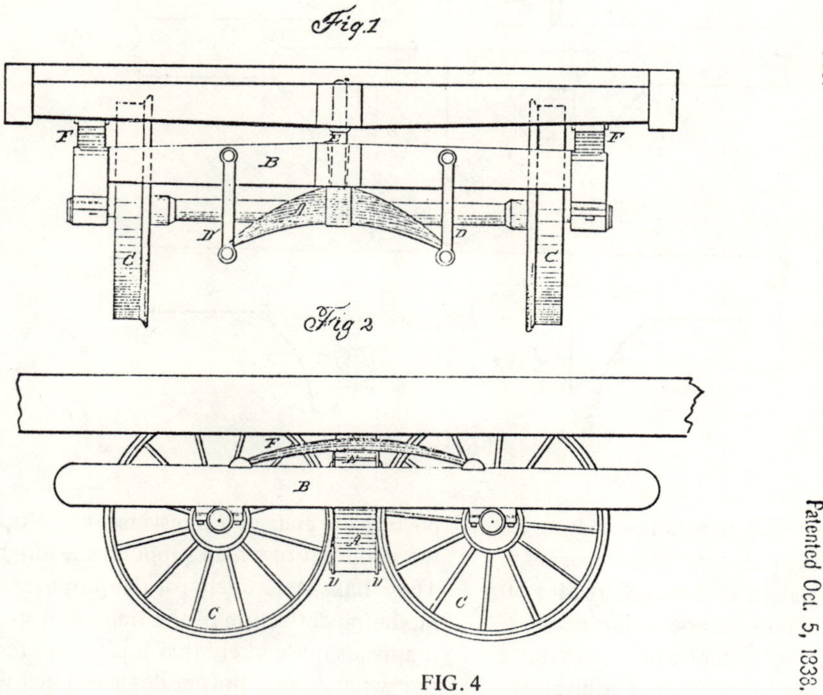

FIG. 4

through the wheelpieces of the truck frame for connection to a spring (not shown) by which the wheelpieces are resiliently supported from the axle boxes.

Cars built by the B&O for its new Washington Branch in 1835 utilized the primary suspension of the Winans' truck described above but, in addition, included a secondary suspension consisting of semi-elliptic springs supported from the ends of the body bolster for supporting the car body. This appears to have been the first truck incorporating a double spring system.

Another form of secondary suspension, although not shown in combination with a primary suspension, was disclosed in R. Imlay Patent No. 389 (Fig. 3) of Sept. 21, 1837, in which the body bolster comprised a semi-elliptic spring supporting the car body at its ends and supported at its center

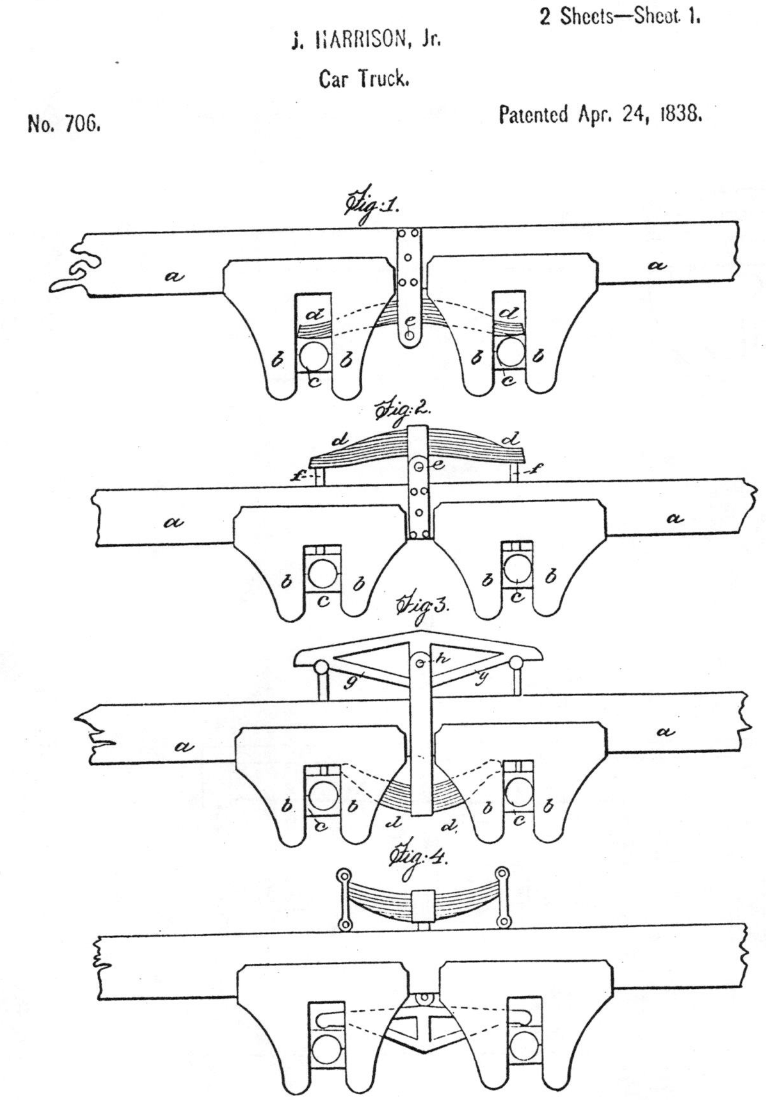

FIG. 5

on the truck frame center transom by mating center plates for pivoting purposes. This appears to have been the first patent to disclose the idea of a center plate comprising a shallow cylindrical pocket on the truck bolster, pivotally receiving a cylindrical boss on a body bolster or equivalent structure. Prior to this, the pivot-forming member had been a single kingbolt passing through the body and truck bolsters, in some cases the body bolster resting directly on the truck bolster and in other cases the body supported at its sides by spring-supported side bearings.

Another early suspension of the car body from the truck frame was disclosed in William A. Davis Patent No. 963 (Fig. 4) of October 5, 1838 which shows a rigid truck frame with swingable links supporting the ends of a transverse semi-elliptic spring, the midpoint of which supports the kingbolt. These links were solely for the purpose of accommodating deflection of the transverse springs, but it appears quite likely that this was a precursor of the swing motion bolster described below.

With a rigid truck frame directly supported on the axles, it would be impossible to equalize the load on each wheel when abnormal track conditions would occur, in which low spots on one of the track rails might be opposite high spots on the other. To compensate for this condition, Joseph Harrison, Jr. obtained Patent No. 706 (Fig. 5) on April 24, 1838. This patent discloses four different means for equalizing the load on the individual wheels of a four-wheel truck. Figure 4 of the

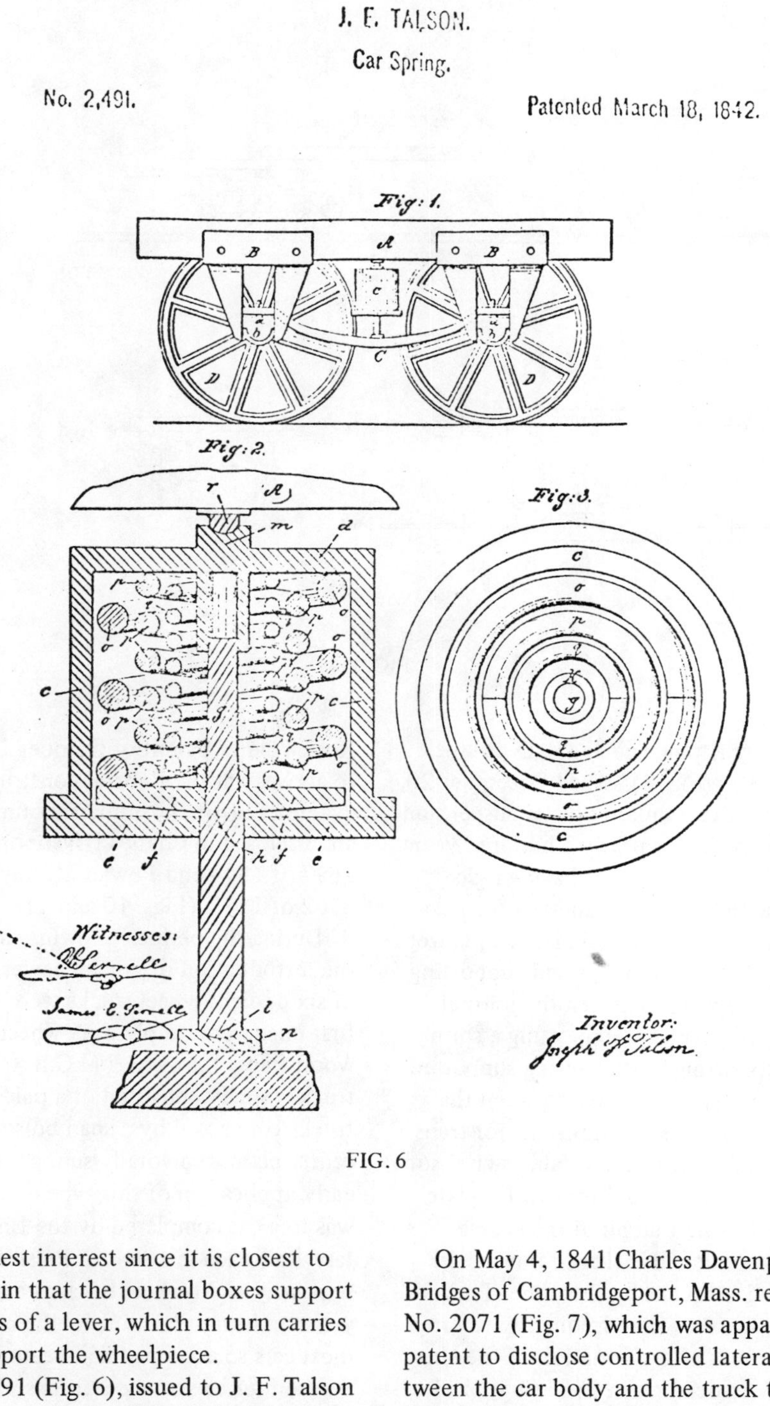

FIG. 6

patent is of greatest interest since it is closest to modern practice in that the journal boxes support the opposite ends of a lever, which in turn carries the means to support the wheelpiece.

Patent No. 2491 (Fig. 6), issued to J. F. Talson in 1842, was apparently the first disclosure of a drop equalizer with its ends resting on the journal boxes and coil springs seated between its ends for supporting the truck frame. It has been impossible to ascertain when the first trucks with equalizers were placed in passenger train service. However, photographs of passenger trains purportedly taken in the middle 1850's show trucks with drop equalizers.

On May 4, 1841 Charles Davenport and Albert Bridges of Cambridgeport, Mass. received Patent No. 2071 (Fig. 7), which was apparently the first patent to disclose controlled lateral movement between the car body and the truck to dissipate lateral shocks received by the wheels from the track by supporting the truck bolster from the truck frame through swing hangers. Although Davenport and Bridges had first taught the desirability of lateral motion and the use of swing hangers to achieve it, the first disclosure of inclined swing hangers supporting a spring plank, with transversely extending full-elliptic springs seated on the spring plank to support the bolster,

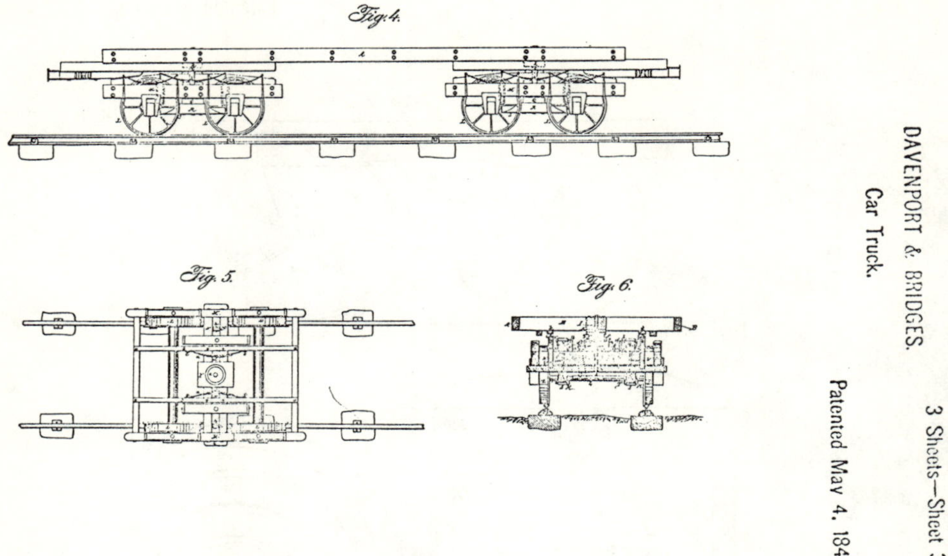

FIG. 7

was made by Henry Kipple and Jacob D. Bullock in Patent No. 26,502 (Fig. 8), issued December 20, 1859. This became the standard bolster suspension for passenger car trucks for nearly a hundred years.

By 1860 trucks were in service having rigid wooden frames with bolted-on pedestals for positioning the axles in the horizontal plane, equalizer beams seated on the journal boxes and supporting the truck frame by springs between the journal boxes, inclined swing hangers supporting a spring plank, a body-supporting bolster spring-supported on the spring plank, chafing plates between the bolster and truck frame center transoms for transmitting longitudinal forces, and a stable swivel support of the car body on the bolster (center plate and side bearings). Note that all of the six elements of design mentioned earlier in the article had been included in trucks by this time. Trucks combining these features were frequently referred to as "M.C.B." trucks after formation of the Master Car Builders' Association in 1867. A typical truck of the 1860's is shown in Fig. 9.

From 1860 to 1904 efforts in the railway passenger car truck field were devoted largely to refinement of the features enumerated above and increasing the strength of trucks because of substantial increases in car body weights and train speeds.

Long before the development of the so-called M.C.B. truck, trucks of all-metal construction had been patented. In fact, the truck described in the Ross Winans' patent of October 1, 1834, referred to above, was of all-metal construction. Two other examples of trucks primarily of metal construction are disclosed in Charles Davenport's Patent No. 3697 of 1844 and Fowler M. Ray's Patent No. 3962 of 1845. (Figs. 10 and 11).

During the period following the Civil War, with the introduction of much heavier cars, a number of six or eight-wheel trucks were introduced. The first ones were simply a six-wheel version of the wood frame four-wheel M.C.B. truck. Eight-wheel trucks usually consisted of a pair of four-wheel trucks connected by a span bolster containing a center plate to pivotally support the car body. An early application of this type of eight-wheel truck was to a car completed by the United States Military Railroads in 1864 for President Lincoln. Subsequently eight-wheel trucks were used for a few years on heavy cars, but by the middle 1870's most cars so equipped had been modified to incorporate four or six-wheel trucks. One such six-wheel truck is shown in Patent No. 77,858 (Fig. 12) issued to C. F. Allen in 1868.

In 1869 Patent No. 86,416 (Fig. 13) was issued to James Kirkley and Hugh Gray of Chicago for a six-wheel truck corresponding closely to the so-called M.C.B. design but in which most of the members, except for a few small castings such as the pedestals, were iron plates riveted together.

Despite the invention of a number of all metal

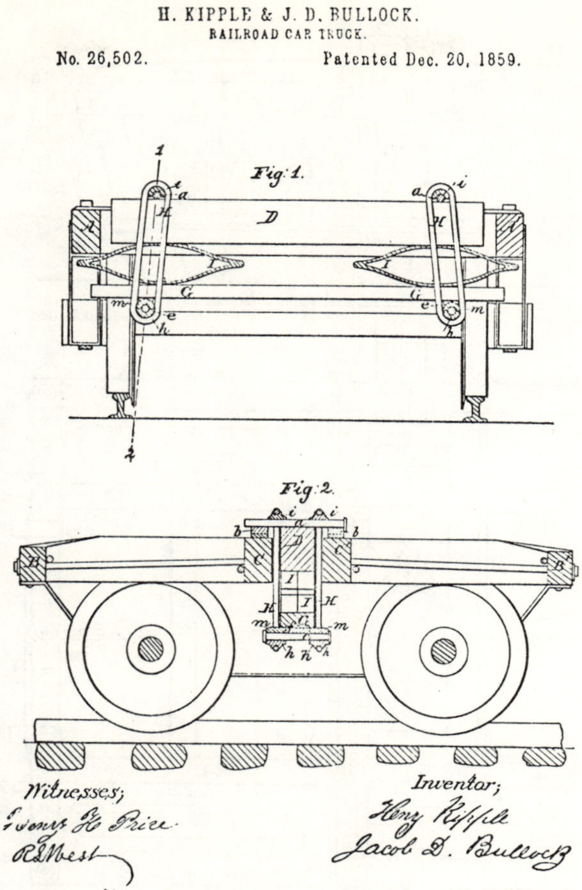

FIG. 8

trucks, as noted above, when the all-wood frame became inadequate to support the extremely heavy cars built during the latter part of the nineteenth century, the basic wood frame structure was retained but was reinforced with trusses made, from bars, rods and plates, and the vertical surfaces of the wheelpieces and transoms were sheathed with heavy metal plates to achieve a much stronger truck frame than the all-wood constructions.

In 1889 Edward W. M. Hughes of the Fox Solid Pressed Steel Company of Chicago obtained Patent No. 408,034 (Fig. 14) on a truck frame in which the wheelpieces, pedestals and end and center transoms were pressed from a single steel blank. The resultant one-piece frame was of inverted channel cross section throughout. It is believed that this design was the earliest approach to a one-piece frame, but was not used extensively despite its advantageous one-piece construction.

Notwithstanding late nineteenth century all-metal truck developments, at the turn of the century the conventional passenger car truck was of composite wood and metal construction.

In 1904, the Commonwealth Steel Company produced the first trucks with one-piece cast steel frames, under Harry M. Pflager and Clarence Howard, Patent No. 774,492 (Fig. 15), which was issued November 8, 1904 for a six-wheel truck having a one-piece cast steel frame with I-section wheelpieces and transoms and integral pedestals having T-section legs. The bolster, including its transverse end beams and central bridge member, was also a one-piece casting, as were the spring planks. Cast single equalizers having bifurcated ends for use with the integral pedestals were also provided. Essentially, the truck shown in the patent was of M.C.B. configuration. Although the patent disclosed a six-wheel truck, the Commonwealth Steel Company also manufactured four-wheel M.C.B. trucks of similar construction.

As the years progressed, the Commonwealth Steel Company continued to improve and refine the design of cast steel truck structures, at an early date substituting box-section and inwardly facing channel members for the I-section wheelpieces

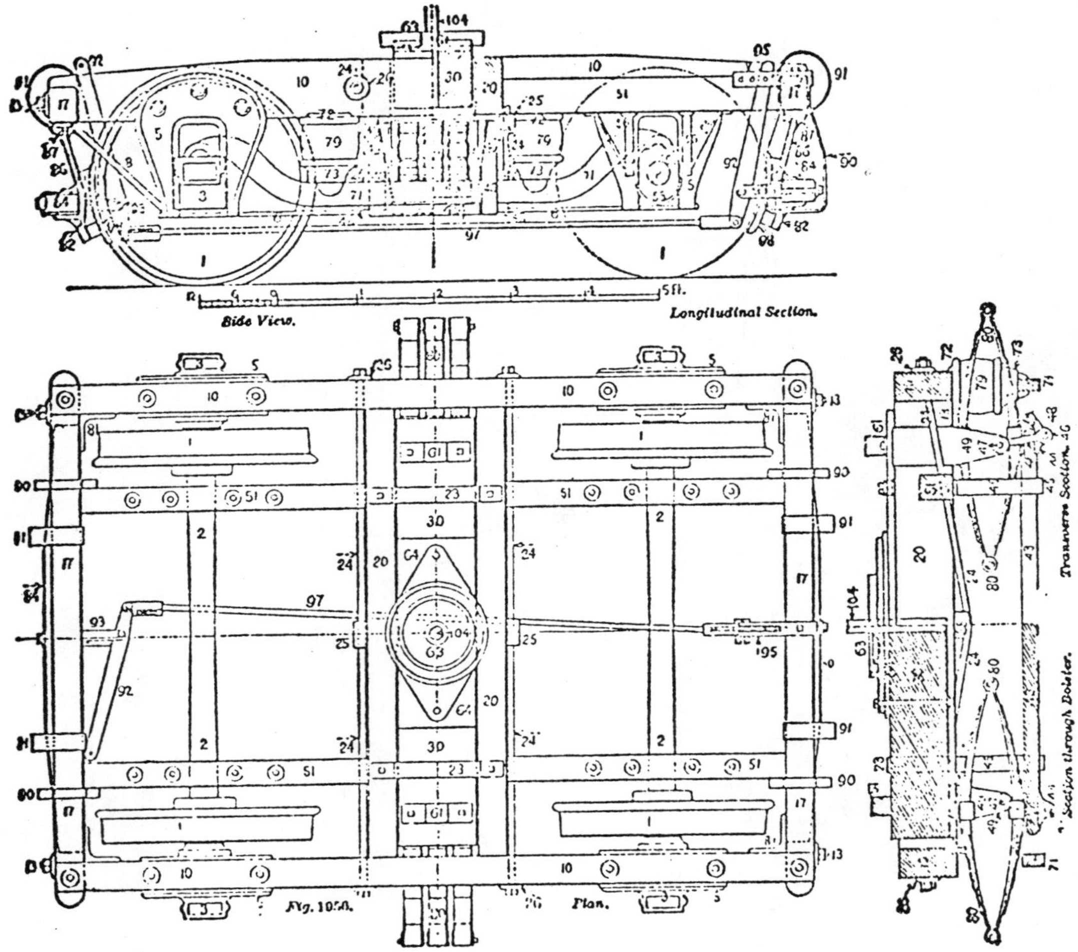

FIG. 9

and end transoms, and separate pedestals for the original integral pedestals to permit the use of conventional single equalizers. In the middle 1920's truck frames for single equalizer trucks were again cast with integral pedestals, this development being made possible by H. M. Pflager's invention of the removable pedestal filler block covered by Patent 1,584,133 (Fig. 16) issued May 11, 1926. Throughout this period, of course, many one-piece cast frames had incorporated integral pedestals for use with double equalizers.

During this period of refinement of trucks with one-piece cast steel frames, some designs deviating from the M.C.B. features were originated. A notable departure was the fabricated steel truck shown in Fig. 17 developed by William F. Kiesel, Jr. of the Pennsylvania Railroad in 1905 (Patent No. 800,921, issued October 3, 1905) and used on the first series of Pennsylvania P-70 coaches. This truck utilized wheelpieces each formed by a pair of inwardly facing channels with cast steel pedestals riveted between them, eliminated equalizers by providing springs over the journal boxes, and eliminated a spring plank and center transoms by supporting the swing hangers from the wheelpieces and supporting a separate elliptic spring carrier at each side on the lower ends of the swing hangers. For centering the bolster and controlling its lateral motion with respect to the truck frame, transversely extending coil springs mounted in the wheelpieces were arranged to act against upstanding abutments on the bolster inboard of the wheelpieces and side bearing abutments on the extended ends of the bolster outboard of wheelpieces. Kiesel also patented a similar six-wheel version of this truck (Patent No. 796,733, issued August 8, 1905). Many of these trucks were ultimately replaced by cast steel equalized trucks manufactured by Commonwealth.

Prior to World War I, most of the principal pas-

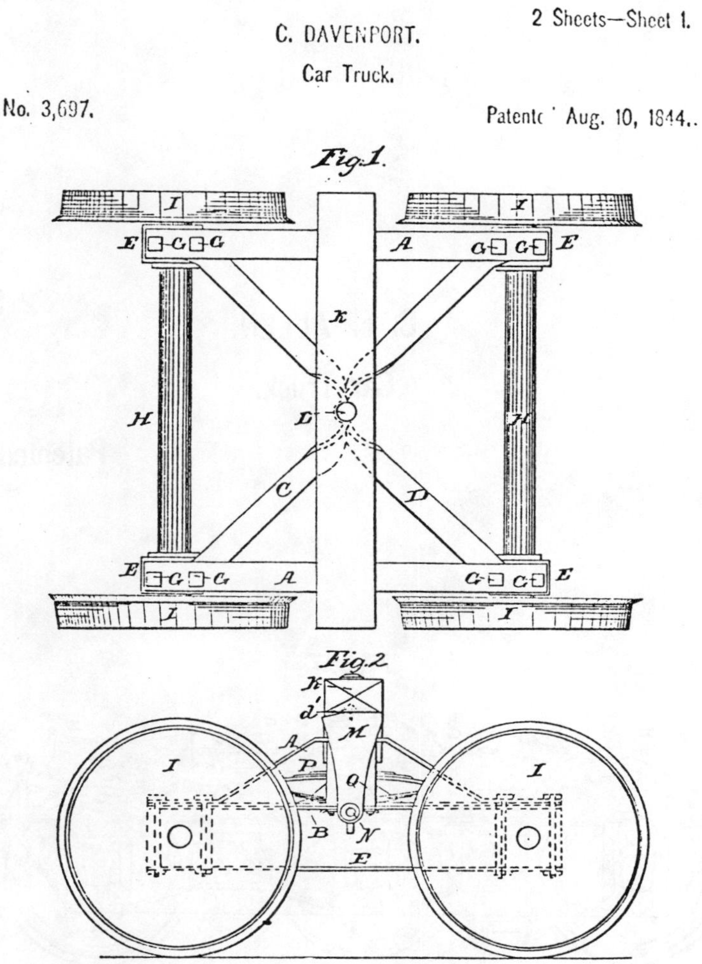
2 Sheets—Sheet 1.
C. DAVENPORT.
Car Truck.
No. 3,697.
Patentc ' Aug. 10, 1844..
Fig.1.
Fig.2

FIG. 10

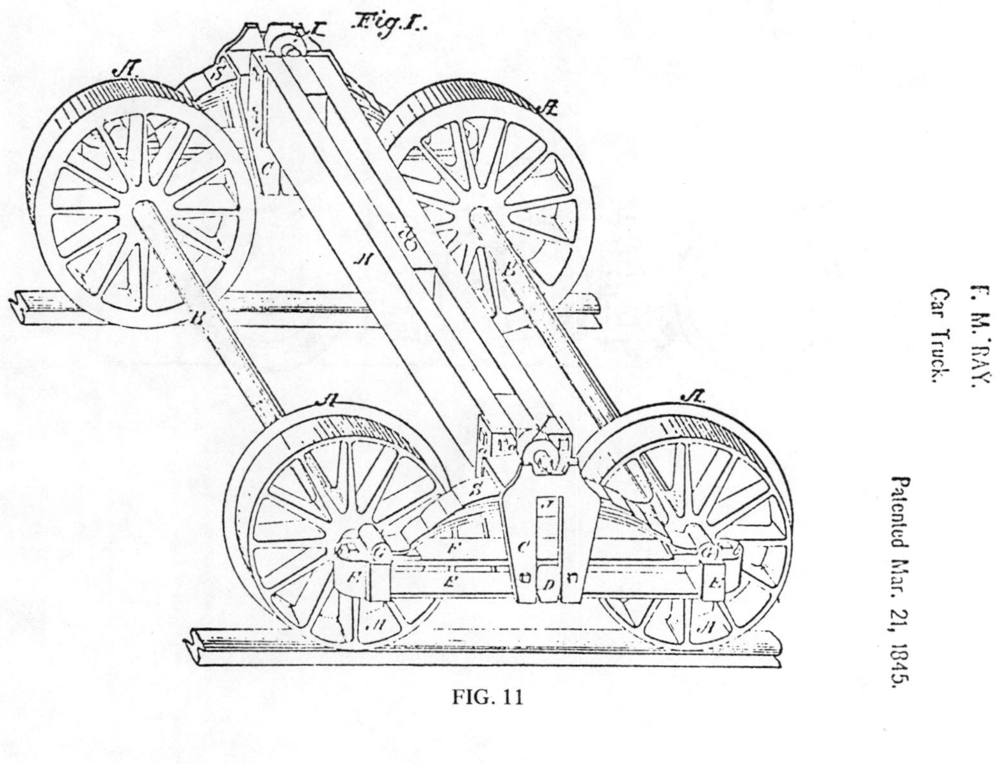
Fig.1.
F. M. RAY.
Car Truck.
Patented Mar. 21, 1845.

FIG. 11

2 Sheets—Sheet

C. F. ALLEN.

Car Truck.

No. 77,858. Patented May 12, 1868

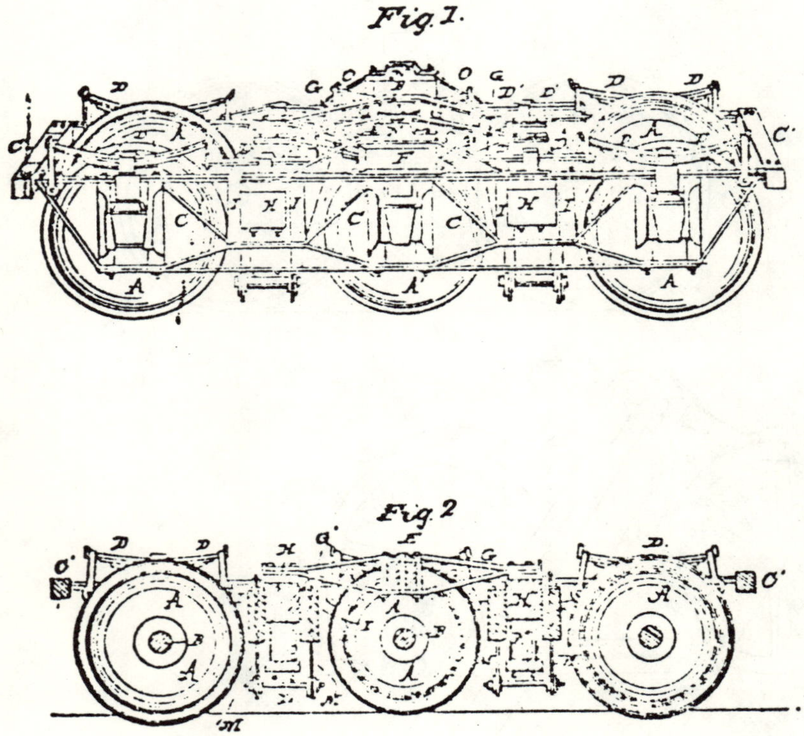

FIG. 12

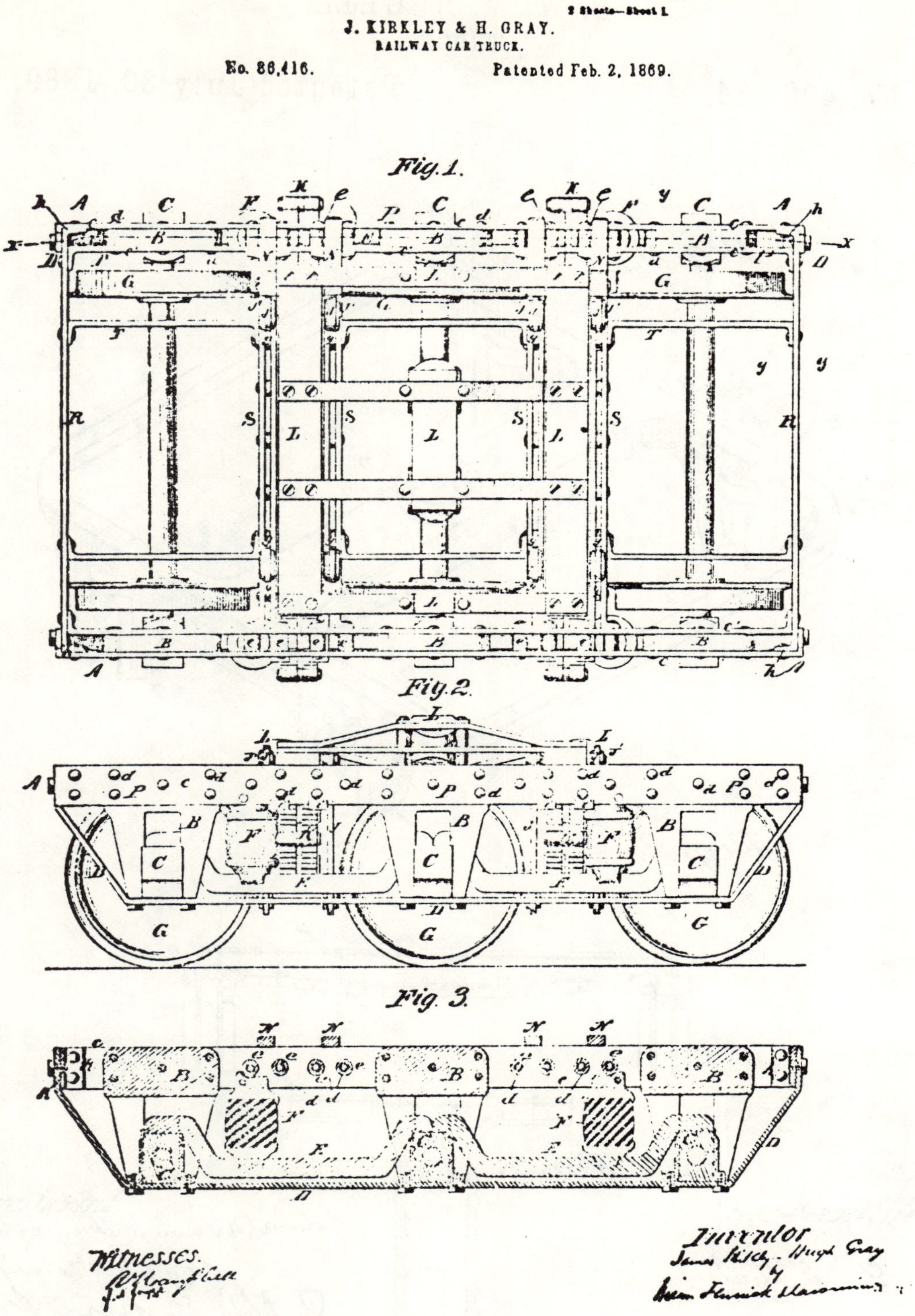
2 Sheets—Sheet 1.
J. KIRKLEY & H. GRAY.
RAILWAY CAR TRUCK.
No. 86,416.
Patented Feb. 2, 1869.
Fig. 1.
Fig. 2.
Fig. 3.
Witnesses.
Inventor

FIG. 13

(No Model.) 2 Sheets—Sheet 1.

E. W. M. HUGHES.

TRUCK.

No. 408,034. Patented July 30, 1889.

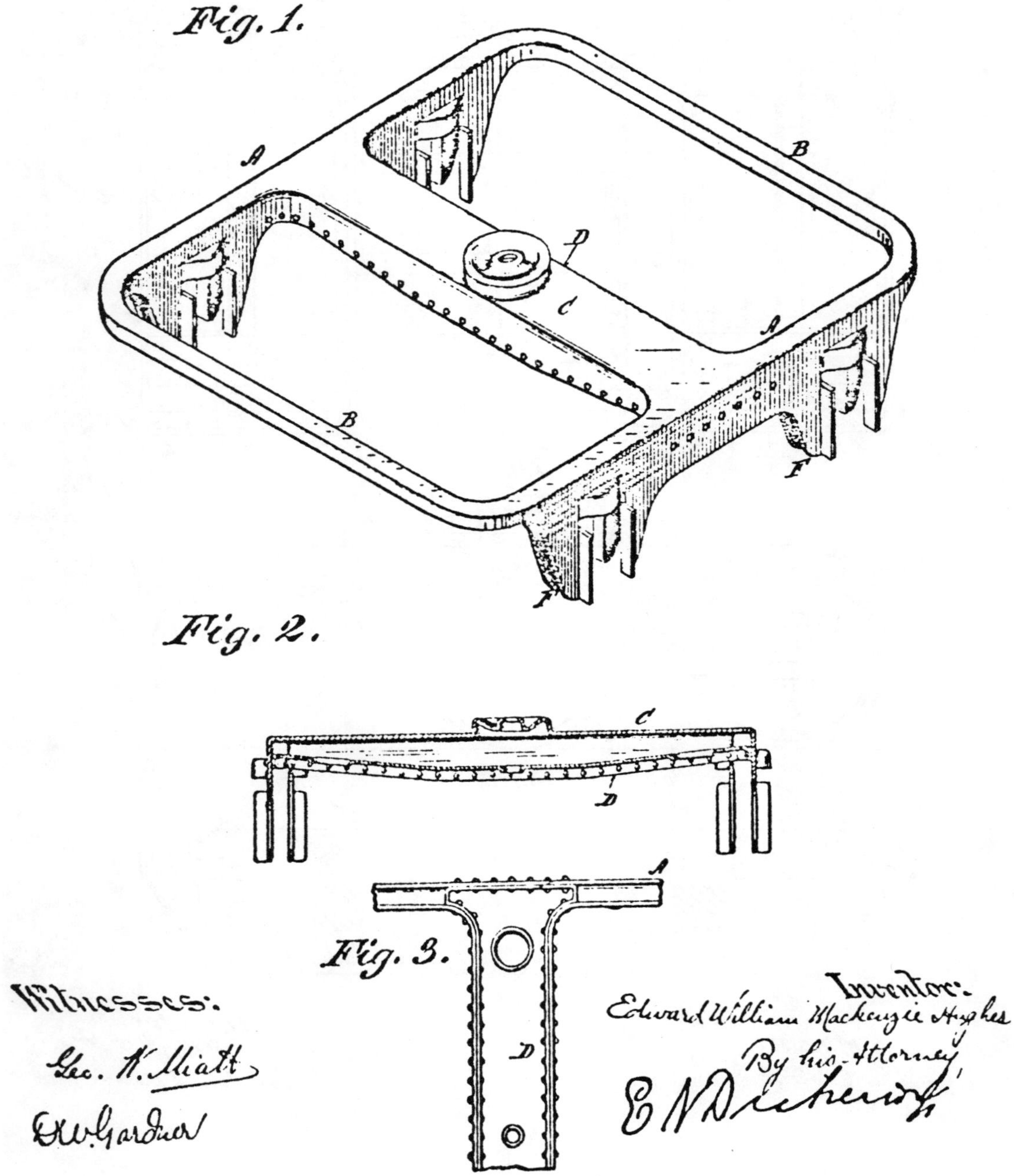

FIG. 14

No. 774,492. PATENTED NOV. 8, 1904.

H. M. PFLAGER & C. H. HOWARD.

CAR TRUCK.

APPLICATION FILED JULY 11, 1904.

NO MODEL. 5 SHEETS—SHEET 1.

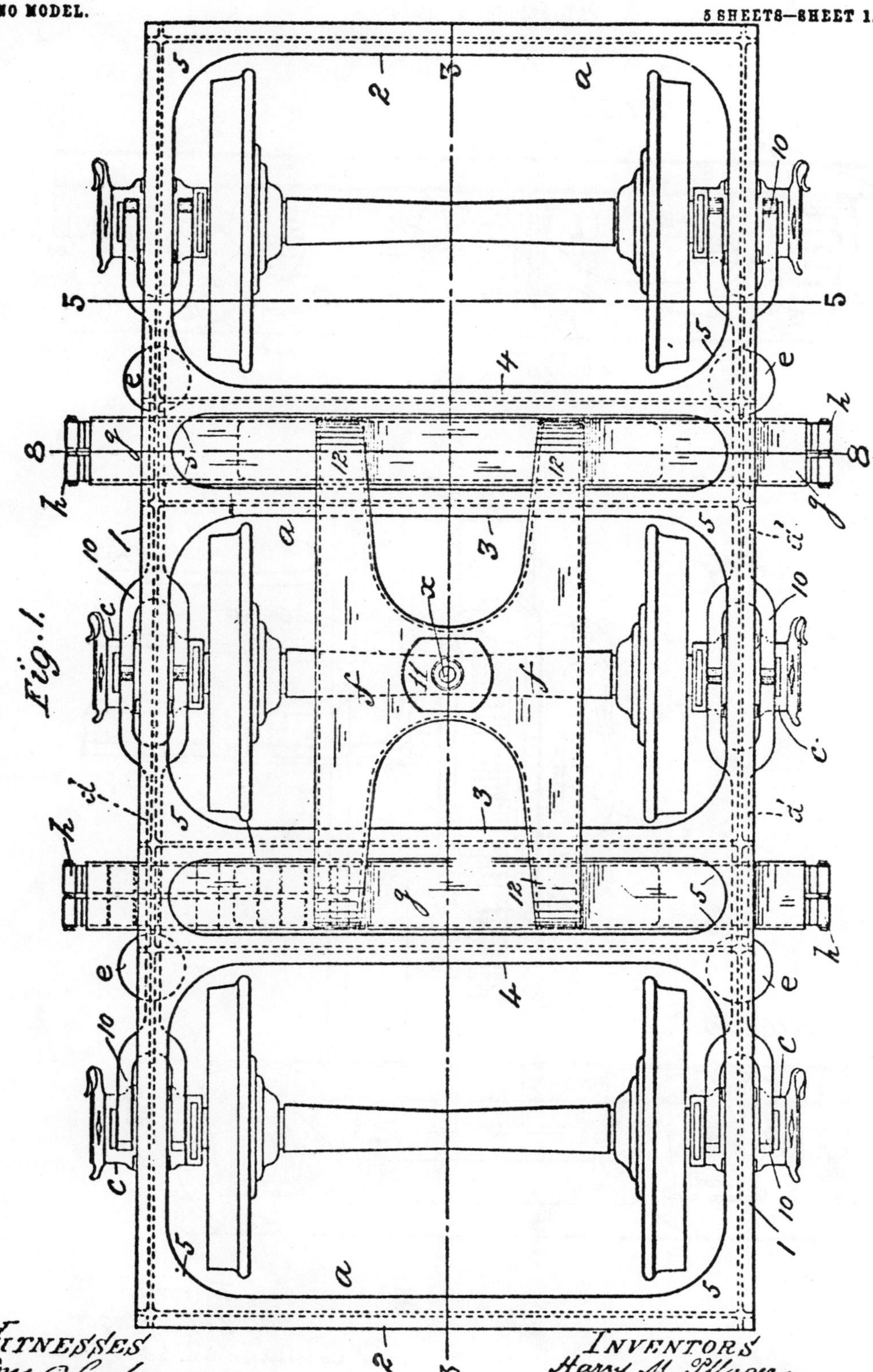

WITNESSES

Wm. Chapman

J. W. Denton.

INVENTORS

Harry M. Pflager

Clarence H. Howard

By Edward W. Furrell

their Atty

FIG. 15

May 11, 1926. 1,584,133

H. M. PFLAGER

CAR TRUCK AND METHOD OF ASSEMBLING THE SAME

Filed June 7, 1924

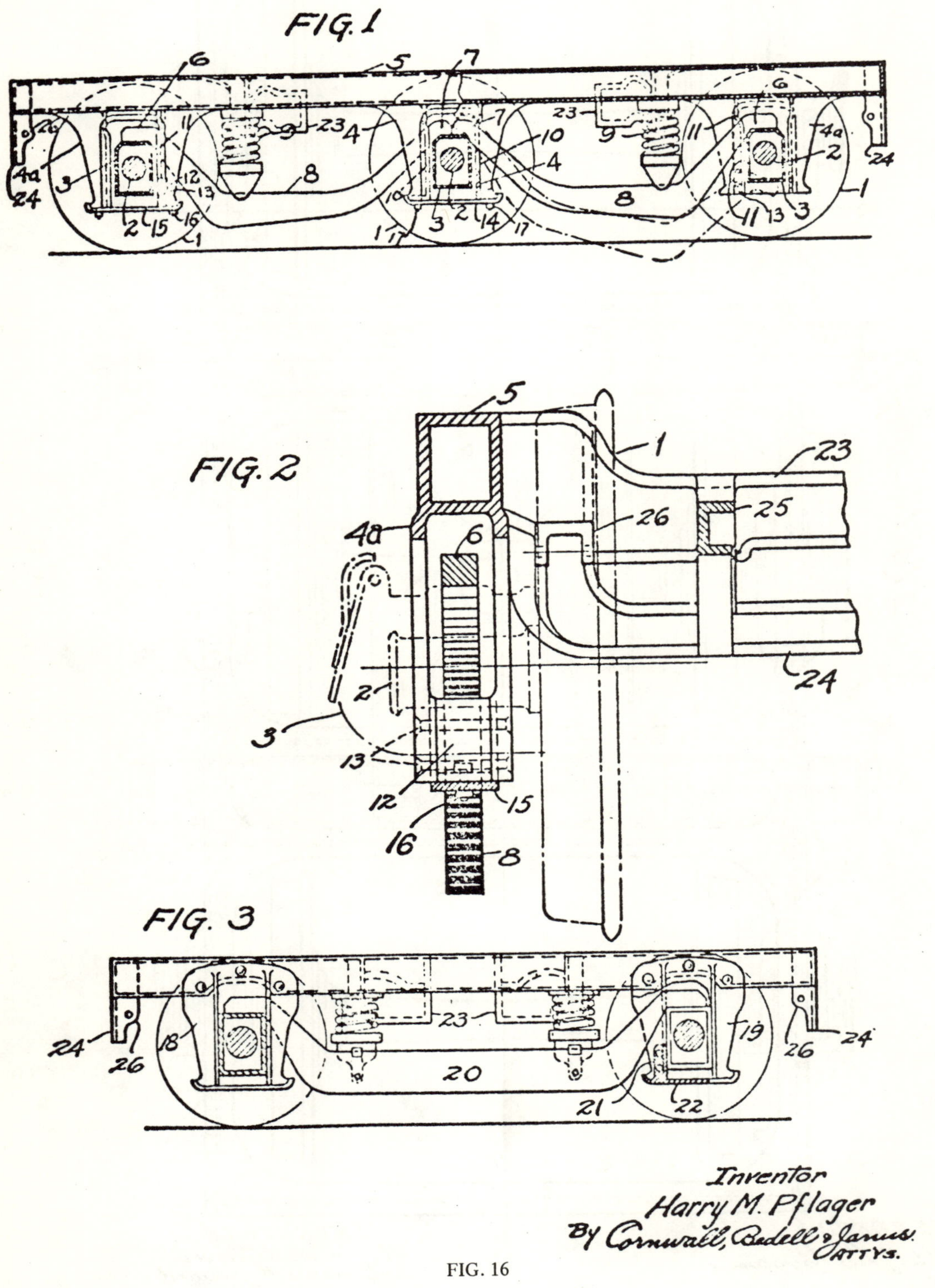

FIG. 16

No. 800,921. PATENTED OCT. 3, 1905.

W. F. KIESEL, Jr.

CAR TRUCK.

APPLICATION FILED JULY 11, 1905.

3 SHEETS—SHEET 2.

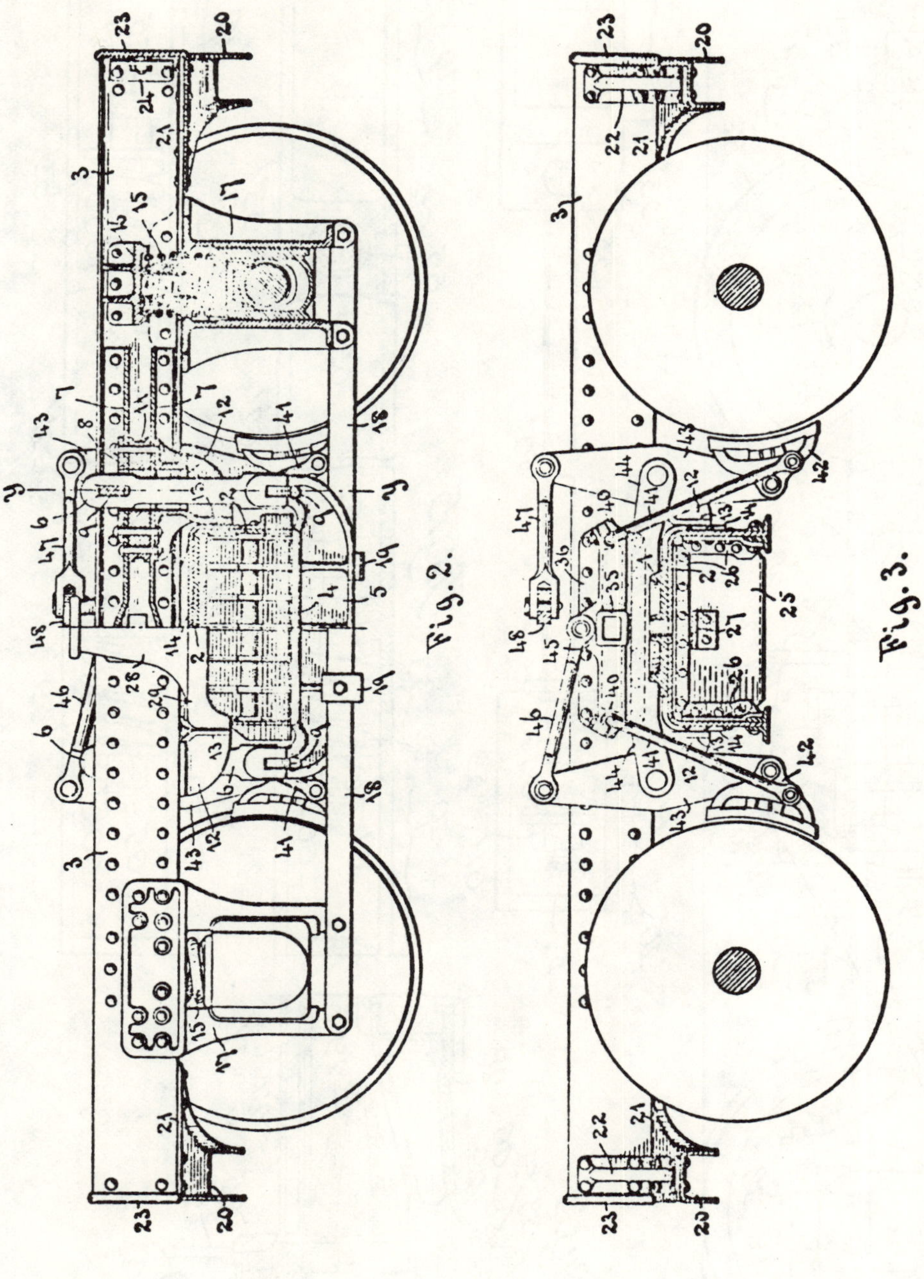

FIG. 17

WITNESSES

J H O'Brien

L. V. Stoltzfus

INVENTOR

William F. Kiesel, Jr.

BY

Eugene Diven

ATTORNEY

Dec. 10, 1935. H. D. TAYLOR 2,023,715

CAR TRUCK

Filed May 16, 1932 2 Sheets-Sheet 2

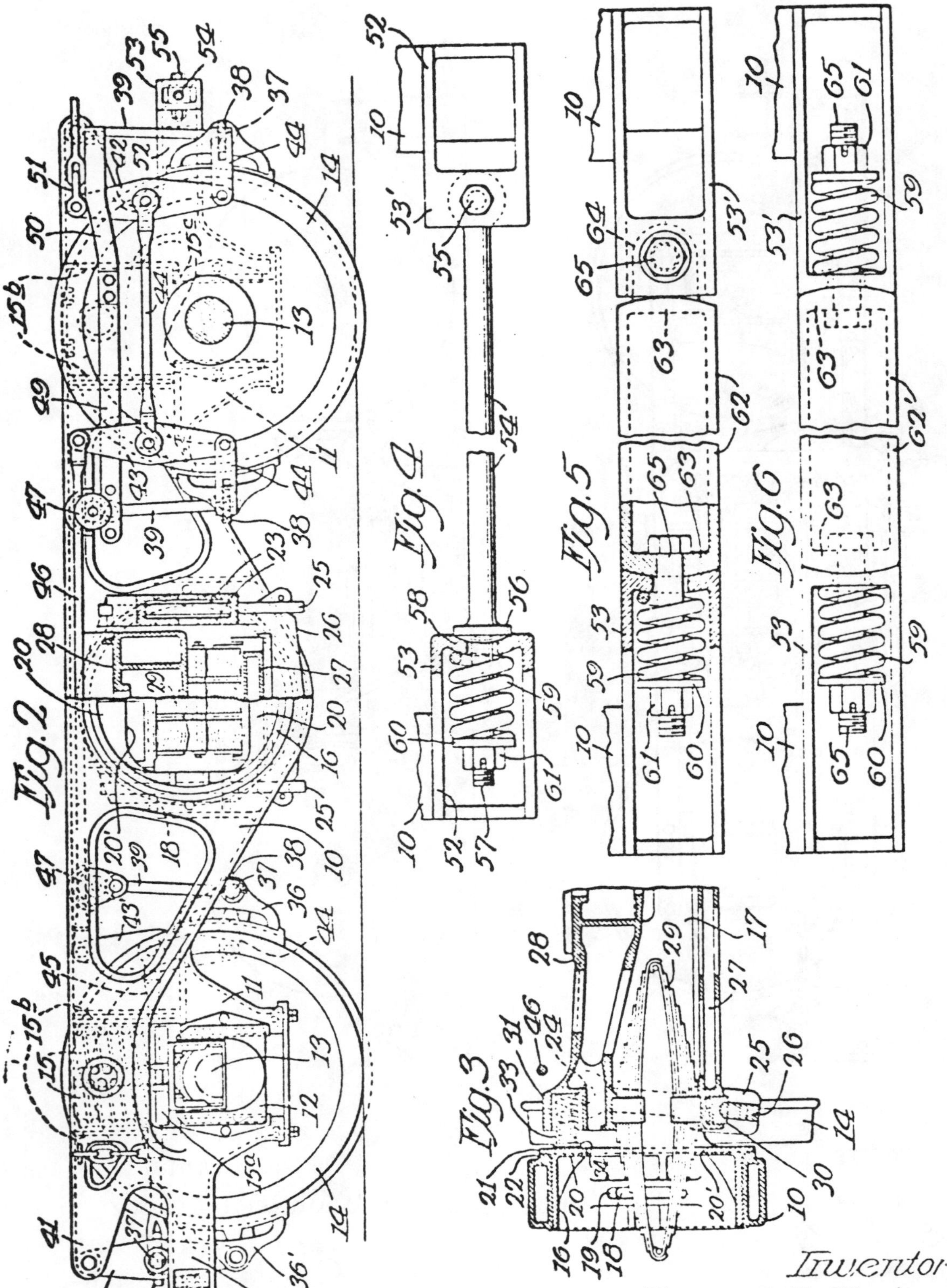

Inventor
Howard D. Taylor
By F... Claff, Soans & Pond Attys.

FIG. 18

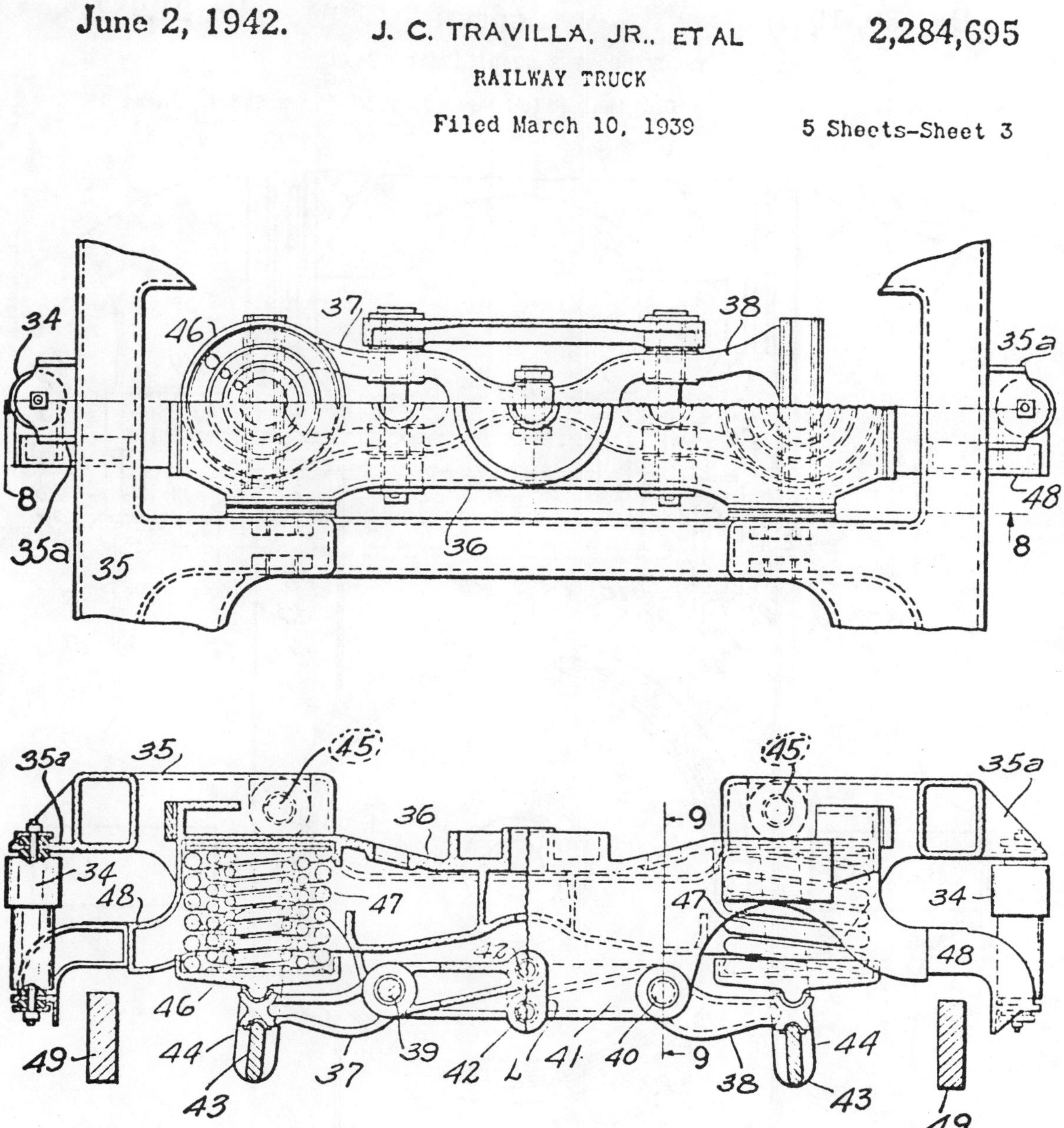

FIG. 19

senger car builders, including Pullman, American Car and Foundry, Barney and Smith, Harlan and Hollingsworth, Pressed Steel Car, Standard Steel Car and St. Louis Car, fabricated M.C.B. trucks, but thereafter few such fabricated trucks were made as the vastly superior Commonwealth one-piece cast steel frame trucks were applied to the majority of passenger cars. Standard Motor Truck Co., an affiliate of Standard Steel Car, also produced M.C.B. type trucks with pressed steel side frames characterized by the elimination of the inner pedestal legs at each corner and the rigid connection of the double equalizers to the journal boxes. American Car and Foundry and Pressed Steel Car also produced trucks designed by Lewis B. Stillwell with truss side frames, journal box springs and sprung swing motion bolsters.

An interesting attempt to achieve a fully equalized swing motion truck with a double spring system but without a rigid frame and separate equalizers was made by Howard D. Taylor of the Flexible Truck Corporation, who obtained Patent No. 2,023,715 (Fig. 18) in 1935 on a truck with separate cast steel side frames, each with pedestals and a circular central window in which a transom casting was mounted. Swing hangers, spring plank,

Dec. 30, 1941. V. L. GREEN Re. 21,987

TRUCK BOLSTER STABILIZING MEANS

Original Filed Nov. 3, 1938 2 Sheets-Sheet 1

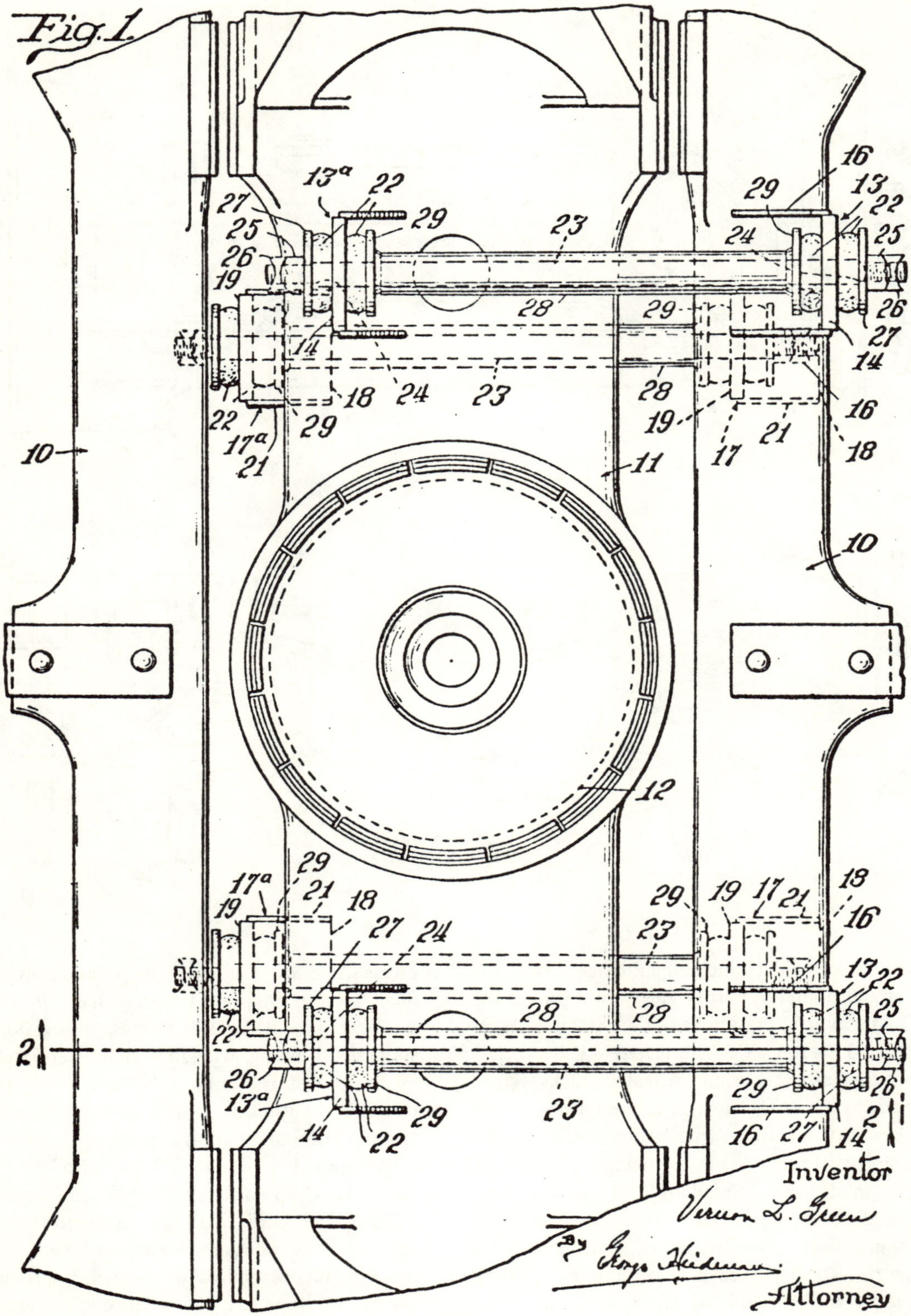

FIG. 20

Jan. 13, 1948. H. M. PFLAGER 2,434,287

RAILWAY TRUCK

Filed March 13, 1944 5 Sheets-Sheet 1

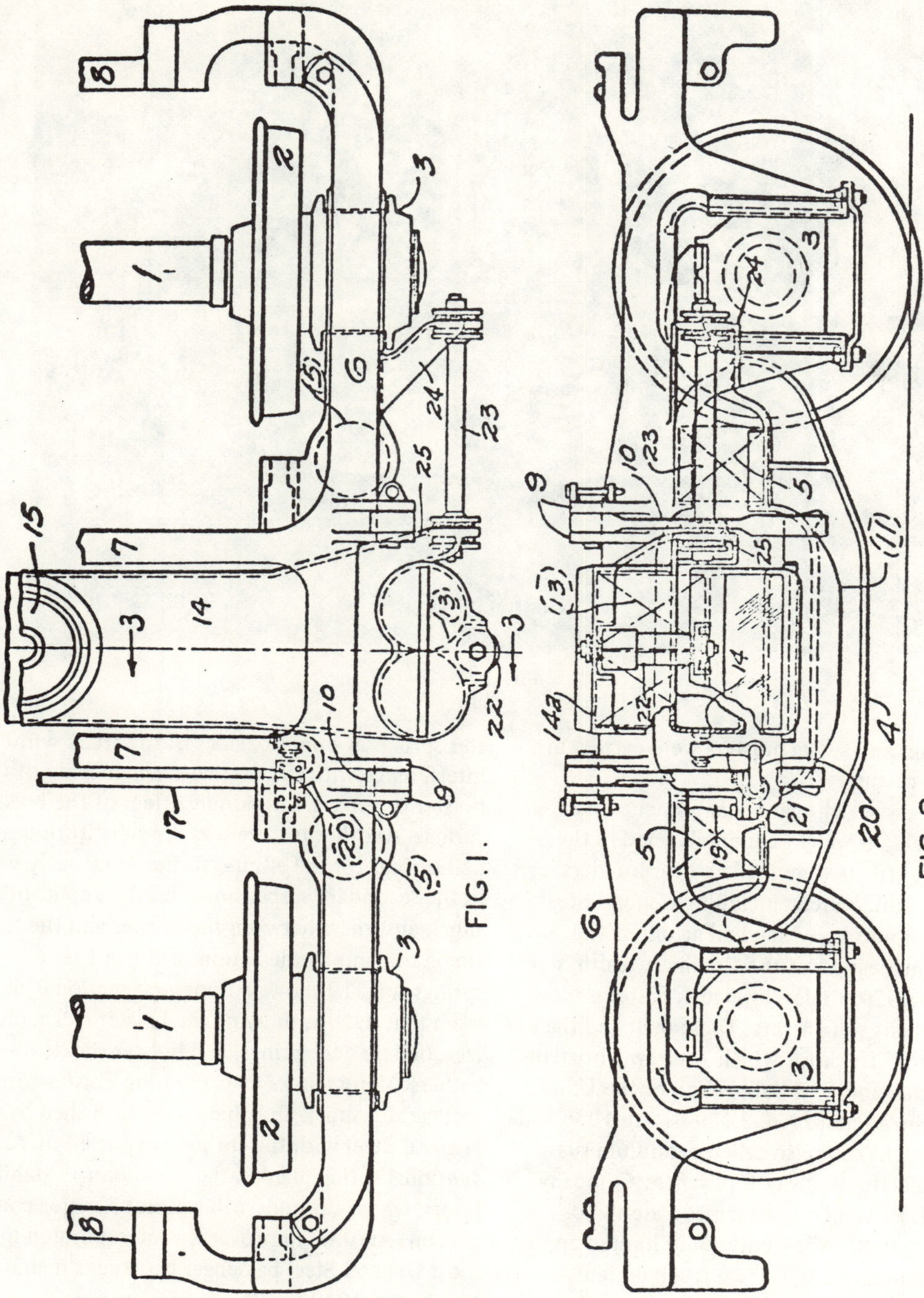

FIG. 21

INVENTOR:
HARRY M. PFLAGER
BY Rodney Bedell
ATTORNEY

FIG. 22

bolster springs and swing bolsters were carried in the transom casting.

As the heavy weight car era approached its close and the light weight car era dawned in the early 1930's, efforts were made by car builders and truck manufacturers, including General Steel Castings Corporation, which had acquired and succeeded to the business of the Commonwealth Steel Company in 1929, to develop truck features more suitable for light weight cars. These included the "triple bolster" truck for which Andrew Christianson of Pullman and Otto Jabelmann of the Union Pacific obtained Patent No. 2,129,910 in 1938 and the development of the so-called pendulum suspension by Pacific Railway Equipment Company on which W. E. Van Dorn obtained Patent No. 2,217,033 in 1940. Apparently only limited applications were made of these two truck designs.

Among the features developed by General Steel for light weight car trucks was the lever type bolster roll stabilizer, on which J. C. Travilla received Patent No. 2,284,695 on June 2, 1942. The purpose of this device was to eliminate objectionable body roll by causing equal depression of the bolster springs at opposite sides of the truck without interference with the normal swing motion of the bolster or vertical cushioning effect of the bolster springs. Another feature was the substitution of coil springs for the elliptic springs previously used in the secondary suspension, and the application of shock absorbers between the bolster and the truck frame to dampen the action of the bolster coil springs. Fig. 19 shows a transverse sectional view from that patent, showing the bolster roll stabilizer, bolster coil springs, and bolster shock absorbers. Many trucks incorporating bolster coil springs and shock absorbers were furnished by General Steel with the springs supported on a conventional spring plank rather than on roll stabilizer levers, where car body roll was not a major consideration. Another improvement incorporated in most General Steel passenger car trucks manufactured after 1940 was the bolster anchor, as shown in V. L. Green Patent 2,219,360, issued October 29, 1940 (Fig. 20), which eliminated frictionally engaged buffing, or chafing, plates between the bolster and the truck frame. The longitudinally extending anchor links, connected at one end to the

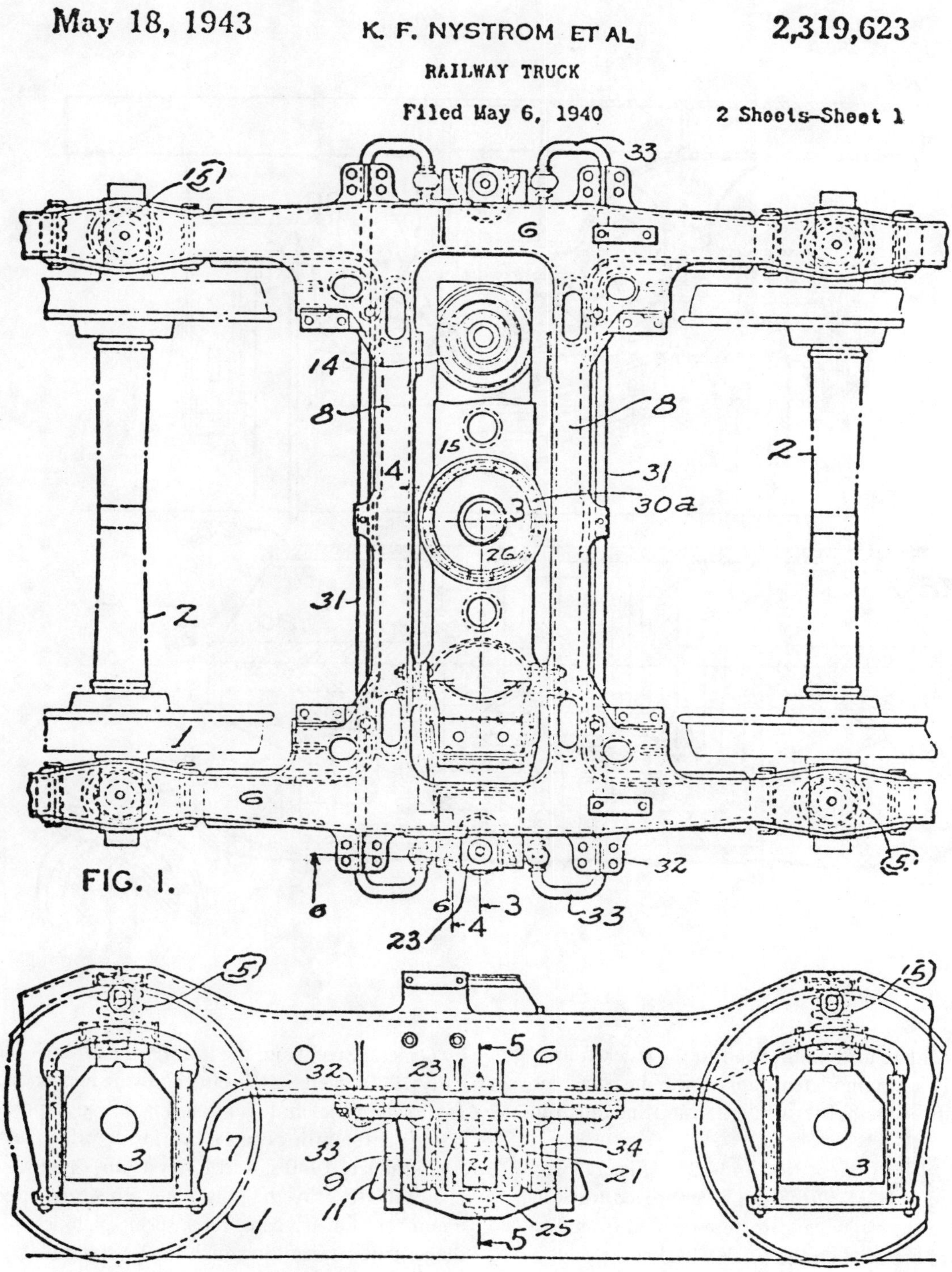
May 18, 1943
K. F. NYSTROM ET AL
2,319,623
RAILWAY TRUCK
Filed May 6, 1940
2 Sheets-Sheet 1
FIG. 1.

FIG. 23

Oct. 13, 1953 J. C. TRAVILLA **2,655,117**

RAILWAY VEHICLE BODY AND TRUCK CENTRAL BEARING

Filed Oct. 10, 1949

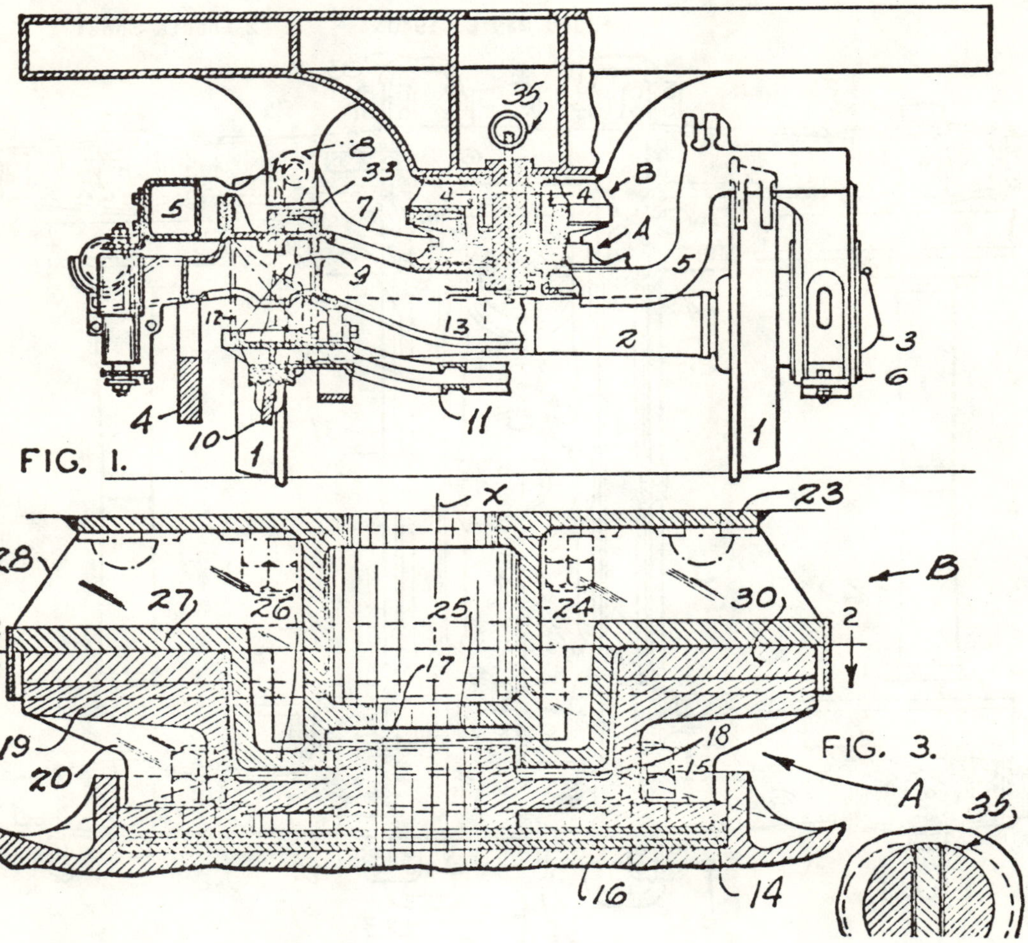

FIG. 24

bolster and at the other end to the truck frame, freely accommodated transverse and vertical movements of the bolster while transmitting longitudinal forces.

In 1948, Patent No. 2,434,287 (Fig. 21) was issued to H. M. Pflager for an improved truck in which the swing hangers were pivoted from outboard brackets on the truck wheelpieces, thus permitting the bolster coil springs to be positioned laterally outboard of the wheelpieces. The increased spring spacing minimized objectionable body roll, while permitting the use of lower rate bolster coil springs without the need for separate roll stabilizers. Most passenger car trucks furnished by General Steel from the late 1940's through the early 1960's were of the outside swing hanger type as shown in the photograph in Fig. 22, incorporating the anti-hunting central bearing described below.

In the late 1940's, Karl F. Nystrom, Chief Mechanical Officer of the Milwaukee Road, patented a number of swing-hangerless, spring-plankless passenger truck designs in which the bolster coil springs rested on brackets extending laterally outboard from the wheelpieces, lateral movement of the bolster being accommodated through lateral shear in the bolster coil springs. Truck designs jointly developed by Nystrom and General Steel and manufactured by the latter included this fea-

FIG. 25

FIG. 26

March 17, 1959 R. L. LICH 2,877,719

RAILWAY VEHICLE STRUCTURE

Filed June 25, 1953 2 Sheets-Sheet 2

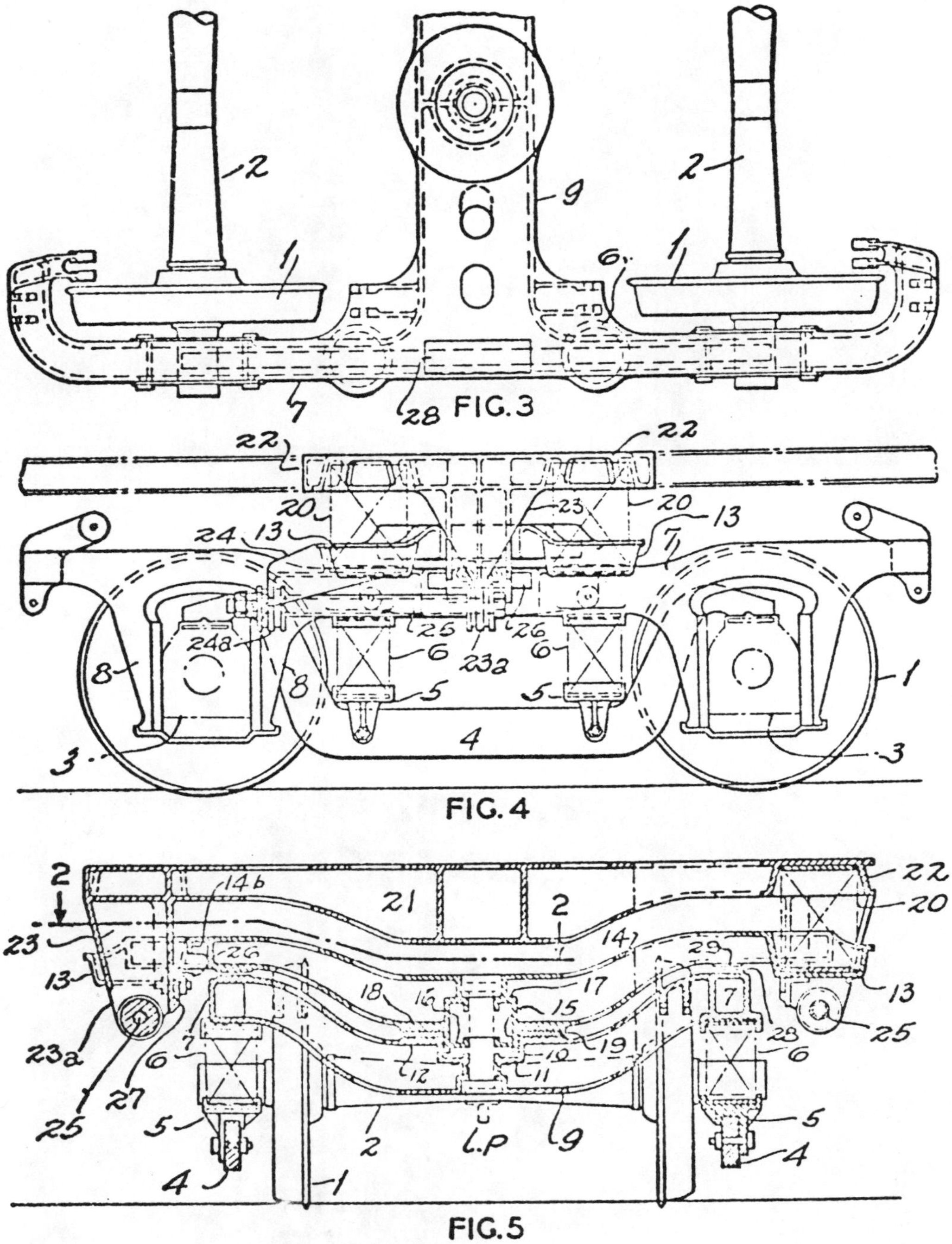

INVENTOR.
Richard L. Lich
BY Rodney Bedell
Atty.

FIG. 27

FIG. 28

ture, along with torsion bar roll stabilizers disclosed in Nystrom Patent No. 2,319,623, which was issued May 18, 1943 (Fig. 23). The trucks also used bolster anchors, and had the journal boxes secured to cast equalizers with longitudinal anchors connecting the equalizer and journal box assemblies to the truck frame to eliminate pedestal friction (Nystrom and Green Patents Nos. 2,323,348, 2,334,024 and 2,347,362). Examples of trucks having these features are illustrated in the 1949-1951 *Car Builders' Cyclopedia,* p. 1026 and 1032

In the early 1950's, General Steel developed a device for opposing truck hunting, i.e., objectionable oscillation of the truck about its swivel axis. This was achieved by the large diameter central bearing invented by James C. Travilla, who obtained Patent No. 2,655,117 on October 13, 1953 (Fig. 24) in which the vertical pivot-forming surfaces were substantially the same diameter as those of a conventional truck center plate, with a larger diameter friction pad between opposed horizontal annular surfaces surrounding the vertical surfaces, to carry the vertical load. This device was very successful in reducing or eliminating truck hunting and has been applied to many passenger car trucks of General Steel design since its invention.

A significant development of the mid-1950's was the introduction of inboard bearing trucks on rapid transit cars, along with the application of package tread brakes. The inboard bearing arrangement offered substantial cost and weight savings along with easier inspection and maintenance of the brake equipment, which was located on the outside of the truck frame. Fig. 25 shows a photograph of one such truck used on the Cleveland Transit System. Another important development of that same period was the introduction of air springs on rapid transit cars. Figure 26 shows the Boston main line transit truck equipped with air springs assisted by coil springs inside the air bellows. This is thought to be the first use of air springs on a quantity production order in the United States.

In 1953, Richard L. Lich of General Steel filed a patent application, which was later issued as Patent No. 2,877,719 (Fig. 27) on a truck which departed substantially from the conventional configuration in that the pivot was placed down with-

FIG. 29

in the truck, instead of between car body and truck, with the bolster springs directly supporting the car body. These springs were yieldable in shear to accommodate lateral motion. In 1957 General Steel developed an inboard bearing equalized truck covered by Lich Patent No. 2,988,019 in which the secondary suspension consisted of combined air and coil springs. This truck (Fig. 28) was applied to several MU cars built by the Budd Company for the Pennsylvania Railroad in the late 1950's and represented the first production order for that type of truck. It has since been used most often with a straight air secondary suspension, marketed extensively under the name "General 70" for rapid transit and commuter cars. One example is the truck shown in Fig. 29 applied to commuter cars built by General Electric.

In 1959 Walter B. Dean of the Budd Company obtained Patent No. 2,908,230 (Fig. 30) on a four-wheel passenger truck having separate side frames clamped directly to the inside axle bearings, each of the side frames being formed with a central inboard extension terminating in a concave vertical axis semi-cylindrical bearing surface. The truck bolster was supported on the side frames in longitudinal alignment with the axle-bearing centers and was pivotally connected to the truck framing formed by the side frames by a depending pivot boss received between the cooperating semi-cylindrical surfaces of the side frame extensions. Upright pneumatic springs seated on the ends of the bolster directly supported the car underframe. A substantial number of trucks with this general configuration have been placed in revenue service.

Subsequently, Rockwell developed designs for and manufactured a number of trucks exemplified by G. B. Sundby Patent No. 3,313,245 issued April 11, 1967 (Fig. 31), having separate side frames clamped to inboard axle bearings, each side frame having a stub transom secured by a spherical pivot to the opposite side frame. Upright springs seated on the side frames and supporting the bolster provide the suspension for these trucks.

In both of the preceding truck forms, wheel load equalization is accomplished through articulation of the truck side frames.

A significant development on transit equipment in the 1970's has been the use of rubber chevron primary suspension at the journal boxes in combination with the General 70 truck secondary suspension. One example is the truck used under the Boeing "State-of-the-Art" cars, shown in Fig. 32.

Generally speaking, a typical transit truck of today would use inboard bearings, air springs,

Oct. 13, 1959 W. B. DEAN 2,908,230

RAILWAY CAR TRUCK

Filed Jan. 30, 1956 5 Sheets-Sheet 1

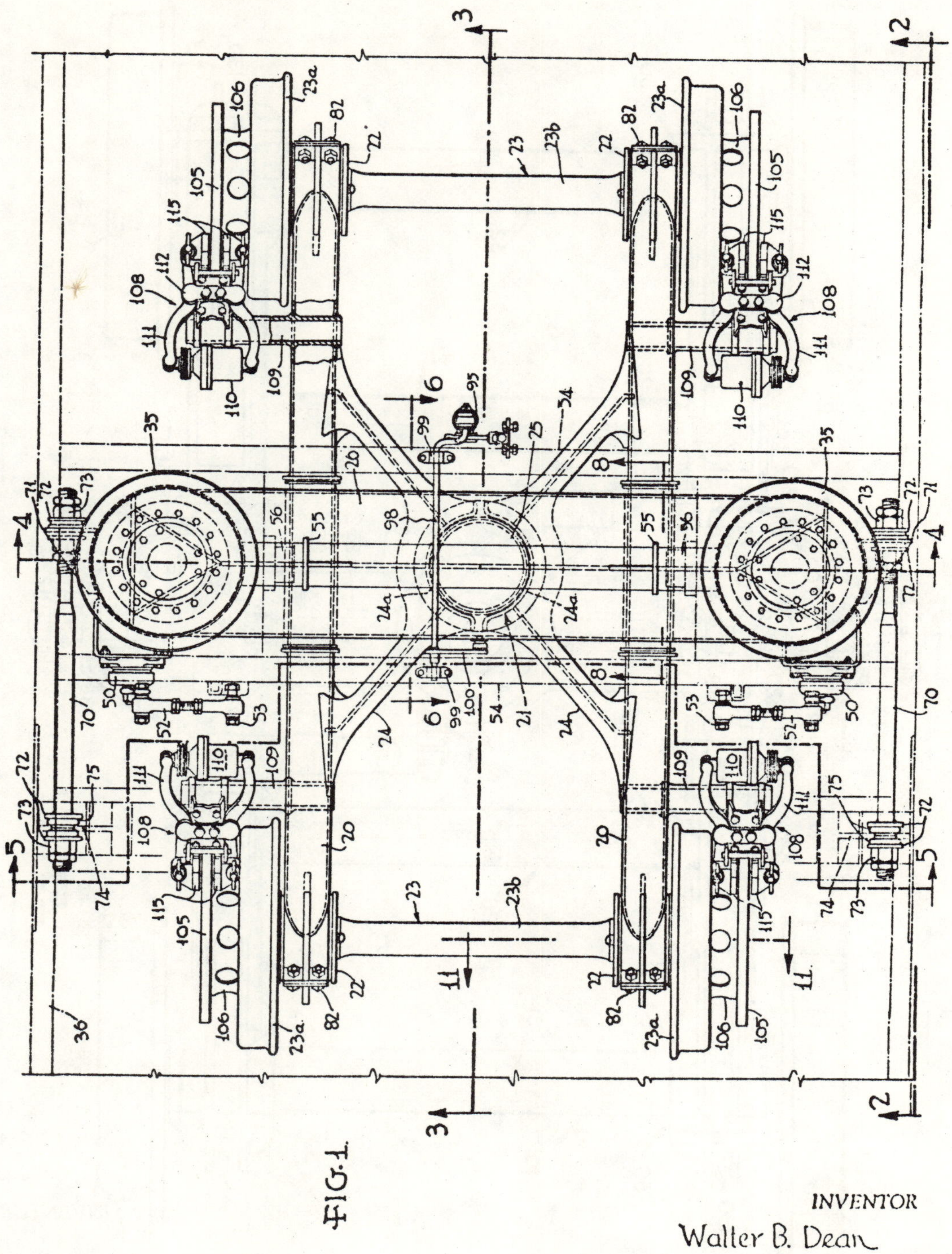

INVENTOR
Walter B. Dean

BY Wm. R. Glisson
ATTORNEY

FIG. 30

April 11, 1967 G. B. SUNDBY 3,313,245

RAILWAY TRUCKS

Filed Aug. 10, 1964 5 Sheets-Sheet 1

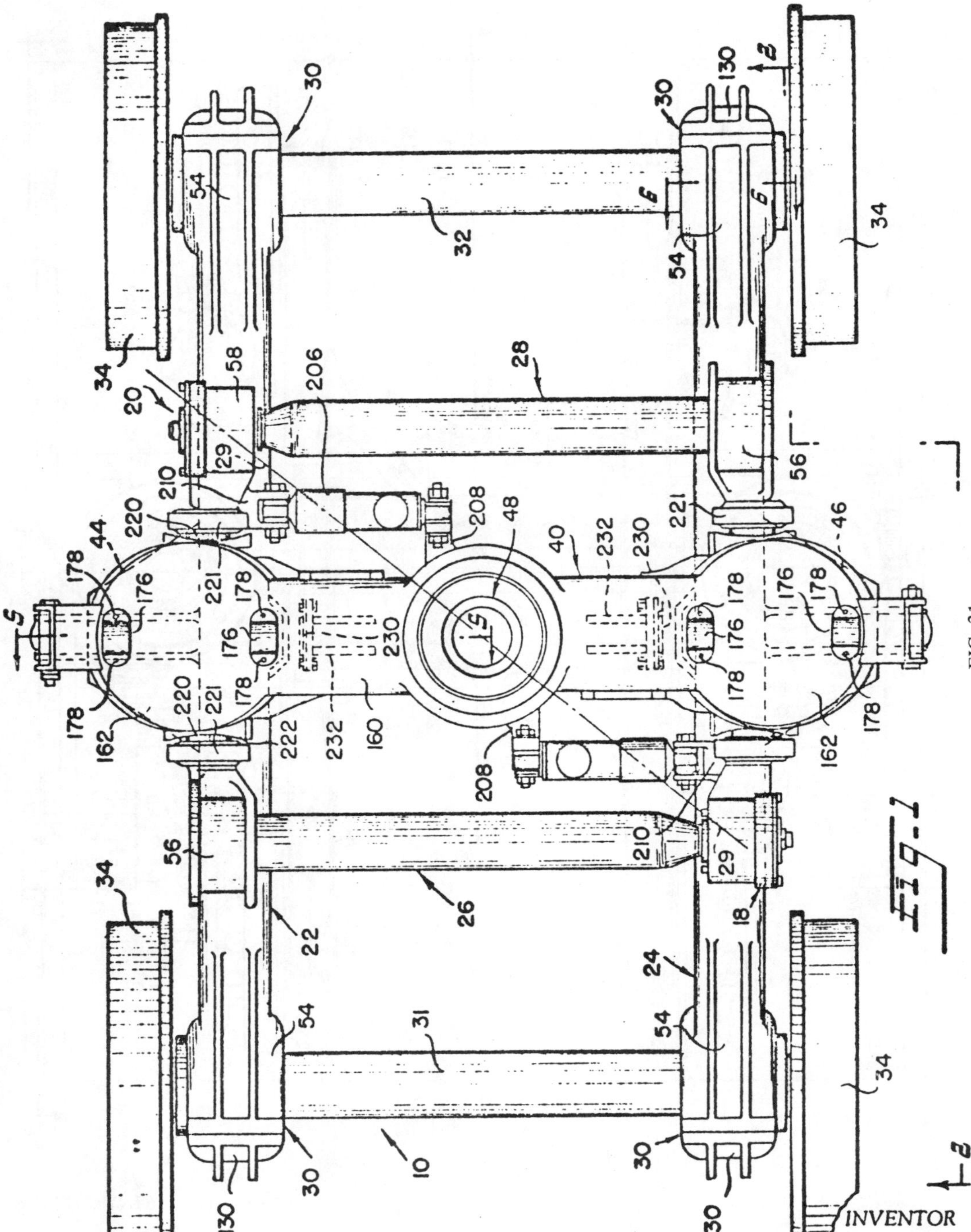

FIG. 31

INVENTOR

GUSTAV B. SUNDBY

BY

Strauch, Nolan & Neale

ATTORNEYS

FIG. 32

package or disc brakes, and a close-coupled parallel motor drive arrangement to achieve a balance between light weight, low cost, good riding quality and low maintenance. One exception to the inboard bearing trend is found in the New York City Transit Authority, which employs outside journal trucks for standardization of the largest fleet of transit cars in the industry. From the photograph of one of the NYCTA trucks, Fig. 33, it can be seen that the standard propulsion package occupies virtually all the space between the wheels and cannot fit into an inside journal arrangement.

Notwithstanding all of the great contributions of the past, the process of truck development still goes on. For example, exciting new ideas are being generated in the field of radial axle truck technology to improve curving performance of rail vehicles. The radial axle feature may someday displace one of the basic principles of truck design, that of keeping the axles parallel. Time will tell if that is to be the case.

The above account merely touches on the highlights of truck development in this country. No attempt has been made to single out which of these developments was of greatest importance, or which individuals or companies made the greatest contribution. The authors will admit to a vantage point influenced by their access to the records of a truck designer and producer, General Steel Industries. However, we believe an impartial research effort, from the vantage point of others, would disclose a story unfolding very much as outlined above.

FIG. 33

II Temperature Comfort in Rail Transportation Vehicles

O. M. CHIESL

Ever since the horse drawn carriages first served as public transportation vehicles, there has been an urgent need to meet comfort requirements of the passengers riding in them. Mass transportation vehicles in competition with air conditioned automobiles must provide similar comfort levels.

Horse drawn rail vehicles riding on rails in the early 1800's not only furnished a smoother ride than the uneven cobblestones, but increased miles per hour by more than a mile and a half. When you travel in today's comfortable trains, it becomes hard to picture those passengers of earlier days sitting on wooden benches, dressed in heavy overcoats, feet nested in straw trying to stay warm. Those who were lucky might find a wood burning pot bellied stove. Sitting too close to it they roasted, and farther away they froze. These early commuters would wait for cars with stoves identifying them by the smoke stacks rising from the roofs.

By 1890 cable, electric powered and steam locomotive drawn cars operated on the city streets and elevated tracks. At one time Chicago had more miles of cable operated cars than San Francisco. Steam locomotives hauled wooden, gas lighted coaches over elevated tracks in New York, Chicago and Boston. Locomotives were tried at street level, but they frightened the horses and were discontinued.

To heat street cars and railroad passenger cars during this period, steam taken from the locomotive was passed through a trainline underneath each car, through a tee, a shut-off valve, and up into the car. Steam entered rows of 2-in. pipe lined along the walls. Although this system distributed heat to each end of the car, it still had shortcomings. Temperatures fluctuated between hot and cold, depending upon how close you sat to where the steam came into the car. And there was always the possibility that these high pressure steam pipes would burst, posing a threat to passenger safety.

About 1905 the first system for heating passenger cars was introduced by Egbert H. Gold Company. Egbert Gold, who was formerly with the Gold Car Heating and Lighting Company of New York, came to Chicago to form his own company. After a succession of names – Chicago Car Heating, Vapor Car Heating, Vapor Heating – it became what is known today as the Vapor Corporation.

It was the early Gold patents that revolutionized passenger car heating and removed the element of danger. A pressure regulator was used to reduce high pressure steam from the trainline before it was admitted into the heating pipes of the car. The pressure regulator, controlled by an expandable diaphragm, was set to admit steam into the car at 212 F and release condensate under the car to the atmosphere at 200 F.

This meant that high pressure steam from the locomotive could be maintained in the trainline at sufficient pressure to reach the last cars in the train without endangering passengers.

By 1907, special arrangements of cut-out valves permitted more exact regulation of car temperatures by using only 25-50 percent during mild weather and 75-100 percent of the total capacity of the system in extreme weather conditions. The third rail and trolley line brought electric lighting to el trains and street cars early in the 1900's, but electric heat was not installed in many of them until later. Some of the metropolitan cars in Chicago had stove heat up into the twenties.

Cars that were motorized and wired for heating and lighting were also revised to operate on the surface lines as well as the elevated tracks. But they were soon revised again, when it was found that the electric heating required more current than was normally available. Once again, other forms of heating and ventilating systems were employed. Hot air and hot water stove heating systems were used for a while until the power load was no problem. Then the cars were changed back again to electric heating. Cars with no heat at all were called "summer cars" and put into service during periods of moderate weather.

Railroad passenger trains were going through somewhat similar transitions. As early as 1856 a heating and ventilating system was developed that brought fresh air in at the top of the car. Forced in, as the train moved, this cold air was passed through a plenum to a hot air stove. Warmed air was distributed into the car through ducts in the ceiling. Cold floor air was removed through suction heads at the corners of the car, delivered to the stove and redistributed.

In 1870, a hot water stove heating system was introduced that circulated hot water from the stove through pipes in the car. This never achieved much general success. When fire damage caused additional loss of life during train wrecks, public protest was aroused against the use of any type of stove inside the transportation vehicles. The first railroad cars to be heated by steam from the locomotive were on New York's Manhattan Railroad in 1883. Fin type radiation replaced the steam pipe in the 1930's. One foot of fin had the heating efficiency equal to four linear feet of pipe.

It was also during this period that the world famous PCC type cars were first introduced. In an attempt to regain riders, the Electric Railway-Presidents' Conference Committee (PCC) developed a model that brought about changes not only for cars of that period but for cars of the future as well. Several cars similar to the PCC type street cars were built and put into service during the 1940's. The heating source was furnished by motoring and dynamic braking. Heat from the resistor strips during these two phases was blown by fans and directed through ducts into the interior of the car. This fan, used to heat the car, also helped to cool and dissipate heat from the resistor elements, making it possible to use lighter grids.

Thermostatically controlled, motor actuated dampers mixed fresh outside air with the air warmed by starting and stopping. If the heated air was not required, dampers would reject this air to the outside. Damper position also regulated the speed of the fans. The large volume of air from these fans also created a greater pressure inside the car that kept out cold drafts and dust.

Just the starting and stopping at each station was usually enough to generate enough heat to warm up the car during normal winter conditions. Auxiliary electric heater strips provided comfortable temperatures for passengers when the weather was extremely cold. These strip heaters were thermostatically controlled. Fans delivered the warmed air through ducts from the ceiling into the car.

At times when the train stops were not frequent enough to heat the car with air blown over the resistor strips, auxiliary heat would be brought in. Manually opening roof dampers and windows provided cooling during the summer.

Various types and kinds of insulation and floor coverings were tried not only to keep the heat in and the cold out but to dampen truck and track noises. More comfortable seat materials and seating arrangements were used. Lighting and windows were improved. Several safety features were added.

Two early forms of rail passenger car cooling employed exactly the opposite means to get results. One system used ice and the other steam. In the ice-activated "air conditioning" system, water was sprayed over blocks of ice that were carried in bunkers under the cars. This water was then pumped through the cooling coils in the overhead ceiling ducts. Fans blowing over these coils forced the cool air through the ducts down into the car.

In the 1940's and 1950's, Santa Fe used steam to cool the cars on their Chief, Super Chiefs, and El Capitan as it sped from Chicago across the western desert to Los Angeles. This steam-ejector type of "air conditioning" required one-third more steam in the summer to cool the cars than it did in the winter months to heat them. Steam provided by the steam generators located at the back of the 4-6 diesel locomotive units flowed through the trainline under the car to the air conditioning unit where it passed through a venturi-type nozzle in the evaporating tank which reduced the pressure in the tank to below atmospheric pressure. The near vacuum formed allowed warm water returning from the cooling coils to boil at 50 degrees and lose its heat. This cooled water was then pumped back into the radiators in the ceiling ducts where blower fans circulated cooled air down into the car.

The electro-mechanical type air conditioning system is most generally used today. In this system an electric motor driven compressor provides suction on the evaporator, compresses the Freon gas and forces it through a condenser where it is liquified. In this liquid form the Freon is boiled in the evaporator and absorbs heat from the air surrounding it. The evaporator itself serves as the heat exchange unit. The air, after it is cooled and has its moisture removed, is then forced by the blower fans through the circulation ducts into the car.

Another more recent approach optimizes air distribution, mixes recirculated air and fresh treated air and delivers it into the car at the window sill. Special nozzles and venturi-type ducts increase velocity and create a circular flow that provides air movement from ceiling to floor. This air distribution system is used on Amtrak's French Turbo-Train as well as their new Turbo-Liners built in the United States.

"Heating, Cooling and Ventilating" has generally been the term used for what happens that brings about temperature changes in transportation vehicles. Fully "air conditioned" transit cars as we know them today did not appear until the 1950's. Some sources have railroaders thinking about air conditioning as early as 1854. However, the first air conditioned railroad car to go into actual service was on Santa Fe's California Limited in 1914. The Baltimore and Ohio takes credit for air conditioning the first entire train put into service in May of 1931. The first passenger automobile to feature "factory air" was a Packard that was shown in November of 1939. Air conditioned inter-city buses followed in the 1940's.

And that was only the beginning. The new era arrived with dramatic swiftness and seems definitely here to stay. Today people expect some form of air conditioning in their homes, offices, supermarts, department stores, theatres and in almost every form of transportation.

There are several factors involved that must be considered when developing air comfort systems for transit vehicles. Trains speeding between stations are subject to shifting conditions of wind wipe and sunload.

Doors are continuously being opened and closed in transit and railroad suburban service as passengers enter and leave. The increased number of people riding in the cars during peak periods build up heat loads that surpass those of normal conditions. Humidity, whether high or low, has to be recognized. Emerging from a subway tunnel into the sunlight can create changes inside the car that has to be instantly dealt with by the thermostatic control system. Some thought was even given to the possibility of air conditioning the tunnels, but the estimated costs were prohibitive even for those days.

Today's modern metropolitan transit cars are equipped with highly sophisticated air comfort control systems. These systems are fully automatic and are usually activated from switches on the motorman's console.

Now, instead of steam cutoff valves, a network of electro-mechanical and integrated circuitry provides fully automatic control of heating, cooling, ventilating, lighting and public address system functions.

If you look into the heating and cooling systems of the latest New York cars instead of straw or rows of steam pipes, you will find three parallel-connected banks, 4000 W each, of electric heating elements that furnish floor heat. Another 12,000 W are available from the overhead heating ducts. Temperature sensors at either end of the car can bring on the floor and overhead heat when inside car temperatures are below 59 degrees. Above 59 degrees, blower fans at either end of the car start the ventilating cycles. These fans operate at low speed whenever the floor heat is on and at high speed during the ventilating or cooling cycles. Cooling units are now located at each end of the car.

You cannot open windows in these modern transit or mainline passenger cars. When temperatures reach 71 degrees at one end of the car, first stage cooling is brought in at that end. It takes 72 degrees to start the cooling at the other end. The overhead heat comes on during these initial stages of cooling and stays on until the duct temperatures reach 74 degrees. Full capacity cooling is initiated at both ends of the car when these temperatures reach 76 degrees.

In 1977 Amtrak was still refurbishing steam heated cars. Their new bi-level and Amfleet cars were electrically heated and air conditioned. And they were converting some of their sleeping cars to coaches and Amlounges.

Although most of the new railway passenger cars constructed in the United States are expected to be electrically heated, some foreign manufacturers are still building steam heated cars.

Research testing and development is essential to a company that wants to maintain its position as a leading supplier to the Mass Transportation market. Designing, developing and testing of products and systems for this market is important. Continuing programs must be devoted to environmental air comfort control systems for public transportation vehicles. Air comfort control, one of the most important human individual needs, is at its best when not noticed by the passengers.

Designing comfort conditioning systems for modern high speed passenger vehicles requires engineering attention to several factors. Car floors and ceilings are constantly exposed to extremes of outside temperature. Because passengers sit close to outside walls of the car, air must be circulated around them. Outside air must be tempered and made comfortable. This must be done whether the air is hot, wet summer air or cold, dry winter air. The amount of current available on the cars to power heating and cooling units is important. The length and shape of the car is another factor that must be reckoned with. Accurate mercury tube thermostats are a vital part of the temperature control system. Several of them, placed strategically in the car, sense the changing temperature conditions and through the temperature control panels automatically effect proper heating and cooling.

Steam Heating Attachment for Locomotive

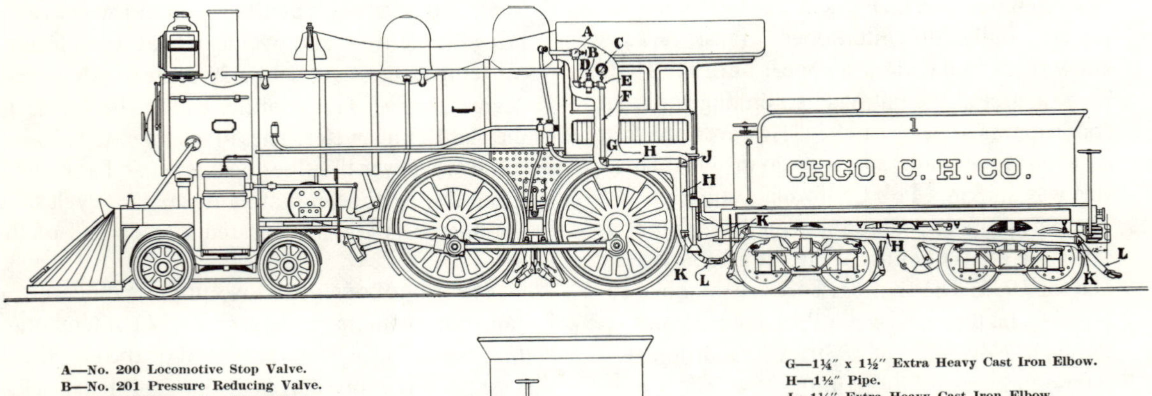

A—No. 200 Locomotive Stop Valve.
B—No. 201 Pressure Reducing Valve.
C—No. 202 Locomotive Steam Gauge.
D—1″ Extra Heavy Cast Iron Elbow.
E—1″ x 1¼″ Extra Heavy Cast Iron Elbow.
F—1¼″ Extra Heavy Pipe.

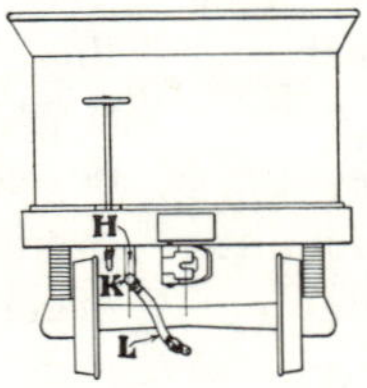

G—1¼″ x 1½″ Extra Heavy Cast Iron Elbow.
H—1½″ Pipe.
J—1½″ Extra Heavy Cast Iron Elbow.
K—1½″ Extra Heavy 45° Cast Iron Elbow.
L—Steam Hose and Coupling Complete.

Metallic connections between engine and tank will be furnished if desired.

FIG. 1.

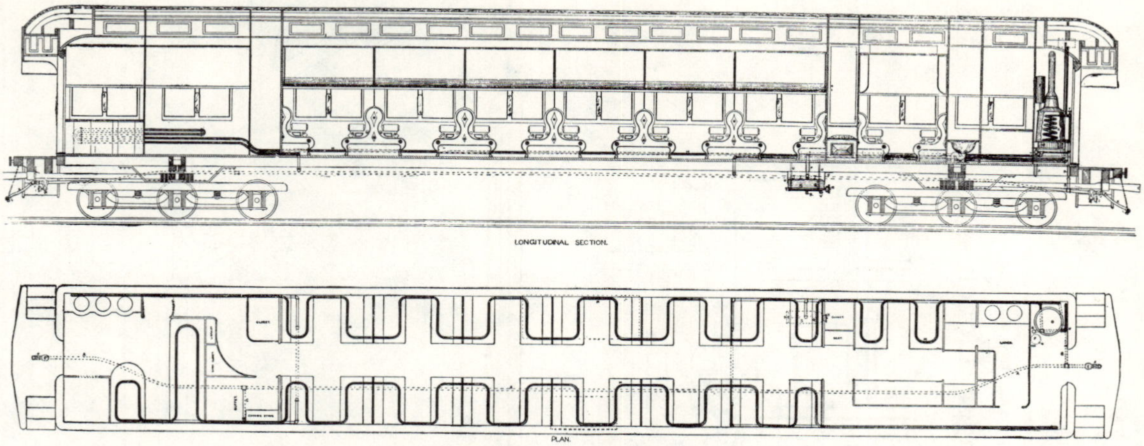

GOLD'S SAFETY STEAM STOVE SYSTEM.

FIG. 2.

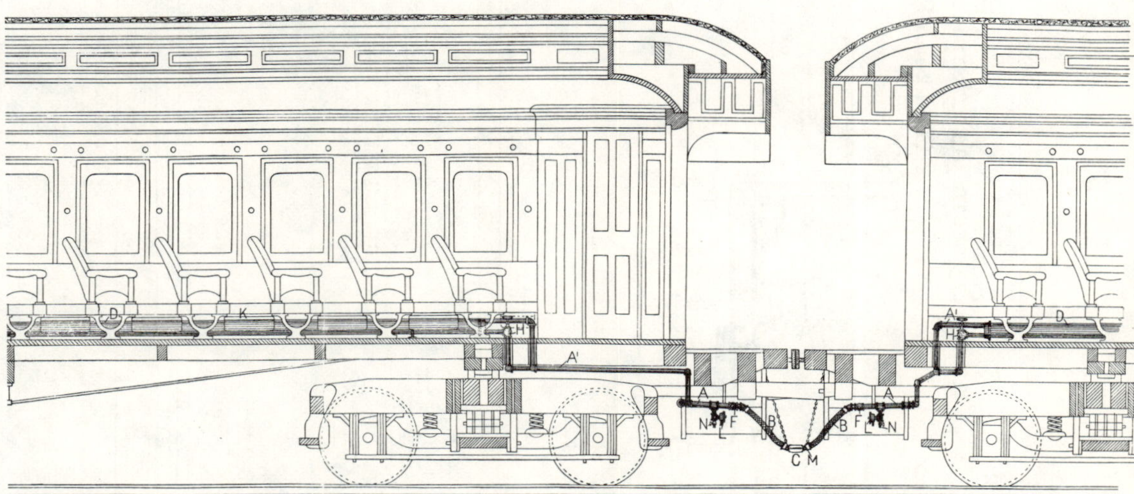

FIG. 3.

Vapor System as Applied to Standard Passenger Coach

(Two 2-inch Radiating Pipes on each side.)

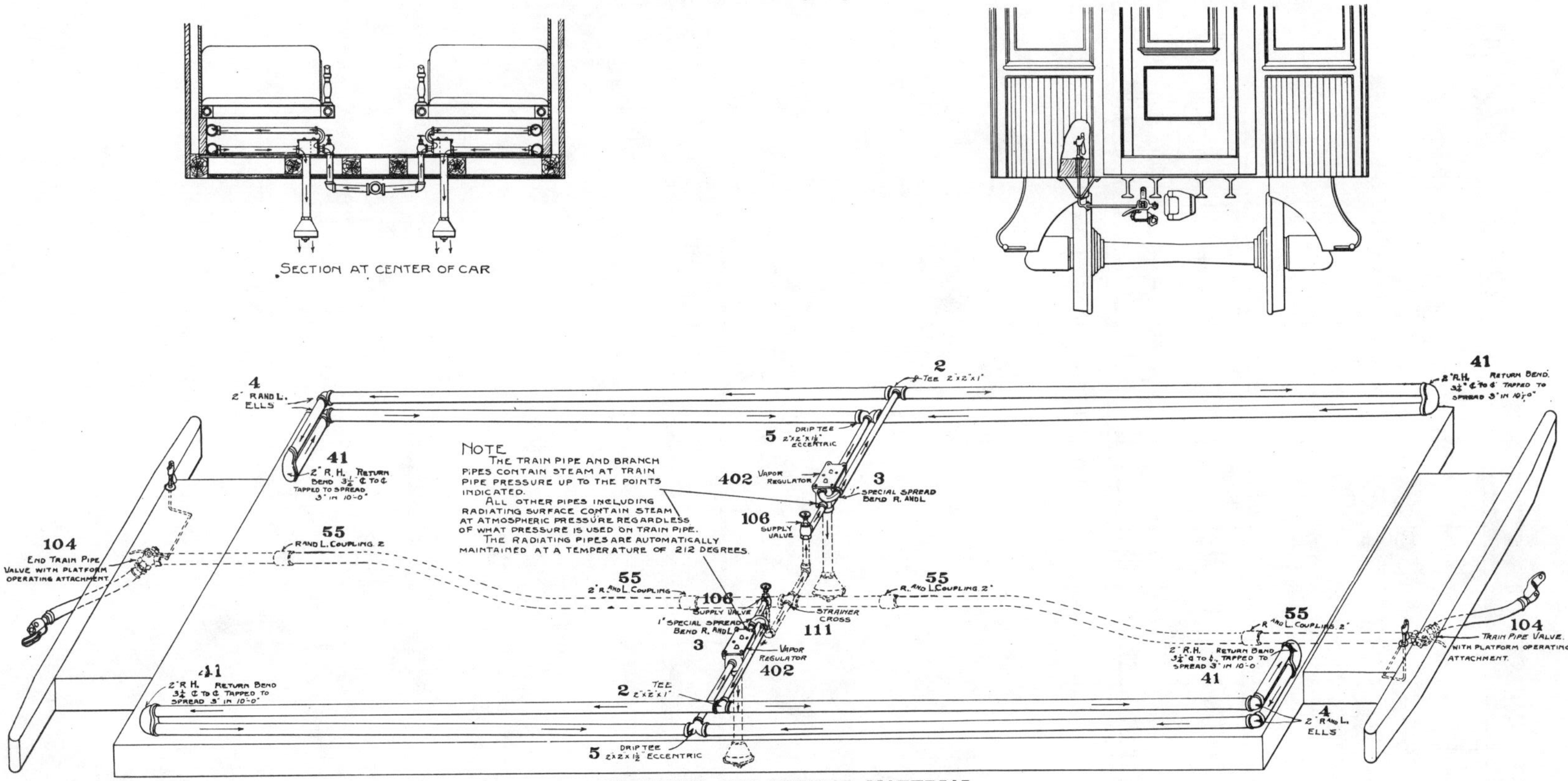

PATENTED AND SPECIAL MATERIAL.

FIG. 4.

Coaches—Chair Cars
Five-pipe Vapor System. Multiple Regulation.

Specially Adapted for Use in Steel Passenger Cars

The two vapor cut-out valves on each side are connected in parallel. The cut-out valve feeding the two-pipe coil on each side is **specially constructed** to permit of drainage when car is left without steam in cold weather and with the cut-out valve in the open position. Either 1½ or 1¼-inch pipes may be used for these coils.

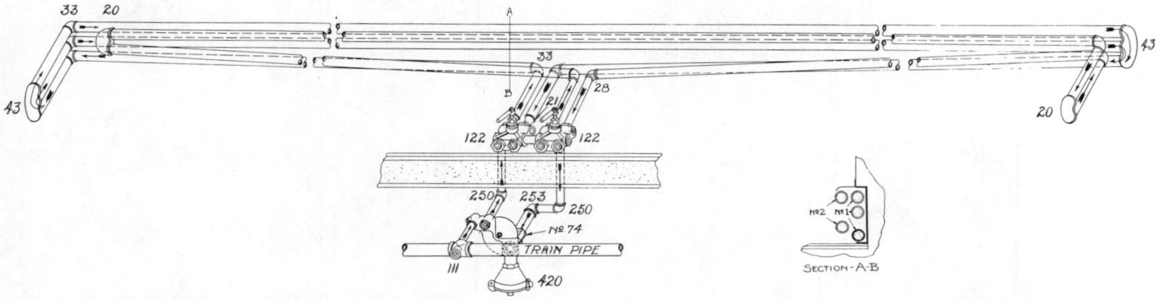

A separately controlled coil of three pipes and a separately controlled coil of two pipes are used on each side of the car.

This arrangement of heating surface and cut-out valves permits of the exact regulation of the temperature in the car in the following proportions of the total heating capacity of the system: 20%, 30%, 40%, 50%, 60%, 70%, 80% or 100%

The above degrees of regulation have been found necessary and very desirable in a steel passenger car because a car of metal construction is particularly susceptible to outside weather conditions, the temperature inside the car responding very promptly to changes in outside temperature and to outside conditions as to sunlight, darkness, etc.

FIG. 5.

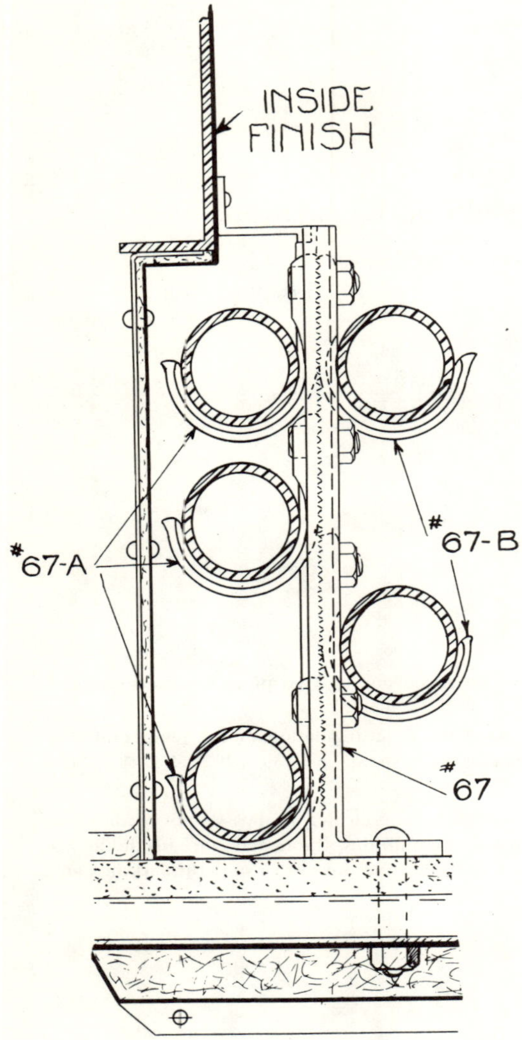

This cut shows the adjustable pipe stands as used for supporting the pipes. It is equally necessary to hold the pipes down; therefore between every two pipe stands that support the pipes, a pipe stand should be used with the pipe clamps reversed as shown.

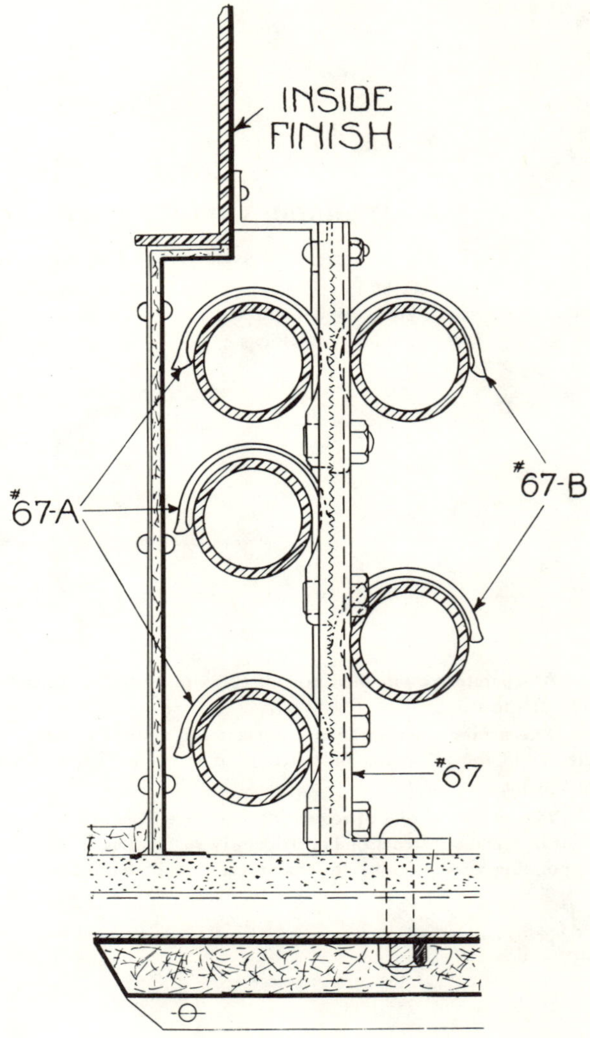

Adjustable pipe stands with clamps reversed to hold pipes down.

FIG. 6.

General Scheme Showing Operation of the Vapor System

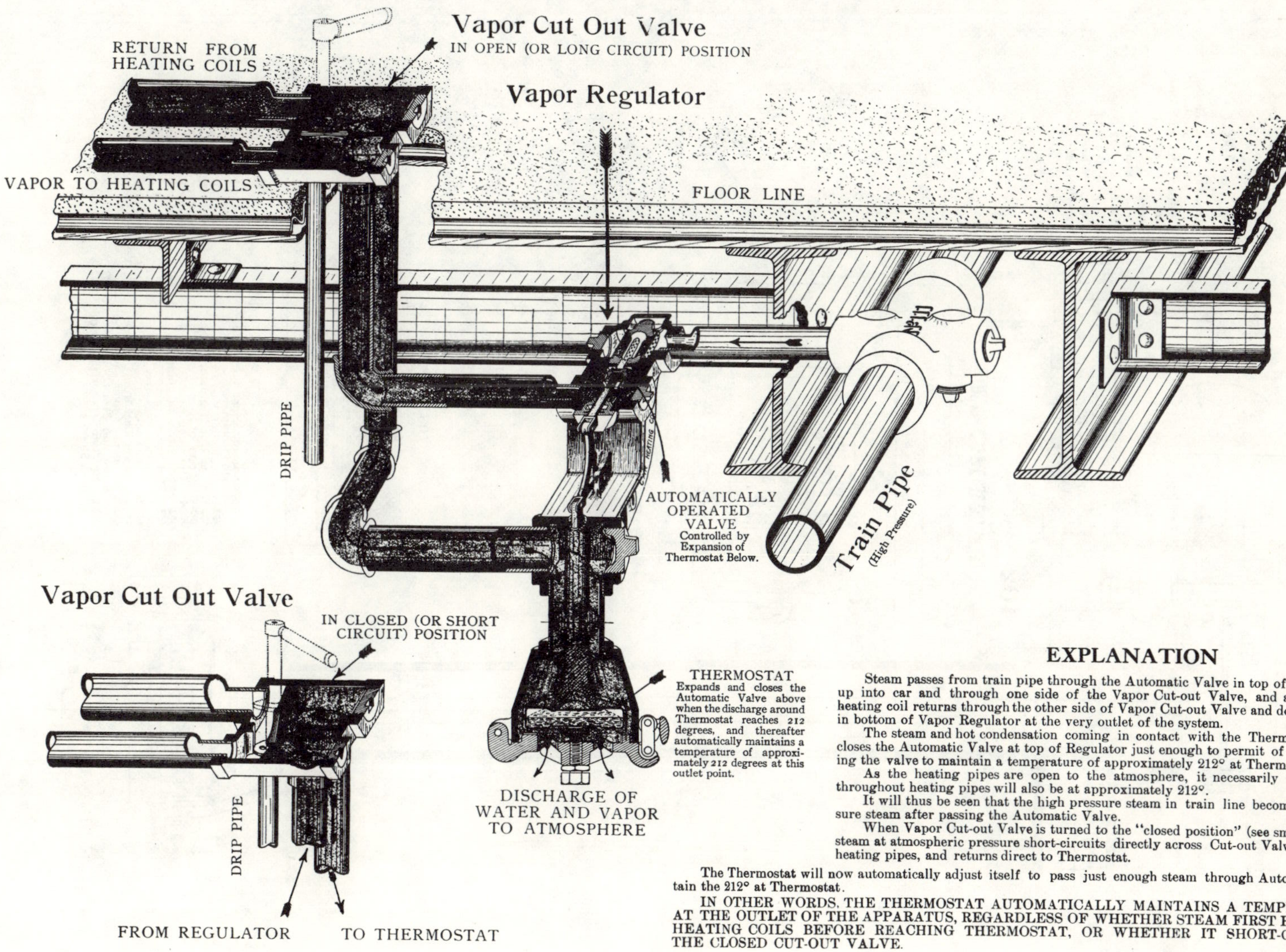

EXPLANATION

Steam passes from train pipe through the Automatic Valve in top of Vapor Regulator and up into car and through one side of the Vapor Cut-out Valve, and after passing through heating coil returns through the other side of Vapor Cut-out Valve and down to the Thermostat in bottom of Vapor Regulator at the very outlet of the system.

The steam and hot condensation coming in contact with the Thermostat expands it and closes the Automatic Valve at top of Regulator just enough to permit of sufficient steam passing the valve to maintain a temperature of approximately 212° at Thermostat.

As the heating pipes are open to the atmosphere, it necessarily follows that the steam throughout heating pipes will also be at approximately 212°.

It will thus be seen that the high pressure steam in train line becomes atmospheric pressure steam after passing the Automatic Valve.

When Vapor Cut-out Valve is turned to the "closed position" (see small cut) the vapor or steam at atmospheric pressure short-circuits directly across Cut-out Valve, instead of entering heating pipes, and returns direct to Thermostat.

The Thermostat will now automatically adjust itself to pass just enough steam through Automatic Valve to maintain the 212° at Thermostat.

IN OTHER WORDS, THE THERMOSTAT AUTOMATICALLY MAINTAINS A TEMPERATURE OF 212° AT THE OUTLET OF THE APPARATUS, REGARDLESS OF WHETHER STEAM FIRST PASSES THROUGH HEATING COILS BEFORE REACHING THERMOSTAT, OR WHETHER IT SHORT-CIRCUITS ACROSS THE CLOSED CUT-OUT VALVE.

FIG. 7.

Old Style Vapor Regulator
Vapor Regulator No. 402—(Sectional View)

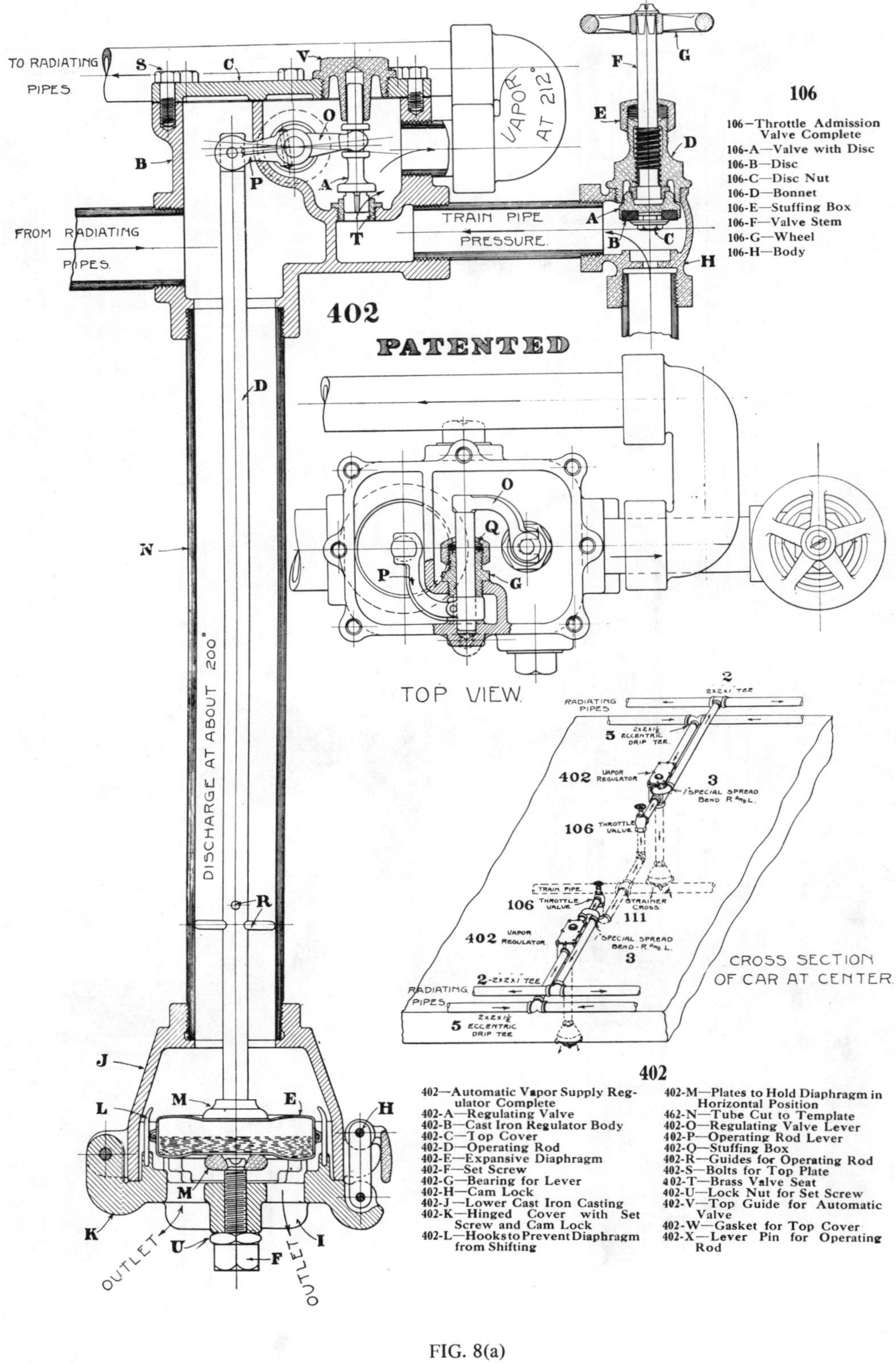

106

106—Throttle Admission Valve Complete
106-A—Valve with Disc
106-B—Disc
106-C—Disc Nut
106-D—Bonnet
106-E—Stuffing Box
106-F—Valve Stem
106-G—Wheel
106-H—Body

402

402—Automatic Vapor Supply Regulator Complete
402-A—Regulating Valve
402-B—Cast Iron Regulator Body
402-C—Top Cover
402-D—Operating Rod
402-E—Expansive Diaphragm
402-F—Set Screw
402-G—Bearing for Lever
402-H—Cam Lock
402-J—Lower Cast Iron Casting
402-K—Hinged Cover with Set Screw and Cam Lock
402-L—Hooks to Prevent Diaphragm from Shifting
402-M—Plates to Hold Diaphragm in Horizontal Position
462-N—Tube Cut to Template
402-O—Regulating Valve Lever
402-P—Operating Rod Lever
402-Q—Stuffing Box
402-R—Guides for Operating Rod
402-S—Bolts for Top Plate
402-T—Brass Valve Seat
402-U—Lock Nut for Set Screw
402-V—Top Guide for Automatic Valve
402-W—Gasket for Top Cover
402-X—Lever Pin for Operating Rod

FIG. 8(a)

Old Style Vapor Regulator—Vapor Regulator No. 402

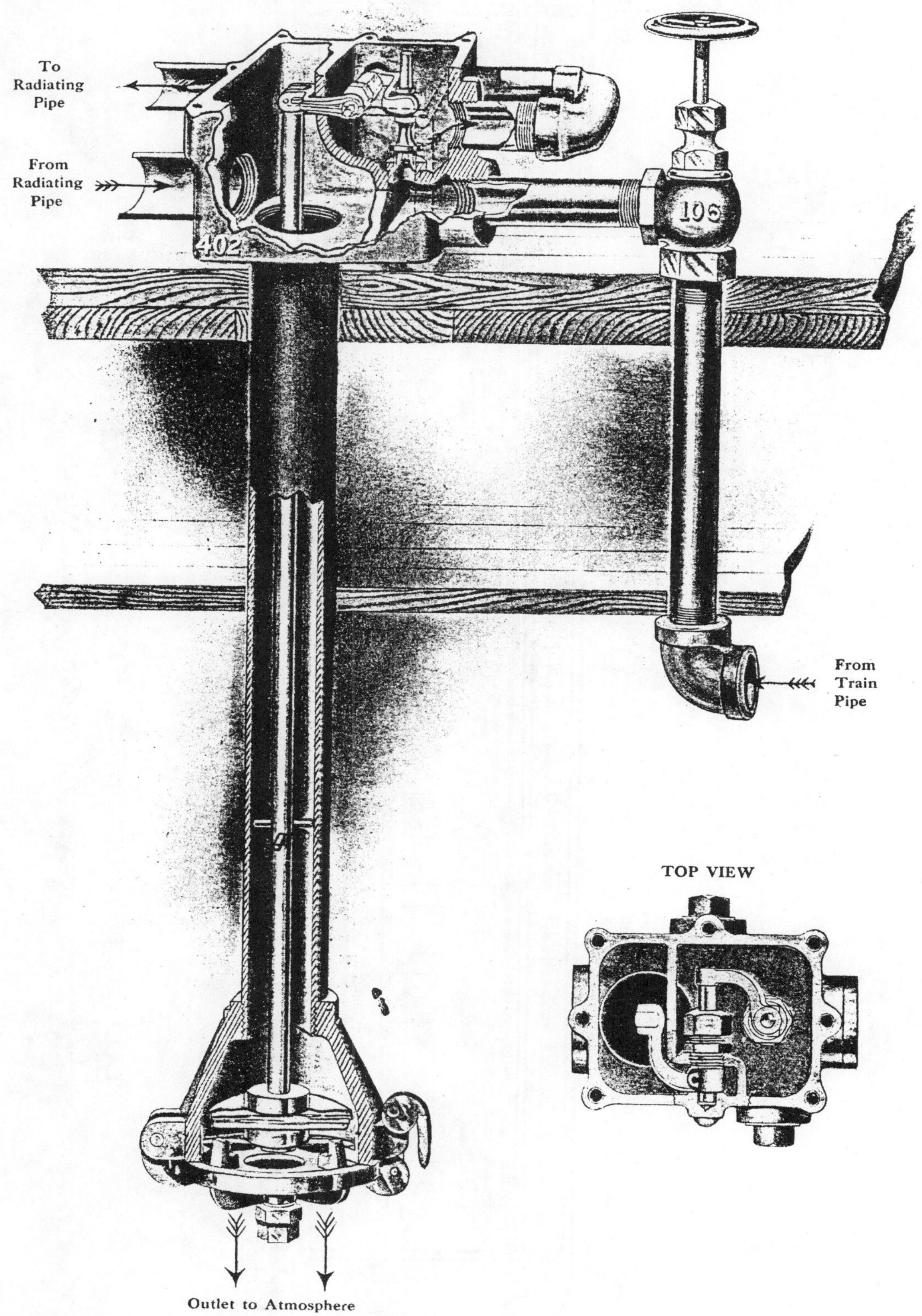

FIG. 8(b)

ELEVATED CAR EQUIPMENT.

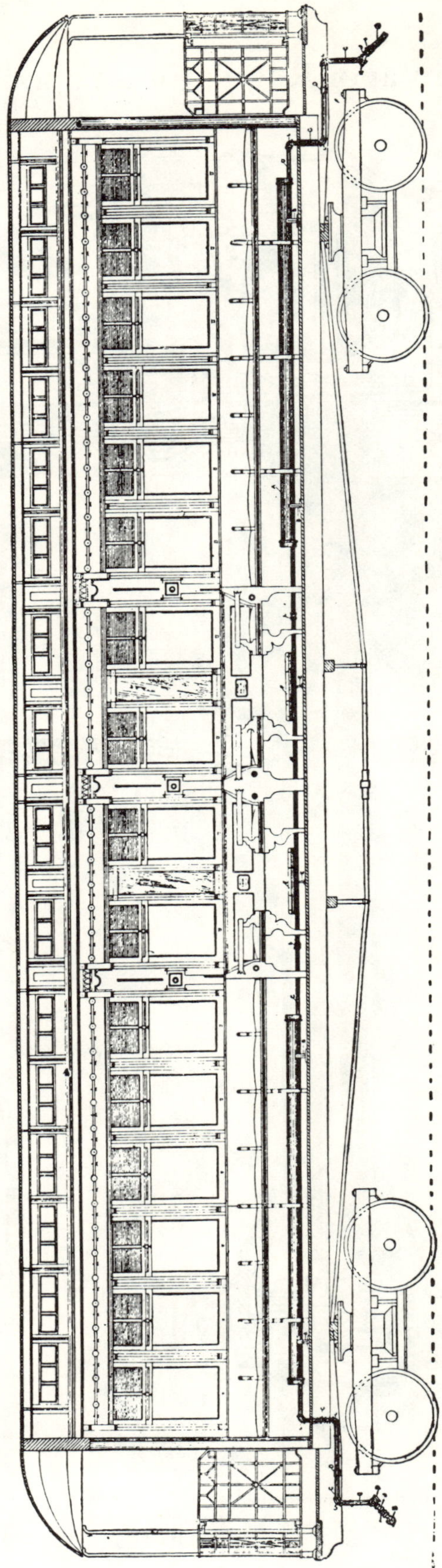

LONGITUDINAL SECTION OF A COACH OF THE MANHATTAN RAILWAY CO.

FIG. 9.

FIG. 10 HORSE CAR WITH "HUMAN BODY HEAT" ONLY.

FIG. 11. ELECTRIC INTERURBAN CAR WITH RESISTANCE HEATERS POWERED FROM TROLLEY CURRENT.

FIG. 12. CABLE CAR TRAINS ON THE BLUE ISLAND ROUTE WERE COMPOSED OF THE CABLE GRIP CAR AND A TRAILER CAR. THIS TRAIN IS AWAITING ITS DEPARTURE TIME AT THE BLUE ISLAND-LEAVITT TERMINAL OF THE ROUTE. WHEN THIS PICTURE WAS TAKEN, IN 1895, THE CABLE CAR SYSTEM IN CHICAGO INCLUDED 86 MILES OF TRACK. WINTER ENCLOSED CARS HAD COAL OR WOOD STOVE HEAT.

CAPITAL TRANSIT CO. 1938-50 CARS.

FIG. 13. PCC STREET CAR USING FORCED AIR HEAT FROM "WASTE HEAT" OF POWER AND DYNAMIC BREAK RESISTORS

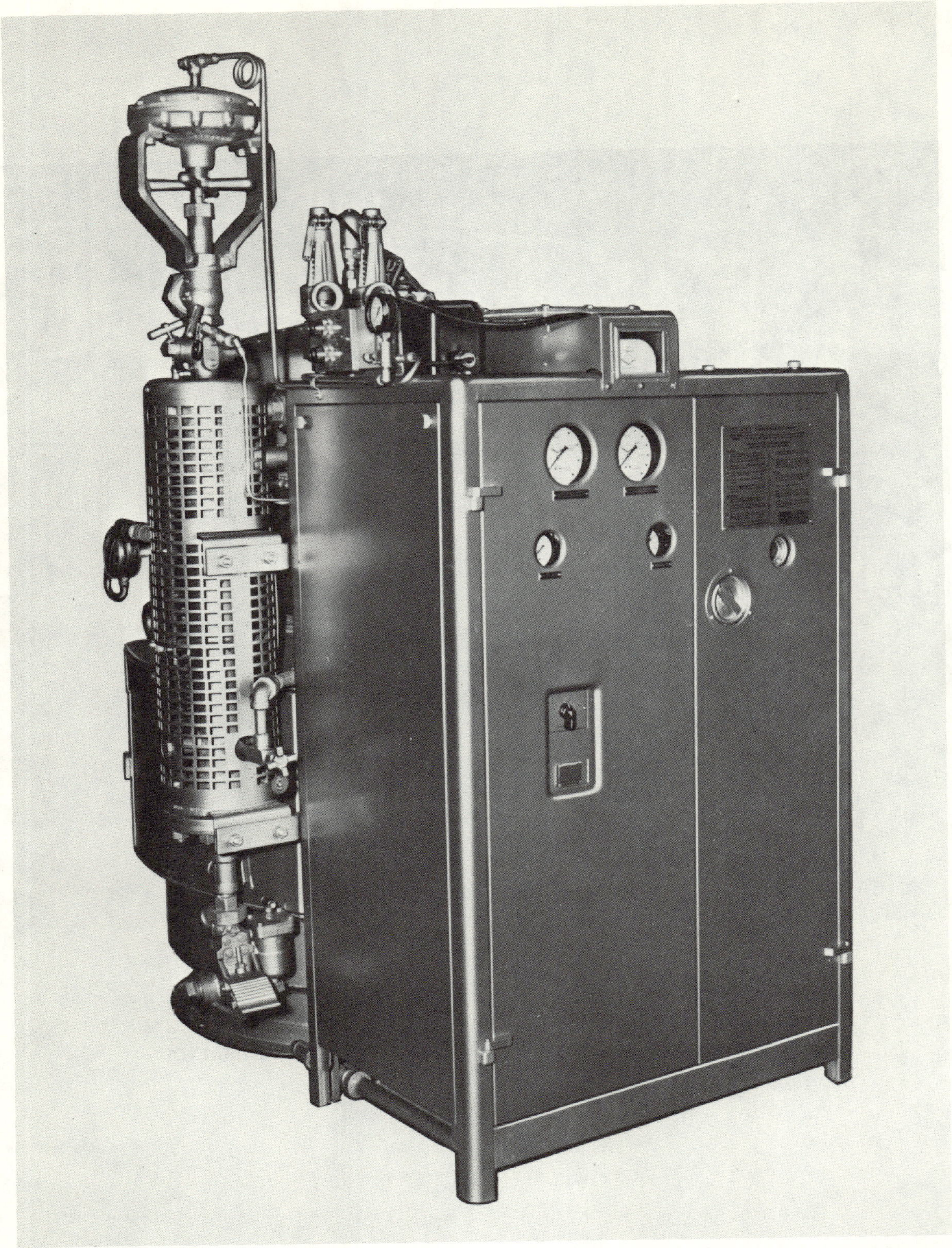

FIG. 14 OK-4625 TYPE VAPOR STEAM GENERATOR CURRENTLY USED ON AMTRAK PASSENGER LOCOMOTIVES

FIG. 15 TYPE OF FLASH BOILER USED ON EARLY BURLINGTON ZEPHYRS

FIG. 16 FIRST TWIN CITIES ZEPHYRS OPERATING AT 90 – 100 + MPH HEATED BY STEAM FLASH BOILERS IN WINTER TEMPERATURES OF -20 – -90 F

12 Wrought Steel Passenger Car and Locomotive Wheel Development

O. B. CAPPS

In the year 1805, Richard Trevithick of England designed what is considered the true forerunner of all steam locomotives. He also invented the forerunner of today's wheel, which was flanged and designed to travel over an iron strap rail. Then in 1812, Matthew Murray designed the first locomotive with flanged wheels for commercial railroad work.

The U. S. entered the railroad business in the year 1827 with the granting of the charter for the Baltimore & Ohio Railroad. Ross Winans, an Assistant Engineer of the B & O, reversed the wheel flanges on locomotives and cars, placing them on the inside of the rails. This enabled motive power and rolling equipment to hold the rail much better. In 1831, he produced the first eight-wheeled passenger coach which was larger and more comfortable and consequently competitive.

Locomotive wheels up to the year 1826 were made of cast iron. At about this time, wrought iron tires were introduced. These met with favor for both locomotive wheels and passenger cars. In 1829, a patent was granted on a conical or tapered tread, as shown by Fig. 1, which begins to look like the present day flange and tread contour.

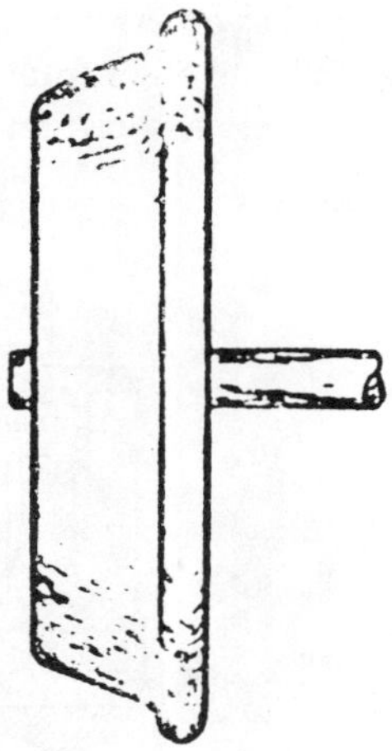

FIG. 1 JAS. WRIGHT'S PAT., 1829

After the chartering of the B & O, railroads in the U. S. expanded rapidly. Operating conditions on grades and increased weight of equipment made constant demands on wheel technology. Wrought iron tires, popular in England, wore out rapidly in American service and were flattened from the use of sand in stopping and constant sliding on grades. There was a return to the use of chilled iron tires, which was the hardest form of iron then known. One-piece cast iron wheels were developed for car equipment. For the larger diameter locomotive drivers, the hubs and spokes were cast of soft iron and the chilled iron tire was bolted to lugs cast in the rim (Fig. 2). The chilled iron tires were used on locomotives for several years until the weight of the engine and the size of the drivers made it extremely difficult to cast a chilled iron tire to meet the service requirements. Around 1865, steel tires came into use for locomotives and remained as standard until the demise of steam in general on American railroads. Freight car wheels were generally of the cast iron double plate design until the beginning of the twentieth century (Fig.

3). Beginning about 1869, the steel-tired wheel became popular for passenger car use. Apparently, early wheel designers had the idea that it was necessary to provide some kind of cushioning action in the wheel structure and the solid-wheel was considered too rigid for this service. This is evidenced by the use of steel-tired wheels with plates made of compressed paper, teak wood, wood black inserts and even hemp packing between the tire and wheel center. These designs proliferated during the last half of the nineteenth century.

In these early days of railroading, each road operated independently without regard to design or standards or interchangeability. Diameters and contours of wheels and gage of track varied in most cases. The confusion resulting from this disjointed system of railroads became quite evident during the Civil War period. Efficient transportation was greatly hampered due to the lack of standardization. This would involve such items as a standard wheel gage, diameter, flange and tread contours as well as other items such as coupler height, etc. To bring order from this chaos, the Master Car Builders Association of the United States of America was organized in 1867. One of the first subjects was the gaging of wheels because of the varying track gages that existed at the time. The other important consideration concerning wheels was the standardization of wheel diameters and tread and flange contours. It was not until 1886 that a standard flange and tread contour was adopted. The 36 in. diameter wheel gradually became generally accepted for passenger cars and the 33 in. diameter freight car wheel was standardized about 1880.

The 33 in. wheel was probably the most common size in freight service and also was the upper limit of manufacture by the casting processes of the day.

The passenger car wheel of 36 in. diameter was finally standardized with the introduction of the mono-bloc forged and rolled wrought steel wheel. The original wrought steel wheels in this country were produced by the Schoen Wheel Mill in Pittsburgh, Pennsylvania. The first of these wheels were shipped in 1903. Today, the generally accepted standard for locomotive and passenger car use is the wrought steel wheel.

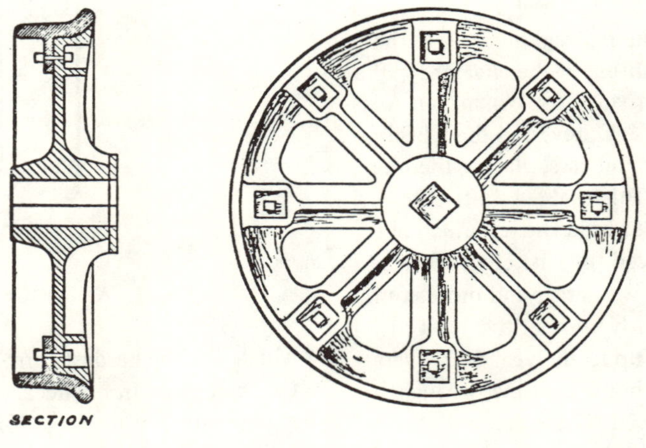

FIG. 2 BUILT UP CAST IRON WHEEL, GEO' HAWKS – 1807

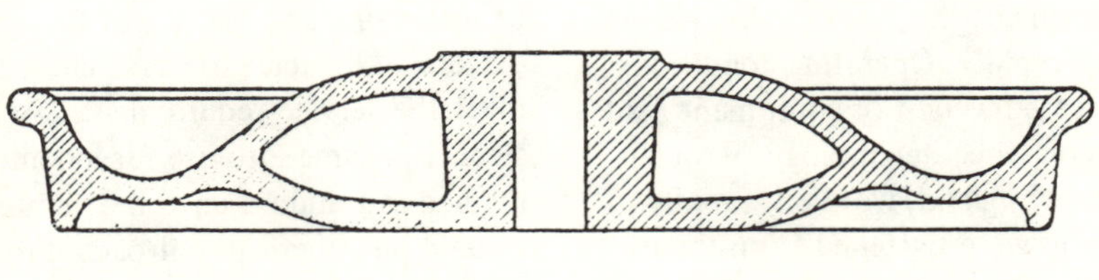

FIG. 3 N. WASHBURN PAT., 1850

Heat treated wrought steel wheels for passenger and locomotive service were introduced in 1930. Today, most wheels for this service are of this type. These wheels are termed multiple wear, meaning the rim is of sufficient thickness to allow the flange and tread to be restored to new contour several times during their service life. These wheels are manufactured to Association of American Railroad standard designs and specifications. Because of the varied designs and services of current motive power and equipment, the wheel diameter might be anywhere from 28 in. to 50 in. These wheels are available in five ranges of carbon content and the range selected usually depends on the nature of service. For example, the lower carbon content affords additional protection against thermal cracking and higher carbon content wheels are more resistant to tread metal breakdown commonly called shelling. In any case, a compromise with the optimum must be made.

Wheels are furnished to railroads under carbon ranges and heat treatments known as Class "U", "L", "A", "B", and "C". A broad description of these various classes is as follows:

	Carbon	B.H.N.	
Class U	0.65-9.80		General freight car service with cars of 70 tons capacity and under, where an untreated wheel is satisfactory. Passenger car and locomotive wheels are not recommended under this class.
Class L	Not over .047	197-277	Passenger car service, primarily rapid transit, with on-tread braking and light wheel loads.
Class A	0.47-0.57	255-321	Passenger car and locomotive service with on-tread braking and moderate wheel loads.
Class B	0.57-0.67	277-341	Freight cars over 70 tons capacity, passenger car and locomotive service with on-tread braking and moderate to high wheel loads.
Class C	0.67-0.77	321-363	Freight cars over 70 tons capacity and locomotive service with high wheel loads and light on tread braking. Passenger car service where off tread brakes are employed.

The Brinell hardness is measured on the front face of the rim with the edge of the impression not less than 3/16 in. from the radius joining face and tread.

Passenger car and locomotive wrought steel wheels manufactured today must comply with specification M-107, the latest revision under jurisdiction of the Association of American Railroads.

Today's service demands on wheels are extremely critical. Because of this, wrought steel wheels are made to meet the highest standards of quality. They are subjected to ultrasonic testing of the rim area in both radial and axial directions as assurance of soundness. The wet method of magnetic particle testing is used on the plate area. This same area is subjected to shot peening under prescribed conditions to impart a beneficial surface compressive stress to increase fatigue strength to the plate area.

The general configuration and nomenclature of the present day wrought steel wheels is illustrated in Fig. 4.

Typical tread and flange contours for passenger cars and locomotives are shown in Fig. 5 (A and B).

Wrought steel passenger car and locomotive wheels are generally manufactured from steel made by any of the following process: open hearth, electric furnace, or basic oxygen process.

MANUFACTURE OF WROUGHT STEEL WHEELS

Processes used by individual manufacturers differ in many details, but all have been developed to produce wheels meeting the requirements of A.A.R. Specifications.

The manufacture of a wrought steel wheel starts with a block of steel of the proper weight to form the wheel. The block may be produced: (a) as an individual ingot; (b) as a block cut from a long ingot; or (c) as a block cut, or sheared from a rolled bloom.

After weighing and inspection, the blocks are made into wheels by a series of hot forming operations, the details of which depend on the practices of individual manufacturers. The essential steps are:

1. The block is slowly and carefully heated to forging temperature and then forged to form a wheel bloom.
2. The bloom is reforged to form a wheel blank of roughly the shape of a wheel, but smaller in diameter.

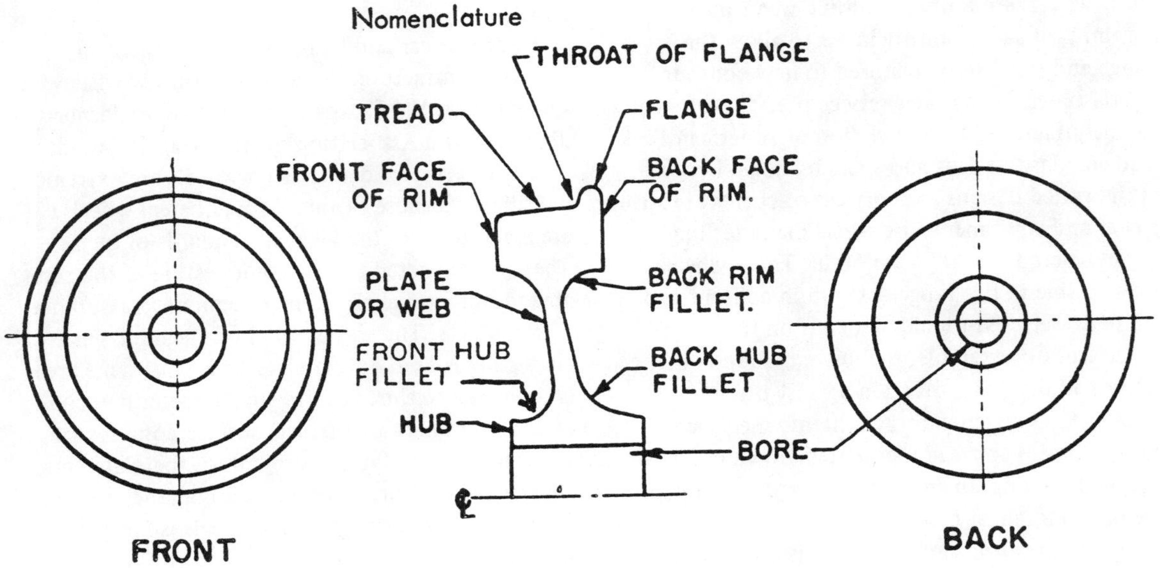

FIG. 4

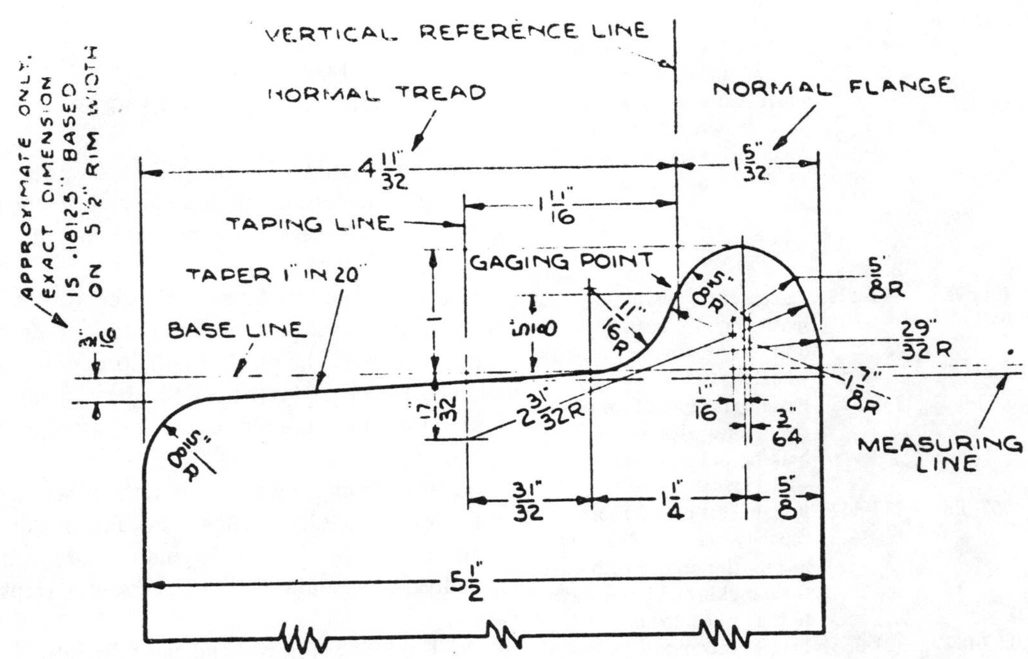

FIG. 5A TREAD AND FLANGE CONTOUR FOR NARROW FLANGE STEEL WHEELS ON CARS, TENDERS, DIESEL AND ELECTRIC LOCOMOTIVES, TRAILER AND ENGINE TRUCK WHEELS

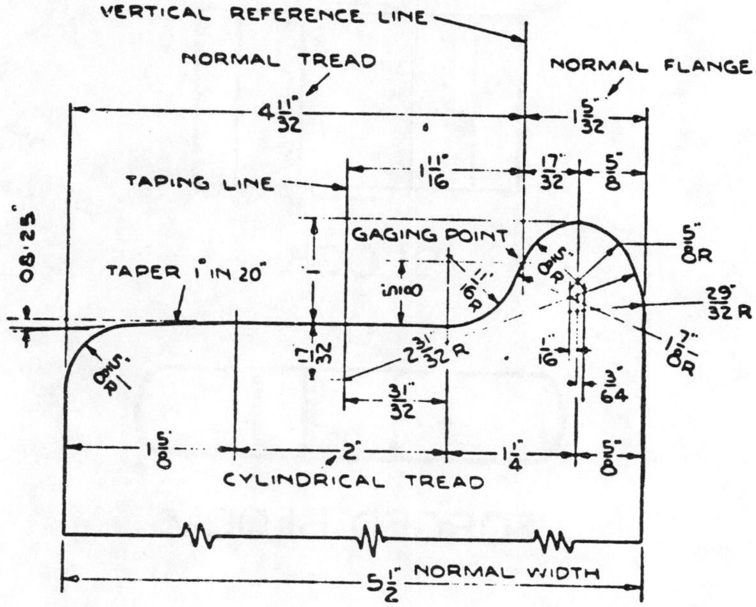

FIG. 5B CYLINDRICAL TREAD CONTOUR
NARROW FLANGE STEEL WHEELS

3. The blank is transferred to the rolling mill and rolled to the desired section and diameter.
4. The rolled blank is transferred to the coning or dishing press, where the plate of the wheel, which was rolled flat, is given its final conical form.

NOTE: These operations are illustrated by the diagrams in Fig. 6. They may be carried out with a single heating of the block, or the bloom may be reheated before blanking and the blank reheated before rolling.

5. The wheels are allowed to cool uniformly under properly controlled conditions.
6. The wheels are heat treated and machined as specified.

The hot rolling operations by which the forged blank is formed into a wheel are illustrated diagrammatically in Figs. 7 and 8.

1. In the vertical type mill, Fig. 7, the wheel blank rotates on a mandrel through the bore, while the web rolls compress the rim against the tread roll, reducing its thickness and increasing its diameter as rolling progresses.
2. In the horizontal type mill, Fig. 8, no mandrel is used. During the rolling operation, the pressure rolls which bear against the tread and flange force the metal to be extruded between the driven edging rolls which bear against the rim faces, inside rim and plate, thus forming the flange and tread contour.

After the final hot forming operation, the rate of cooling is controlled primarily to prevent thermal ruptures and flaking of the steel.

Heat treatment is performed primarily to increase the strength and hardness of the wheel and tread, thus increasing the load carrying capacity and resistance to wear. In the process of heat treatment, the wheels are cooled to a temperature below the critical range, then reheated to a temperature above the critical range, quenched and tempered. The quenching may be performed on the entire wheel or on the rim only. For untreated Class U wheels these heat treating operations are not performed.

Wheels are machined to design dimension and shot peened before inspected.

The processes used by any individual manufacturer for the production of wrought wheels are the same whether the wheels be one wear, two wear, or multiple wear.

Figures 6, 7 and 8 illustrate typical forging and rolling steps and equipment necessary to making the wheel.

The mono-bloc steel wheel is the generally accepted standard today for use in passenger car and locomotive service. Some transit cars have recently

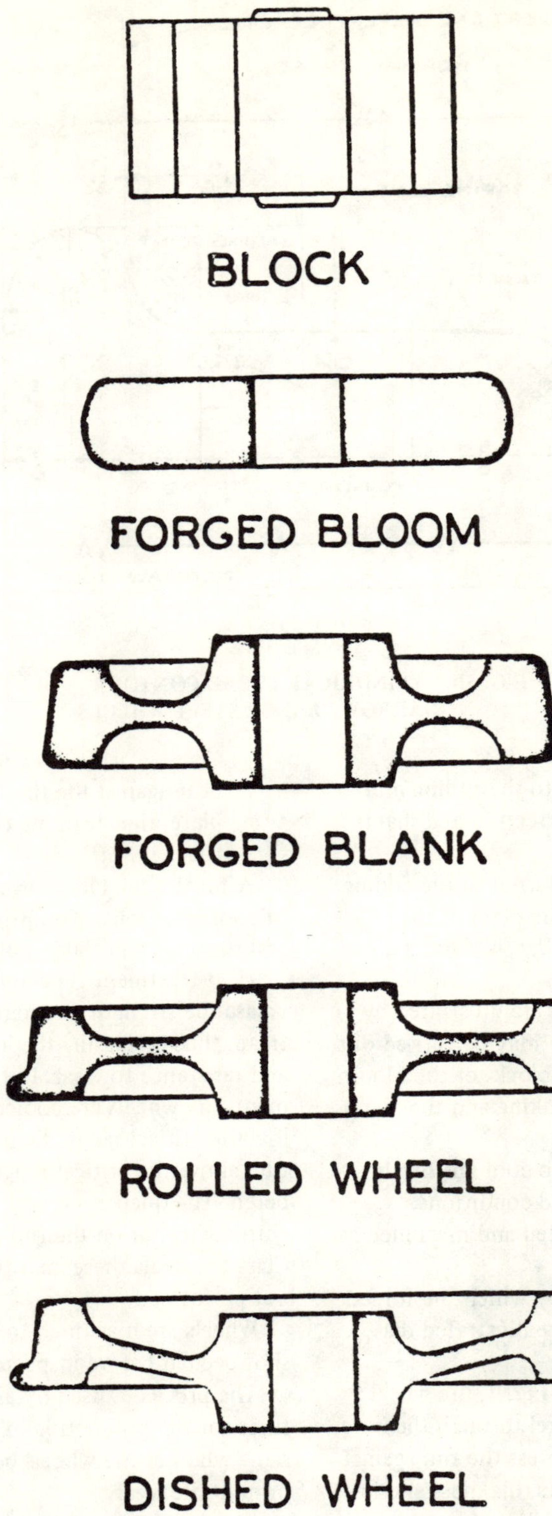

FIG. 6 TYPICAL STEPS IN THE MANUFACTURE OF WROUGHT STEEL WHEELS

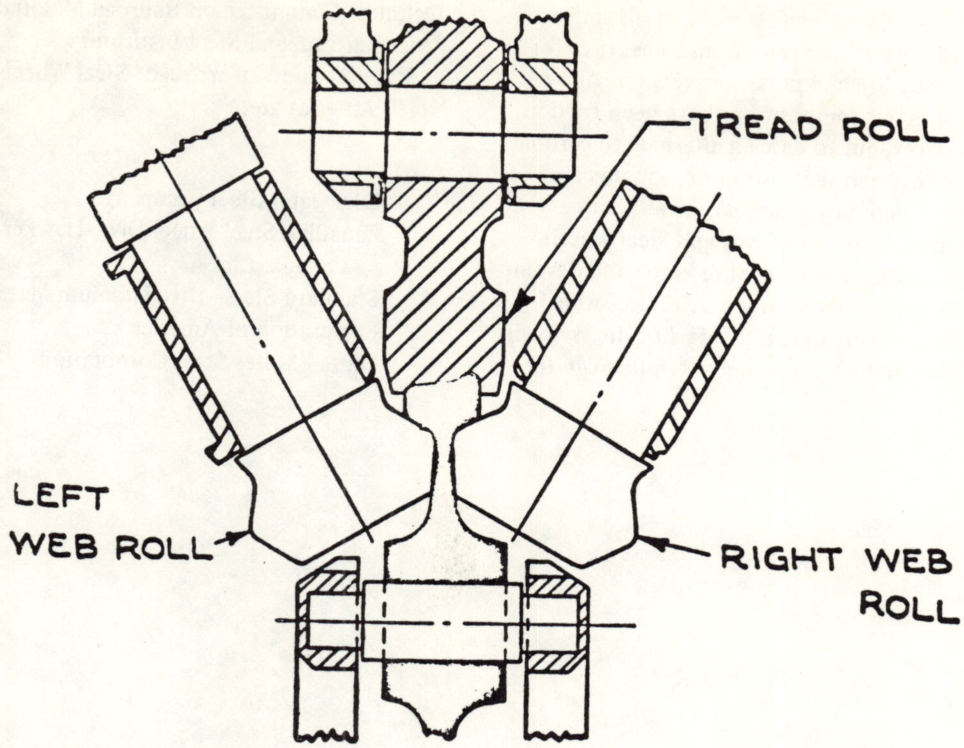

FIG. 7 PLAN OF MILL ROLLING WHEEL VERTICALLY

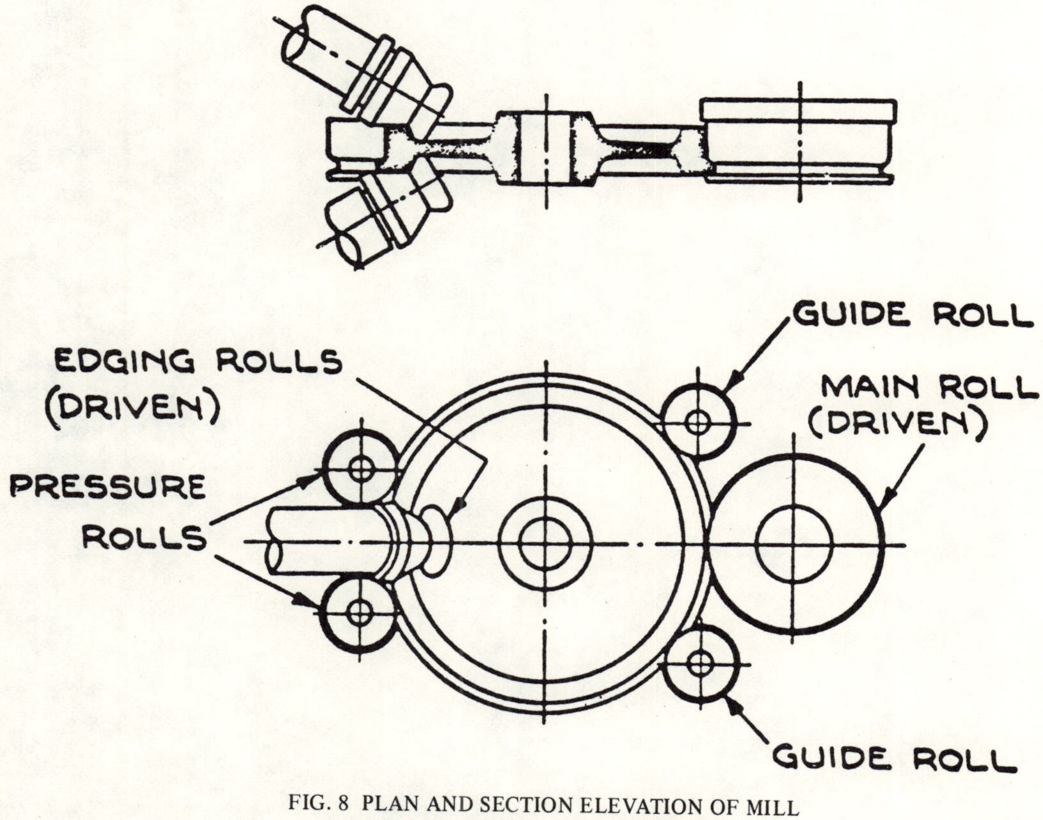

FIG. 8 PLAN AND SECTION ELEVATION OF MILL ROLLING WHEEL HORIZONTALLY

been equipped with wheels made of dissimilar materials, i.e., aluminum centers and steel tires for weight reduction and noise suppression. Alloy wheels of many compositions have been tried in these services, but to date an alternate to carbon steel has not been successful primarily because of the basic economics in cost of manufacture.

The manufacturers of wrought steel wheels have continuing programs of research and development for improvement of wrought steel wheels.

Cooperative efforts on the part of the Wrought Steel Wheel industry are conducted through the Technical Committee on Railroad Materials of the American Iron and Steel Institute.

Manufacturers of Wrought Steel Wheels in North America are:

Edgewater Steel Company
Canadian Steel Wheel Div.—Hawker Siddeley Canada, Ltd.
Standard Steel—Div. Titanium Metals Corporation of America
United States Steel Corporation

13 Cast Steel Freight Car Wheel Development

B. J. ECK

Archeologists have unearthed wheels in Russia and India which are estimated to have been made about 3000 BC and, although wheels have been used in Asia, Africa and Europe for thousands of years, it is noteworthy that America was a wheelless continent until the arrival of the Europeans. Thus, the initial use of horsedrawn railroads occurred in Continental Europe for use in mines in the 1500's. These trains used a board track laid on the ground so the cars did not get stuck in the mud or have difficulty negotiating the rough, rocky, mine pathways. The flangeless, spoked, wooden wheels on these freight cars were similar to those used on regular horse drawn, street type, wagons. Since there were no flanges on these wheels, the cars were kept on the track by board flanges fastened to the wooden rail. To improve the wear and riding characteristics, iron strips were later fastened on the tops of the wooden rails. Some soft iron wheels were used on mine cars in the 1500's in Germany. A completely iron, flanged type rail was first cast in England in 1767. The use of cast iron rails and wheels led to the development in the late 1700's of the tee rail and the double flanged wheel, which was later supplemented by the single flange cast iron wheel.

In North America, the first horse drawn freight trains were used in mines in 1807. With the development of the first practical steam locomotives in England in the early 1800's, and their subsequent use in North America in the 1820's, the soft grey iron wheels proved unsatisfactory because they had a short life and developed flat spots. Fortunately for the railroads, at this same time, a discovery was made in England that if molten iron was solidified rapidly by pouring it against a cold metal surface, a phenomenal change takes place in the metal. This chilled metal, "case hardening" as it was called, became a white crystalline metal which is very hard and wear resistant, whereas the grey iron wheels, in which molten iron was cast against a sand mold and cooled slowly, formed a flake graphite structure which was soft. So experiments were conducted to find a way of making a wheel of this newly discovered chilled iron. By using a metal band or chiller around the tread and flange surface of the mold, a method was found to make a wheel with a hardened, chilled iron, wearing surface and a soft, easily machined, grey iron, hub, and the desired rotundity. The first known comparison service tests of chilled and grey iron wheels were conducted by George Stephenson (the steam locomotive developer) and his associates on one of the early steam engines in England in 1818. Eleven years later, 1829, the first chilled iron wheels were manufactured in America at a foundry in Maryland and used on a railroad hauling coal from the mines. The use of iron wheels rapidly accelerated and in 1943 an all-time record of 3,306,463 chilled iron wheels were shipped from 44 foundries in the United States and Canada.

The first wheels were held on the axle with pins. The first pressure fitting of wheels on axles was accomplished around 1830 with a hand operated screw press. After the invention of the hydraulic press in 1857, the screw presses were gradually replaced.

Many different track gauges were used in North America in the early 1800's. With the advent of the Civil War, the need for interchangeability and a common track gauge became more apparent, and the most common gauge (4 ft. X 8½ in.) was accepted as the standard. This gauge had its origin in the spacing of the wheels on the coal cars in existence in England at the time they were first placed on a rail. These coal carts had been built to accommodate the existing hard roads which were based on the Roman road system built in about 200 B.C. The Roman road spacing, stone slabs 4 ft, 8½ in. apart, in turn had the spacing of the chariots brought to England by the Roman Legions. The Romans had adopted as their standard gauge for chariots, the Pace. The "Roman Pace" is the distance stepped by their army in marching. Today much of the world has adopted this gauge as standard for its railroads.

To handle the many problems which were created by the interchange of traffic, the Master Car Builders Association (MCBA) was organized in 1867. This is the predecessor organization of the Association of American Railroads (AAR). These organizations set up the rules and specifications for freight car wheels. The MCBA's first rules concerning wheels, which were issued in 1877, governed removal of slide flat and worn flange wheels. Other defects were added from time to time.

By 1867, the diameter as well as the exact tread and flange contour of the 33 in. cast iron freight car wheel were defined by the MCBA. The diameter was dictated by the fact that this was the largest diameter the early English foundries could cast. During the years, the car capacity was increased from 10 to 70 tons and changes were made in the wheel contour and casting quality, but the standard interchange cast iron freight car wheel diameter stayed at 33 in. It was after the advent of the steel wheel and during the phasing out of the iron wheel that the heavy, larger diameter, freight car wheels came into use.

During the 1800's, the cast iron wheels were of spoke design and the hubs were split to relieve castings strains. In 1838, a wheel was patented with two parallel plates which extended from the hub to the rim with a hollow space between the plates. This design, modified from time to time, is known as the double plate wheel and was the standard for railroads until superseded in 1928 by the single plate design. During the years, more than 900 patents were taken out for various designs of chilled car wheels. The single plate, chilled iron design, adopted in 1928, reduced the development of cracks in the plate with which the double plate wheel was troubled. Brackets were cast on the plates of chilled iron wheels for heavier cars to strengthen the flanges and rims.

Until the late 1800's the cost of producing steel was much higher than the cost of iron, but as the relative cost of steel production decreased, many methods of manufacturing wrought and cast steel tires and wheels were evaluated and patented. The steel tired, pressed paper center wheels produced by the Allen Paper Car Wheel Company near Chicago, were first placed on Pullman cars in 1871 and many went into passenger service, but there is no record of this type of wheel being used on freight cars. The steel tires for these wheels were manufactured by Krupp in Germany. In addition Krupp manufactured cast steel wheels for locomotives in the late 1800's, but they were not used on freight cars. Cast steel and cast iron centers for locomotives and cast steel wheels for nonrailroad applications were manufactured in North America in the late 1800's and early 1900's.

Composite steel-iron wheels and solid manganese steel wheels manufactured by Taylor Iron and Steel Company of New Jersey are listed in the Car Builders Directory of 1898 but they apparently were not used in freight service.

The first widely accepted cast steel freight car wheels were made in North America in the early 1900's. J.C. Davis originated the idea in 1898 for a centrifugal cast one-wear, one piece, steel wheel. In his process, ground ferro manganese was introduced into the first portion of the low carbon steel as it was added to the spinning mold. A soft hub and harder tread was thus produced in the wheel. The first "Davis" wheels to go into service were cast in 1903 at the American Steel Foundries East St. Louis plant and by 1910 large quantities were being sold. In 1938, this method of manufacturing wheels was discontinued.

National Malleable and Steel Castings Company developed a "Spun Steel Wheel," which was cast in a spinning mold somewhat similar to the Davis wheel. It was used in limited service from 1934 until production was stopped in 1953. In addition, an English firm, Steel, Peech and Tozer, put a small quantity of experimental spun cast steel wheels in freight service in 1955, but they were removed after failures occurred.

As mentioned earlier, for the first 130 years of railroad service in North America, most freight cars had 33 in. chilled iron wheels, but with the advent of the high-powered diesel locomotive, a new concept of freight service was evolved. The new cars were designed to carry greatly increased loads at speeds which were inconceivable for freight service earlier. It became apparent that the chilled iron wheels could not take the increased mechanical forces and braking so that the chilled iron wheel manufacturers sought to develop methods of making steel wheels. Their experience in casting iron wheels led them to develop special casting methods for producing steel wheels. One manufacturer, Griffin Wheel Company, developed the Pressure Pouring Process of forcing liquid steel by air pressure through the bottom opening of graphite molds, as shown in Fig. 1. Another manufacturer, ABEX, developed a process of pouring molten metal from bottom pouring ladles by gravity through a feeding head at the center of a composite mold. Initial approval was given by the AAR in 1951 for service trials of experimental cast steel freight wheels made by these two processes. Unlimited approval was given in March 1, 1957, with the adoption of the AAR cast wheel specification, M-208. During the 1960's, cast wheels were placed in passenger and diesel service and the M-208 freight wheel specification was expanded to include passenger and diesel wheels.

The last iron wheels for railroads were cast in 1963 and today these wheels are prohibited in freight car interchange service. Thus, today, wrought and cast steel, one piece, wheels are the only AAR approved wheels for railroad service in North America.

The advent of the diesel, as discussed earlier, led to heavier cars and in 1961 AAR approval was given for the first experimental 36 in. freight wheels for 100 ton cars. The 28 in. wheels for low deck freight cars and 38 in. for 125 ton cars were approved at about the same time.

Since the braking on passenger cars is more severe than on freight cars, more problems with thermal cracking were experienced in passenger service and, therefore, the lower carbon steel wheels, such as AAR Class L, A, and B, were initially developed for that service, and later applied to freight car wheels. In like manner, the quenching of the steel treads to obtain a harder and more wear resistant rim was initially tried on the lower carbon passenger cars, but as the freight car loading became heavier, the use of quenched wheels to increase life and reduce shelling and other types of rim deterioration came into use. Thus, in freight car service the use of the quenched Class B and C wheels has become common along with the unquenched Class U wheels. The AAR chemistry and hardness specifications for these wheels are as follows:

		Percent	Minimum Hardness	Maximum Hardness
Carbon	Class U	0.65 to 0.80		
	Class L, not over . .	0.47	197 BHN	277 BHN
	Class A	0.47 to 0.57	255 BHN	321 BHN
	Class B	0.57 to 0.67	277 BHN	341 BHN
	Class C	0.67 to 0.77	321 BHN	363 BHN
Manganese		0.60 to 0.85		
Phosphorus, not over		0.05		
Sulfur, not over		0.05		
Silicon, not less than		0.15		

Various laboratory tests have been devised for the purpose of evaluating wheels. One of the early test methods designed to simulate brake shoe heating of wheels consisted of placing a wheel in a specially designed mold and pouring 4-6 in. of molten iron against the surface of the tread, as shown in Fig. 2. The criteria used to evaluate wheels in this test was the absence of cracks in the wheel after a period of time. Another early test was the "drop test," shown in Fig. 3. The next significant development was the use of dynamometers which simulated service braking on full size wheels. A comprehensive article on this subject is the University of Illinois' Engineering Experiment Station Bulletin Series No. 387, published June, 1950, written by H. R. Wetenkamp, "The Effect of Brake Shoe Action on Thermal Cracking and on Failure of Wrought Steel Railway Car Wheels". In the 1950's a method of simulating brake shoe heating was developed which consisted of using an induction coil and high-frequency generator to create a controlled source of heat to simulate a brake application, as shown in Fig. 4. Test stands are often used to subject strain-gaged wheels to lateral and vertical loads. However, one of the most significant technological developments of this century for evaluating wheels is the use of computer mathematical programs to simulate and determine stresses in wheels due to combinations of mechanical and thermal service loads. Some of the most significant papers illustrating the developments in this field are listed below chronologically:

1. Johnsson, S., "Digital Machine Computation of Railway Wheels," International Wheel-Sets Congress, Pergamo, April 7-12, 1963.

2. Riegel, M. S. Levy, S. and Sliter, J. R., "A Computer Program for Determining the Effect of Design Variation on Service Stresses in Railroad Wheels," *Journal of Engineering for Industry,* Nov. 1966.

3. Novak, G. E., and Eck, B. J., "A Three Dimensional Finite Difference Solution for the Thermal Stresses in Railcar Wheels," *Journal of Engineering for Industry, ASME Transactions,* Vol. 91, Series B, Aug. 1969, pp. 891-896.

4. Wetenkamp, H. R. "Thermal Stress Developed in S Plate, Straight Plate and Deep Dish Wheels," *ASME Paper No. 73-RT-1.*

5. Novak, G. E., Greenfield, L. P., and Stone, D. H., "Simulated Operating Stresses in 28-In.-Dia. Wheels," *ASME Paper No. 75-RT-10.*

6. Novak, G. E. and Eck, B. J., "Cyclic Distortions and Stresses in 36-In. Wheels Under Combined Service Loads," *ASME Paper No. 75-WA/RT-5.*

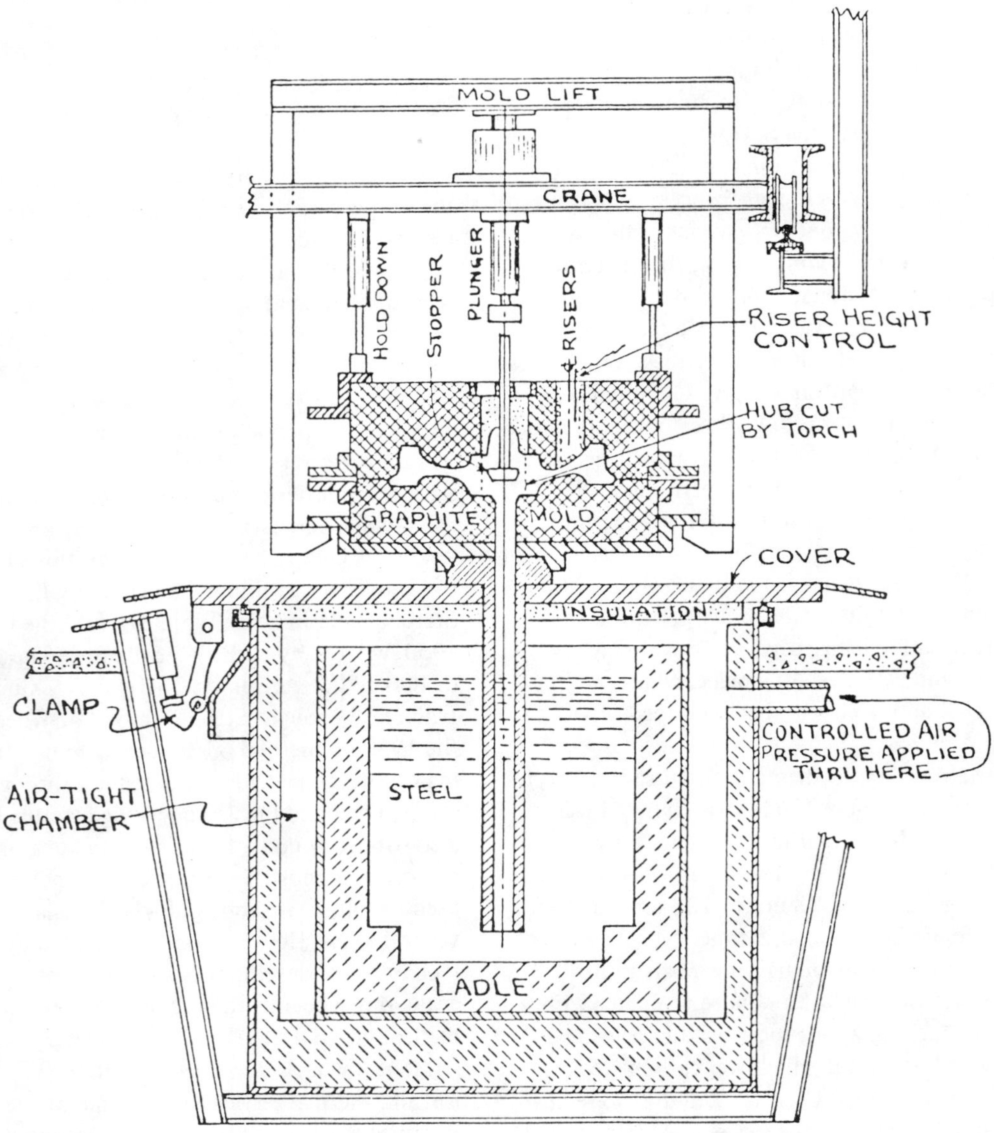

FIG 1. SCHEMATIC ARRANGEMENT FOR PRESSURE POURING CAST STEEL WHEEL IN GRAPHITE MOLD

FIG. 2 WHEELS SELECTED FOR THERMAL TEST WERE LAID FLANGE DOWN AND A CHANNEL WAY FORMED AROUND THE WHEEL. THE CHANNEL WAY WAS FILLED WITH MOLTEN IRON AND THE WHEEL COULD NOT BREAK DURING A PRESCRIBED TIME.

FIG. 3 WHEELS SELECTED FOR DROP TEST WERE LAID FLANGE DOWN ON SUPPORTING LUGS. A TUP FELL STRIKING A BLOW ON THE HUB. AAR SPECIFICATIONS REQUIRED THAT THE WHEEL DID NOT BREAK IN A GIVEN NUMBER OF BLOWS.

FIG. 4 INSTRUMENTED WHEEL AND INDUCTION COIL ASSEMBLY USED FOR SIMULATING BRAKE APPLICATIONS.

14 Technological History of Plain Journal Bearing Development

JOHN ZUPEZ

The first steam railroads in the United States were started about 1829 and by 1869 had expanded to 53,000 miles of track, creating one of the greatest industrial revolutions in the western world. During this period the conversion of existing technology of steam engine, wheel, axle and bearing produced many designs, particularly in the bearing assembly, but it was not until 1869 that the present journal bearing was accepted as a standard. Its history is the topic of this paper. First, some mention should be made of the organizations which contributed to railway journal bearing development.

In 1869, the year of the golden spike, the Master Car Builders' Association was founded, the first of its kind in the world. Its primary function was to establish a code of rules for cars to be acceptable in interchange and to set standards and recommended practices for car construction. In a 1919 consolidation, MCB became a part of the American Railway Association, predecessor to the present Association of American Railroads. The Operation and Maintenance Department, Mechanical Division, became a committee and subcommittee for specific mechanical investigations. About 1950, an AAR Research Center was built on the campus of the Illinois Institute of Technology with various divisions, one being the Mechanical Research Division with a journal bearing testing machine. Divisions for inspection or investigation worked closely with the Mechanical Division and also evaluated new products from the supply industry.

In 1901 car men formed the Chief Interchange Car Inspectors and Car Foremen's Association, presently known as Car Department Officers Association. There is also a supplier committee organization for many products. Finally, ASME input can be traced back before 1920, when the Rail Transportation Division was formed. All of these organizations facilitate the exchange of information, the solution of problems, and the application of new developments.

CAR JOURNAL BEARINGS

The founding of the Car Builders' Association in 1869 brought the adoption of the MCB Standard Journal Bearing, or Car Brass (Fig. 1). In general appearance, with modifications, it is the same as that in use today. The bearing was 3¾ × 7 in. size, steeple back with wedge (or key), dust guard, and oil lubricated with waste packing. (Fig. 2). The bearing was iron-back babbitt lined. The babbitt was mechanically anchored and its bore was not machined. In 1875 bronze, lined or unlined, was made optional and soon after was made standard. The iron-back was tested again in the 1930's and results submitted to the AAR Lubricating Committee. It was approved as an alternate standard but never gained acceptance by railroads. Several times in the 1950s the aluminum-back bearing was approved for test by the AAR but never accepted as an alternate standard for use in freight cars. The bronze-back bearing was preferable because it was easier to obtain a chemically-bonded lining and new bearings could be purchased at the cost of remaking from the customer's scrap bearings. Iron-backs are

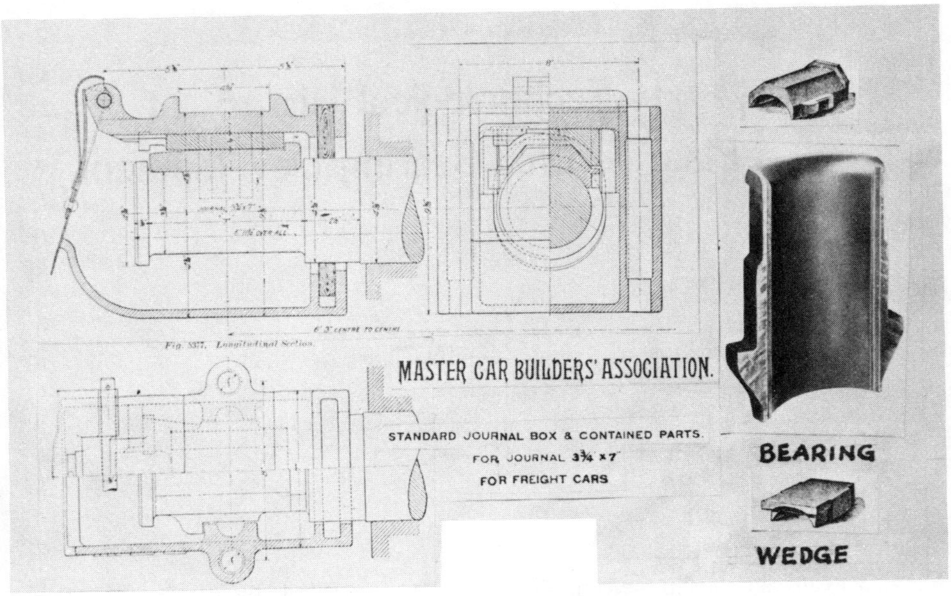

FIG. 1

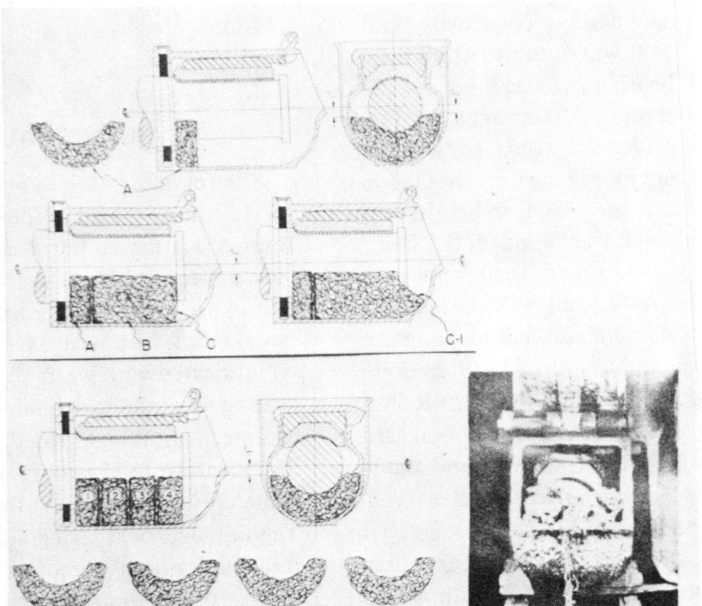

FIG. 2 JOURNAL STABILIZATION

in use in various countries where bronze and babbitt alloy metals are in short supply and subject to pilferage. Metal supply was a problem in the U.S. only very briefly during World War II.

As the capacity of cars was increased, larger sized bearings (measurements are in inches) were added: 3¾ x 7 (1869), 4¼ x 8 (1889), 5 x 9 (1896), 5½ x 10 (1900), 6 x 11 (1914), 6½ x 12 (1920), with the respective rating for two four-wheel trucks at 25, 30, 40, 50, 70 and 100 tons.

Basically, the babbitt-lined, waste-lubricated bearing operated on the principle of hydrodynamic lubrication, in which oil was carried through the bearing by movement of the journal. The bearing bore diameter was 1/16 in. larger than a new journal and up to 7/16 in. larger on some remachined journals. This resulted in a boundary or mixed-film lubrication until soft babbitt could wear into a seat of about 2 in. or more. Some overheating took place during this period. In 1955 a change was made to use three step-size bearings and journals with normal clearance of 0.015 in.

Bearing failures or "hot boxes" in the early days were not a serious problem since freight locomotives seldom exceeded 30 mph and frequently stopped for fuel and water. Also, trains had fewer than 50 cars on the average and train crews had sufficient time to inspect cars for mechanical faults. The average freight car was in transit for about 12 percent of the time and averaged only 16,200 miles per year. Failures were due mostly to fore and aft movement of the axle journal from coupling shocks and emergency brake applications, moving the waste toward the bearing and resulting in waste grabs. This movement also damaged the dust guard.

Around 1930 studies were made of various ways to stabilize the journal. Journal stops, flat-back bearings, and hi-hat bearings were tested and approved, and became mandatory after 1966. Bearings in use today are steeple, flat-back and hi-hat. It should be noted here that stabilization with the latter two bearings was due to eliminating the steeple and increasing the bearing arc about 22 degrees, namely from 112 degrees to 134 degrees on a 6 x 11 in. bearing. See Fig. 3.

Concurrent with journal stabilization, in 1960 it became mandatory to replace waste packing with lubricator pads (Fig. 4). Although waste was superior in feeding oil to the journal, it was a laborious job done in terminal yard repair tracks and would seldom meet the rules in the Lubrication Manual. The lubricating pad idea touched off pro-

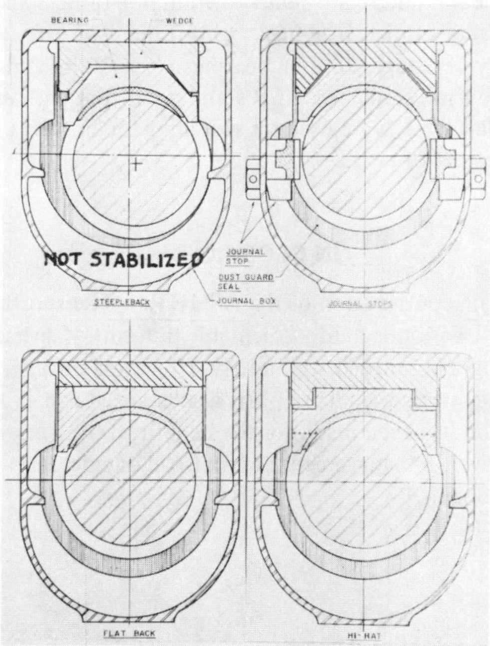

FIG. 3

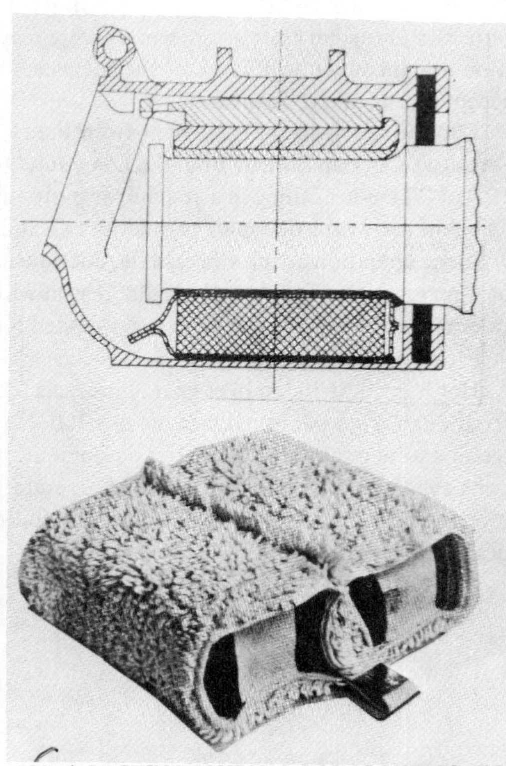

FIG. 4 PACKING JOURNAL BOXES LUBRICATORS

lific inventive research, resulting in many designs submitted for approval. Basically, nearly all were of foam rubber core with chenilled cotton cover. At present the AAR lists about 18 as conditionally approved. Many were approved for tests but never reached the conditionally approved status. Once journals were stabilized it was decided to change the design of the dust guard to a seal and also use a gasket seal for the box lid. Suppliers soon developed seals that were tested at the AAR Research Center and, at present, there are at least 10 of each type in use. This completed improvements to the journal box assembly and resulted in reducing the number of hot box set-outs from 201,276 in 1951 to 27,744 in 1964 (over 7 to 1), with 11 percent of the fleet on roller bearings (Fig. 5). With more cars per train and increased train speeds, less time for inspection and oiling should be offset by hot box detectors and other technical surveillance. This, in turn, can provide a helpful record of plain journal bearing performance until these are entirely replaced by roller bearings. At present about 70 percent of all cars are on roller bearings, which are mandatory on all new cars built.

Worthy to mention is the development of cartridge journal bearings as approved for testing on interchange cars in about 1950; Kargo by GMC Allison Div., Clevite by Clevite Graphite Bronze Co., and National by Abex Corp. (Fig. 6). Results of tests and further tests after changes were made were not encouraging because of the advance acceptance of the roller bearing.

Food for thought is a statement from Fred Peronto's Car Department Briefs, RL&C, July 1970: "There is nothing in a freight car built anywhere in the world that is so vital to its safe and efficient operation as the wheel, axle, and journal bearing assembly. This is perhaps the reason why International Wheelset Congress was founded back in 1963."

Hot box statistics on over-heated journals for freight cars were published starting in 1950. No records were published for captive equipment, such as passenger cars or locomotives which operate mostly under owner control, therefore the following history will be brief.

PASSENGER CARS

These were mostly six-wheel trucks with plain bearings, AAR 5 x 9 in. New cars were built with cartridge type heavy-duty roller bearings in about 1940. Earlier applications were made in about 1919 of both roller and ball bearing designs.

STEAM LOCOMOTIVES

Driving box or crown bearings were high-lead bronze, grease lubricated, and designed by locomotive builders (Fig. 7). Engine and trailer truck bearings were bronze back, babbitt-lined, oil lubricated. Main and side rod bushings were bronze, mostly floating type, turning in a gun-iron outer bushing, grease lubricated. Hennessy lubricators were used by some roads. A plunger pump was actuated by the lateral movement of the axle to supply oil to waste packing. Inspection and lubrication was done by the engine crew at intermediate stops and at terminals. Overheating was usually due to misalignment or improper application.

DIESEL LOCOMOTIVES

Motor support bearings or axle bearings (Fig. 8) bronze babbitt lined, are used on all locomotives. Development in this has been to improve wick feed lubricator and sealing ends to eliminate oil loss and entrance of dirt. Iron-back babbitt lined were used many years ago. Journal bearings 6½ x 12 in. and 7 x 14 in., similar to AAR standard except with extended sides, were used on switching locomotives for a number of years.

IN CLOSING

The contribution of the ASME Rail Transportation Division is hardly calculable in terms of specific changes in railroad technology. However, through its papers and annual reports a substantial contribution has been made for the benefit of both executives in management and railroad engineering personnel.

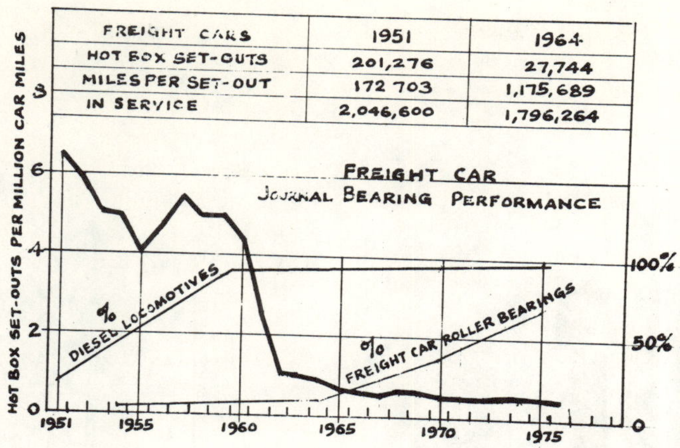

FIG. 5

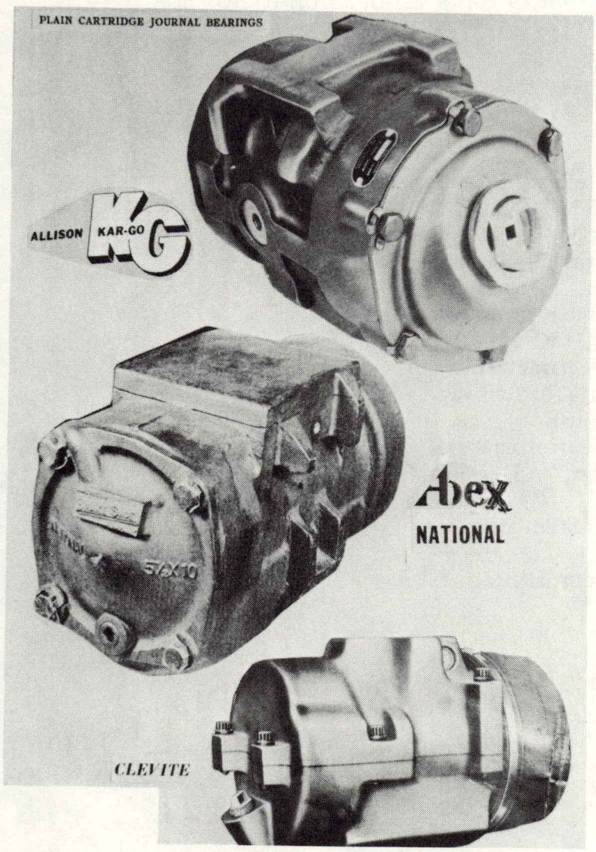

FIG. 6 CARTRIDGE BEARING

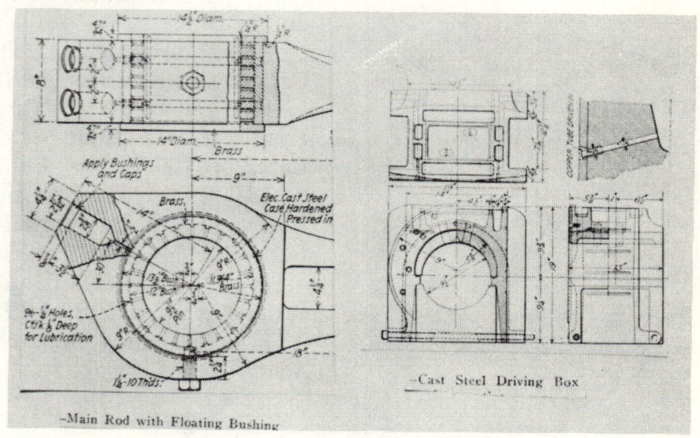

FIG. 7 STEAM

Pair of Motor Support Bearings; bronze back, Babbitt lined. For traction motors of diesel electric locomotives.

FIG. 8 DIESEL

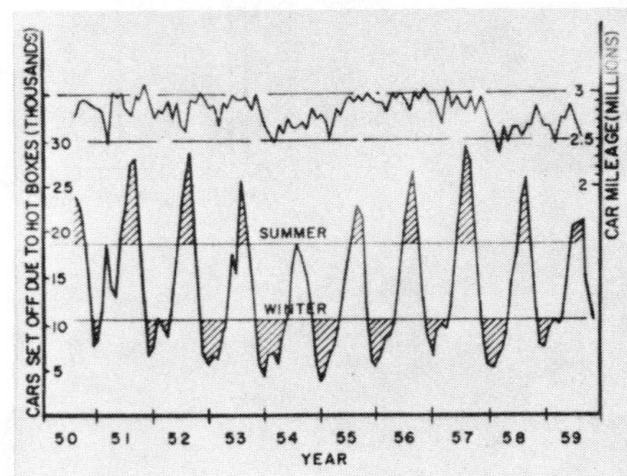

FIG. 9 CARS SET-OFF DUE TO HOT BOXES

REFERENCES

[1] Hawthorne, J. W., "Hot Boxes – Some Fundamental Problems," ASME 53A104.

[2] "Comparative Economics of Journal Bearings," AAR Research Center Report 20,000, November 1955.

[3] Hanson, M. A., "Development in Railway Car Journal Bearings," ASLE 57AM3C-2.

[4] Zupez, John, "Operating Characteristics of Positive Control Flat Back Bearings for Railway Cars," *ASLE Journal,* February 1961.

[5] Zupez, John, "Optimum Performance of Railway Car Solid Journal Bearings," ASLE-66AM-6D1.

[6] "100 Years of Innovation Standardization Progress," *RL&C,* June 1967.

[7] Bierlein, John C., "The Journal Bearing," *Scientific American,* July 1975.

15 Railroad Roller Bearing Development

HARRY M. JONES

Roller bearings were applied to the axle journals on a streetcar of the Brooklyn City Railroad Company in 1894, which was probably the first application of roller bearings to railroad axles in the United States. The bearings were made by the Hyatt Roller Bearing Company (now New Departure-Hyatt Bearings, Division of General Motors Corporation). The inner and outer races and rollers were formed by spirally winding ribbon steel into tight coils. Another similar application was made in 1897.

A Pennsylvania Railroad passenger coach was equipped with ball bearings in 1913, and in 1914 a seven-car train was similarly equipped. However, the life of the ball bearings on these applications was found to be unsatisfactory.

Progress in railroad bearing development was slow until the self-propelled gasoline rail motor

FIG. 1 TIMKEN-EQUIPPED INTERURBAN CAR

FIG. 2 THE TIMKEN-EQUIPPED "OLYMPIAN" ON THE MILWAUKEE ROAD

car was introduced in the early 1920's. Power characteristics of the gasoline engine made it necessary to find a means to lighten the starting loads. The Timken Roller Bearing Company (now The Timken Company) placed test applications of tapered roller bearings on interurban streetcars operating between Cleveland and Canton, Ohio in 1923 (Fig. 1). Experimental application of roller bearings was made to a Wheeling and Lake Erie box car in 1923 and other applications to box and hopper cars in 1925 and 1926.

A passenger car on the Chicago, Milwaukee, St. Paul and Pacific was equipped with SKF spherical roller bearings in 1926.

In 1927, the Pioneer Limited and the Olympian (Fig. 2), crack passenger trains of the Chicago, Milwaukee, St. Paul and Pacific Railroad were completely equipped with roller bearings. The two Pioneer Limited trains operating between Chicago and Minneapolis and accumulating 153,000 miles annually, were the first completely roller bearing equipped passenger trains. Eight Olympian trains operated between Chicago and Seattle with annual mileages of 202,000 miles.

The Wheeling and Lake Erie, the New York Central, and the Baltimore and Ohio also put test cars in service and the demand for roller bearings for passenger equipment began to increase.

Roller bearing applications were designed for steam locomotive trailing axles and tender axles in 1929. The same year, The Timken Company ordered from the American Locomotive Works, a 4-8-4, 355-ton steam locomotive fully equipped on all axles with roller bearings. This locomotive (Fig. 3), number 1111, was named the "Four Aces". Delivered in 1930, it was tested on 14 different railroads, accumulating over 100,000 miles in 18 months. This was followed by development of roller bearing applications for crank pins and crossheads (Fig. 4). The first application of roller bearings to all rods and crank pins of a steam locomotive was in 1934, on a Pacific type locomotive on the Pennsylvania Railroad (Fig. 5).

In 1927 and 1942, two different railroads ran comparative dynamometer tests on complete passenger trains with both roller bearings and plain bearings. Speeds from 10-65 mph on tangent level track showed 10-17 percent less train resistance for roller bearings.

An experimental application of roller bearings was made in 1930 to 100 70-ton hopper cars on the Pennsylvania Railroad (Fig. 6), at which time

THE TIMKEN ROLLER BEARING COMPANY
CANTON, OHIO
LEADING, TRAILER, TENDER TRUCKS, BOOSTER AND ALL DRIVERS ON TIMKEN BEARINGS
BUILT BY AMERICAN LOCOMOTIVE COMPANY

CLASS 484 — ROAD NUMBER 1111

Gauge of Track	Cylinders Diameter	Cylinders Stroke	Driving Wheel Diameter	Wheelbase Driving	Wheelbase Rigid	Wheelbase Engine	Wheelbase Engine & Tender	Firebox Length	Firebox Width	Fuel Kind
4'-8½"	27"	30"	73"	19'-3"	12'-10"	45'-10"	88'	132[illegible]	96¼"	Soft Coal or Oil

Boiler Inside Dia.	Boiler Max. Dia.	Boiler Pressure	Weight in Working Order, Pounds: Leading	Driving	Trailing	Engine	Tender	Maximum Tractive Power: Engine	Booster	Factor of Adhesion: Drivers	Booster
84¼"	96"	235 LBS. 250 LBS.	73,000 65,000	240,000 256,000	110,500 102,500	423,500	299,000	60,000 LBS 63,750 LBS	12,300 LBS 13,100 LBS	4.00 4.01	4.87 4.28

Tubes Number	Tubes Diameter	Tubes Length	Evaporating Surfaces, Square Feet: Tubes	Flues	Firebox	Syphons	Arch Tubes	Total	Superheater Surface Sq. Ft.	Grate Area Sq. Ft.
66 194	2[illegible] 3[illegible]	21' 6"	832	3805	360	105	18	5120	2100	88.3

TENDER TYPE, 12 WHEELED — CAPACITY, WATER, 15000 GALS. — FUEL, 21 TONS
COMPLETED DECEMBER, 1929

FIG. 3 TIMKEN 1111, FOUR ACES, LOCOMOTIVE

FIG. 4 ROLLER BEARING EQUIPPED CRANK PINS

FIG. 5 PENNSYLVANIA LOCOMOTIVE, CLASS K-4-S, WITH TIMKEN BEARINGS ON THE CROSSHEAD AND ALL CRANK PINS: ALSO THE MAIN AND SIDE RODS, PISTON AND PISTON ROD, CRANK PINS AND CROSS HEAD ARE MADE OF TIMKEN HIGH DYNAMIC STEEL. THE RECIPROCATING PARTS WEIGH LESS THAN 1,000 LB. THIS ENGINE IS ALSO EQUIPPED ON ALL AXLES WITH TIMKEN BEARINGS. THE DRIVING AXLE AND TRAILER TRUCK BOXES FIT INTO PLAIN BEARING PEDESTAL OPENING. BEARING AND ROD APPLICATIONS WERE MADE IN APRIL 1934, AND THE LOCOMOTIVE AVERAGES 12,000 MILES PER MONTH. THE LOCOMOTIVE HAULS THE "CINCINNATI LIMITED" AND THE "AMERICAN" BETWEEN COLUMBUS AND PITTSBURGH.

FIG. 6 TEST TRAIN AT TRINWAY

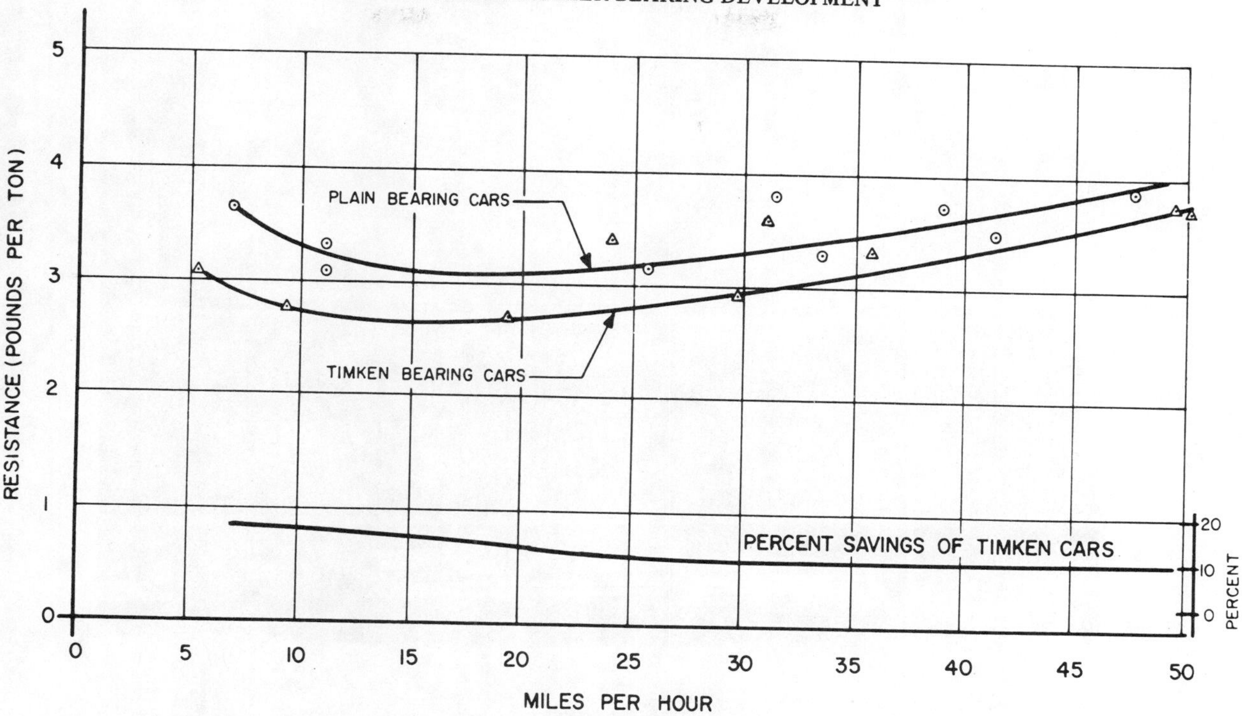

FIG. 7 TRAIN RESISTANCE, EQUILIBRIUM TEST–LOADED CARS, WINTER–1931

complete dynamometer tests were made. The roller bearing trucks were completely furnished by the bearing manufacturer. These were the first tests made on a complete freight train equipped with roller bearings and results clearly demonstrated the advantages of roller bearings for freight cars.

Trains of cars with both roller bearings and plain bearings were hauled enough miles for the bearings to reach equilibrium temperature and then total train resistance was measured over tangent level track at speeds up to 50 mph. The plotted results (Fig. 7) of resistance in pounds per ton versus speed in miles per hour show 10 percent less resistance for roller bearing cars at medium speeds and 9 percent less at 50 mph. The tests indicated no need for change in tonnage ratings between summer and winter months.

Starting tests on a 0.3 percent ascending grade with a 2-10-0 steam locomotive showed that 76 cars could be started with either type bearing by taking slack (starting one car at a time). With stretched trains (no slack), 65 roller bearing cars were started compared to 37 plain bearing cars. The easy starting characteristic of roller bearings could be used to reduce shock in couplers, draft gear, and lading.

Starting resistance of roller bearing cars was not greatly in excess of the low speed rolling resistance while starting resistance for plain bearing cars was 10 or more times the lowest rolling resistance for these cars.

The adoption of roller bearings for freight cars faced the difficult problem of interchange. Roller bearings represented a high initial cost in comparison to the total cost of a freight car and with interchange of cars from one railroad to another, the owning railroad would not realize the full benefit of such an investment. Also, the housing type roller bearing applications required separate spare axle assemblies for each different type of truck frame.

The Union Pacific's Challenger (Fig. 8), a 10-car merchandise freight train was placed in service in 1939, equipped with roller bearings to obtain maxi-speed in delivery of shipments.

The first large application of roller bearings was to 300 50-ton stock cars on the Union Pacific in

FIG. 8 UNION PACIFIC HIGH-SPEED FREIGHT CARS ON TIMKEN BEARINGS.
ONE OF THE TIMKEN-EQUIPPED HIGH-SPEED FREIGHT CARS ON THE UNION PACIFIC. THESE CARS WERE PLACED IN SERVICE IN 1939 AND ARE USED IN HIGH-SPEED MERCHANDISE SERVICE.

FIG. 9 UNION PACIFIC LIVESTOCK CAR 49018 EQUIPPED WITH TIMKEN BEARINGS WHICH FIT INTO AAR STANDARD SIDE FRAMES OF THE INTEGRAL BOX TYPE. AXLES WERE MACHINED FROM WORN FRICTION BEARING AXLES. 800 EXISTING LIVESTOCK CARS WERE PLACED IN SERVICE ON TIMKEN BY THE U.P. DURING 1947 AND 1948.

FIG. 10 ROLLER FREIGHT. TIMKEN BEARING EQUIPPED 70-TON HOPPER CAR FOR THE CHESAPEAKE & OHIO RAILWAY. THIS IS THE SAMPLE CAR FOR A LOT OF 1,000 CARS ALL OF WHICH WILL BE ON TIMKEN BEARINGS. THE WHITE STRIPE UNDER THE CHESAPEAKE & OHIO NAME INDICATES ROLLER BEARING EQUIPPED.

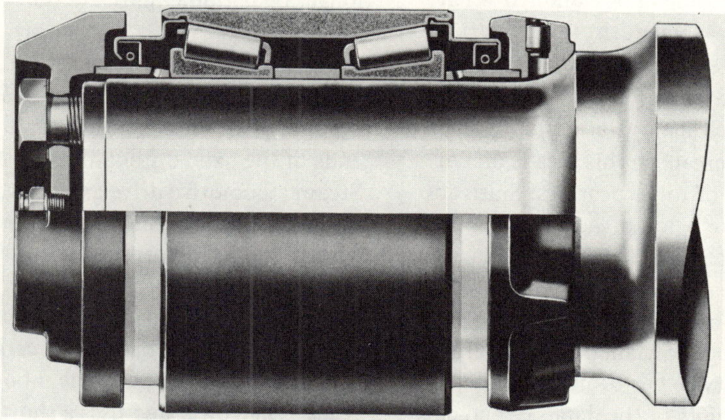

FIG. 11 TIMKEN "AP" BEARING

1947 (Fig. 9). These cars operated between Salt Lake City and Los Angeles with a 58-60 hr running time. After being rebuilt with roller bearings applied, the running time was reduced to less than thirty hours. This enabled shipments to market without an intermediate stop to feed and water the cattle. This service was so effective that another 500 cars were ordered.

Later in 1947, a second large application of roller bearings was made to 1000 new 70-ton hopper cars on the Chesapeake and Ohio (Fig. 10) for hauling coal from mines in West Virginia and Kentucky.

These first applications of roller bearings to freight cars were to cars in captive service where the owning road could realize the return on investment.

Roller bearings became standard equipment on passenger cars and diesel road locomotives. Also, approximately 10,000 car sets of housing type roller bearings were applied to freight cars into the 1950's.

A breakthrough came in 1954 with the introduction by The Timken Company of a new concept in railroad roller bearings. The Timken "AP" (all-purpose) bearing was the first self-contained, sealed bearing unit which became the basis for standardization by the AAR (Association of American Railroads) of freight car journal roller bearings (Fig. 11). There had been no previous standardization of roller bearing applications until 1956, when AAR adopted "Specifications for Approved Journal Roller Bearings Applicable to Interchange Freight Cars." AAR approval was eventually given to twelve different bearing manufacturers for various designs of tapered, cylindrical, and spherical roller bearings. Currently, two U.S. and six foreign manufacturers produce bearings that have AAR approval for railroad use. These are all tapered roller bearing designs except for one cylindrical roller bearing.

The self-contained bearing applied to sphare axle sets does not require a housing for protection of the bearings. The same axle set can be applied to many different truck designs with the use of appropriate frame adapters.

The self-contained bearing applied to spare ed the cost of roller bearing application to freight cars. During the 1950's, many detailed economic studies were made comparing cost of operating a fleet of roller bearing equipped cars versus a fleet of plain bearing equipped cars. Rates of return on investment show the advantage for roller bearings which increases as labor rates increase. The maximum advantage will be obtained when the freight car fleet is 100 percent roller bearing equipped.

The "hotbox" or overheated plain journal bearing has been a troublesome operating problem on the railroads. Improvement in this area is one of the major advantages of roller bearings. Statistics published by the Association of American Railroads can be plotted (Fig. 12) to show the increase in the use of roller bearings as a percentage of the AAR freight car fleet and the corresponding decrease in the percentage of plain bearing equipped cars. A measure of service performance is also shown. At the end points of the data, 39 percent of the fleet on plain bearings accounts for 89 percent of the overheated journals while 61 percent of the fleet on roller bearings accounts for only 11 percent of the overheated journals.

From 1954, roller bearing use on freight cars accelerated. In 1966, roller bearings were made standard by AAR on all new and rebuilt cars of 100 tons capacity and larger. In 1968, they were made standard on all new cars and in 1970 on all rebuilt cars. An estimated 65 percent of the freight car fleet under AAR jurisdiction is now roller bearing equipped.

Development of railroad roller bearings has been a continuous process of improvements in materials and designs. In 1975, the AAR adopted the "No Field Lubrication" bearing design which will be implemented in 1976. This will require roller bearings for interchange freight car use to operate until cars are shopped for wheel turning without the addition of lubricant. There will no longer be periodic lubrication between wheel turnings.

Most freight car, passenger car, and locomotive journal roller bearing applications are outboard mountings where the bearings are located on the ends of the axle, outboard from the wheel hub. Steam locomotive drive axles have inboard mountings where the bearings are located between the wheels due to the drive rods on the outboard side of the wheels. Inboard trucks are used on some passenger cars and particularly on self-propelled cars such as transit or subway cars which are all roller bearing equipped. The inboard passenger car truck is narrower and uses a shorter axle for a significant weight reduction over the outboard truck. Tubular axles are often used for a further

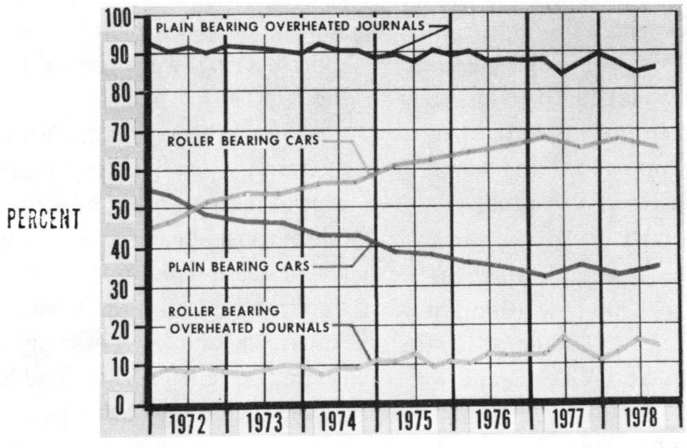

FIG. 12 ROLLER BEARING CARS VS. PLAIN BEARING CARS – AAR SEMIANNUAL CAR SURVEY. ROLLER BEARING VS. PLAIN BEARING OVERHEATED JOURNALS – AAR QUARTERLY OVERHEATED JOURNAL REPORT

weight reduction. Advantages of the inboard truck are reduced unsprung weight and reduced energy requirements for car acceleration due to reduced weight.

In recent years there has been a revival of interest in mass transit with expansion of existing rail transit system and planning and building of new systems. Many rail transit cars are being built which are roller bearing equipped, including roller bearings in the axle mounted gear boxes.

Diesel electric locomotives in the United States, since the first one was built in 1923, have used split sleeve bearings for traction motor suspension on axle hung traction motors. These bearings have done a satisfactory job but require continuous and expert maintenance. One major U.S. locomotive builder in collaboration with a bearing manufacturer has developed a tapered roller bearing suspension application for traction motors on electric and diesel electric locomotives. A test application went into service in 1966 and production applications began in 1973. Another major U.S. locomotive builder has cylindrical roller and tapered roller bearing suspension applications in test.

16 100 Years of Progress in Railway Car Hand Brakes

C. F. ROSELIUS

The first railway cars were built by wagon and carriage works, and had no brakes of any kind, as the trains were stopped by means of the locomotive. When brakes were used they were probably adaptations of the systems found on horse drawn vehicles, such as stage coaches. Foot and hand operated levers forced wooden blocks against the iron tires on the wheels.

Who first thought of wrapping chain on a shaft as a hand brake mechanism is not known. Ship anchors and windlasses for well water buckets employed chain, and possibly these mechanisms suggested the use of chain for hand brakes.

When we think of hand brakes on old time freight cars we usually visualize a horizontal wheel on top of a staff extending above the roof of a box car. A brakeman with a club stuck through the spokes of the wheel to increase his leverage completes the picture. However, the earliest examples of hand brakes, illustrated in old publications, on cars built in 1856, turn out to be vertical wheel hand brakes. Examples of these are shown in Figs. 1 and 2, which depict a 9 ton box car and a 20 ton triple pot iron body hopper car with wood underframe. It is interesting to note that the hand brakes are on the right side of the car; the Master Car Builders' Association not yet having been formed.

Before the advent of power brakes the locomotive, as mentioned earlier, provided the means of both starting and stopping the train, and even though the air brake was introduced in the 1870's, it was a good many years before all cars were equipped with power brakes. To assist the locomotive engineer in stopping the train, and possibly to prevent the extreme slack action occurring in trains not having cars with power brakes, a rather novel idea was proposed in 1880 whereby a lever-operated hand brake located in the caboose enabled the train man in the coupola to assist the engineer in braking the train (as shown in Fig. 3).

Today all interchange freight cars are equipped with air brakes, but by Federal Regulations they must also have a hand brake.

The United States Safety Appliance Standards, administrated by the Federal Railroad Administration of the U.S. Department of Transportation, states in Section 11, "All cars must be equipped with efficient hand brakes." The Railroad Safety Appliance Standards only require that the hand brake be of any efficient design, but must provide the same degree of safety as the design shown on Plate A (Fig. 4).

All types of railroad cars were equipped with this hand brake. It consisted of a vertical shaft with a 16 in. diameter hand wheel at the top supported on the bottom by a metal stirrup or brake shaft step. A piece of chain, bolted to the bottom of the shaft, extended to the brake rod. A ratchet and pawl manipulated by the foot was located on the brake step for holding the brake in the applied position.

As the brakeman turned the hand wheel, most likely using a brake club to increase his leverage, the chain wound up on the shaft. The chain could wrap on itself or "ball up" and often slipped just when the brakeman was exerting his greatest effort,

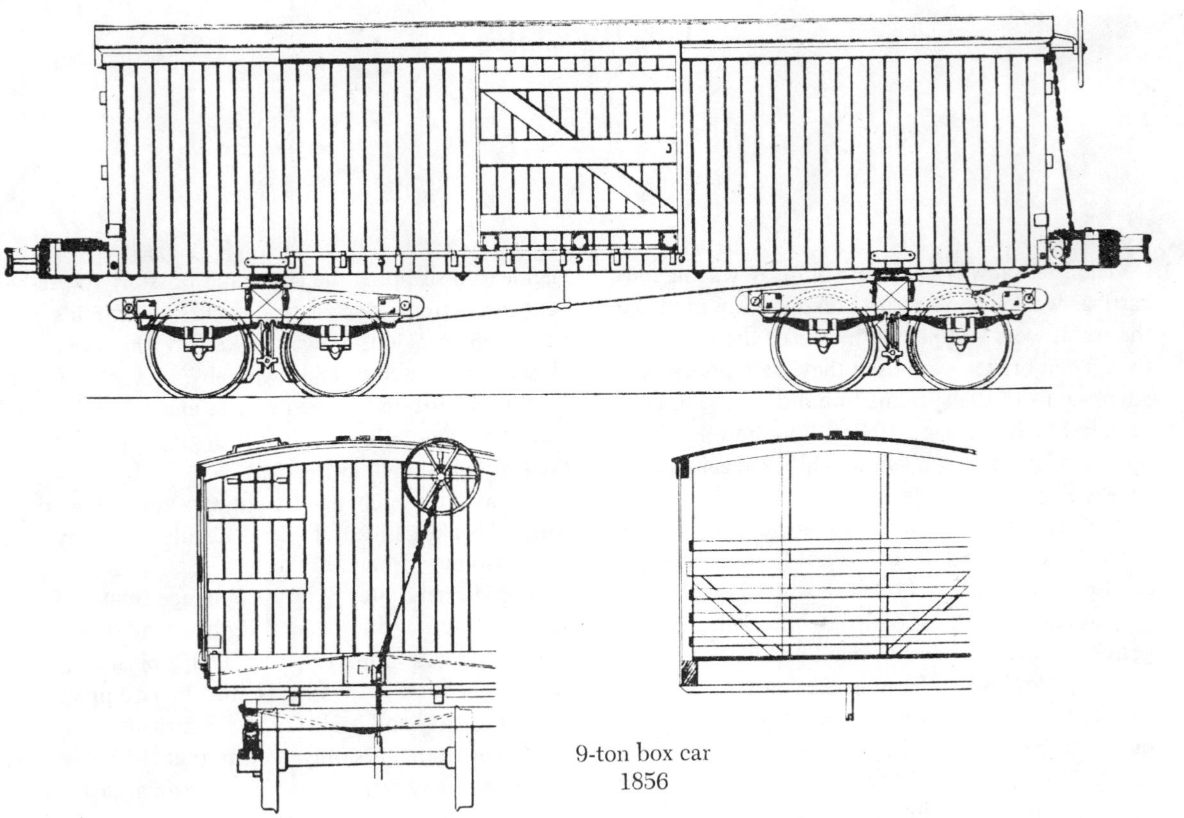

FIG. 1 "WINDLASS" TYPE VERTICAL WHEEL HAND BRAKE ON 9 TON CAPACITY BOX CAR – 1856 VINTAGE

Forerunner of the second largest class of freight cars, this 20-ton, triple-pot, iron body hopper car, with wood underframe, as built by the B&O Railroad over a period of twenty years, starting in 1856.

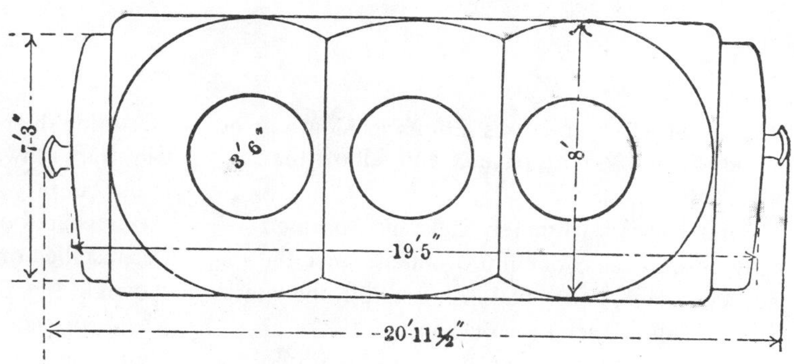

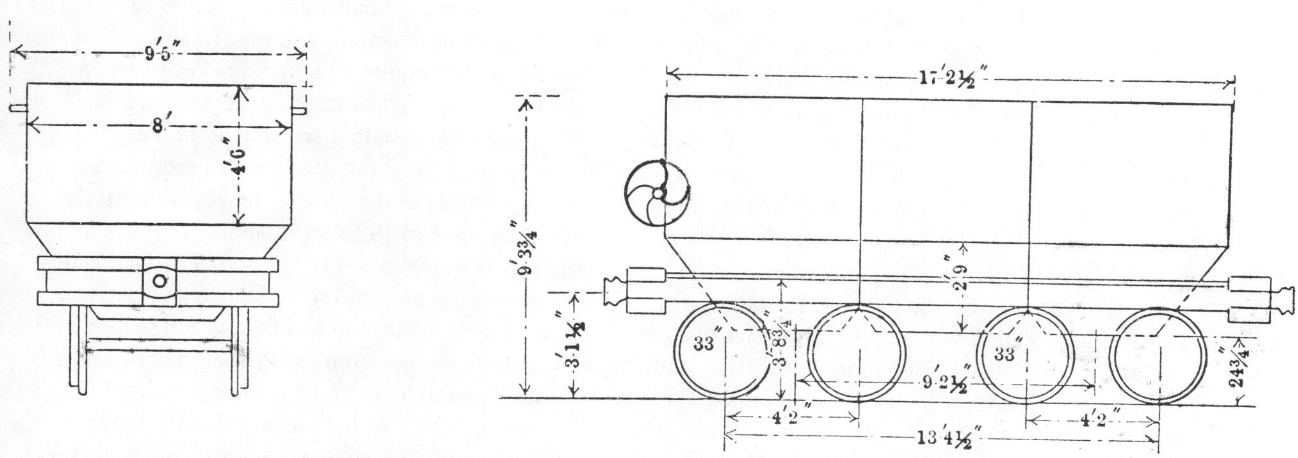

FIG. 2 VERTICAL WHEEL SIDE MOUNT HANDBRAKE ON 20 TON BALTIMORE & OHIO, WROUGHT IRON BODY TRIPLE HOPPER 1856-1876

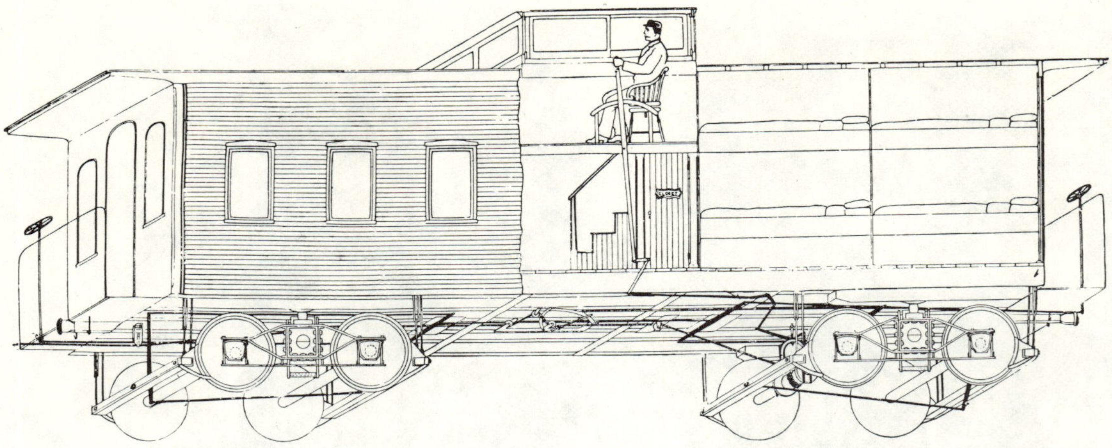

Train man, in caboose cupola, operates with engineer in brake system designed, 1880. Shows crew sleeping quarters.

FIG. 3 CABOOSE WITH CUPOLA OPERATED HAND BRAKE CIRCA 1880 WITH CREW OR STOCKMEN (DROVER) SLEEPING FACILITIES. SIMILAR IN PRINCIPLE TO EUROPEAN BRAKE VAN CARS OF THE PERIOD.

which resulted in numerous serious accidents as the operator could lose his balance and fall off the car (Figs. 5 and 6).

Hand brakes, fortunately, have incorporated many novel mechanical improvements since the "good old days," but "Plate A" is still in the latest edition of the Federal Regulations.

Part of this development has been the result of the free enterprise system, including our patent laws, and part from regulations and specifications of the Association of American Railroads.

Sometime about 1900 some ingenious individual came up with the idea of winding the chain around a decreasing radius single-plane drum located at the bottom of the vertical shaft. This was the first quick take-up device, and likewise the first power brake. The next step in the development was an improvement made by the use of a decreasing spiral cam located at the bottom of the shaft. This combined "take-up" and force multiplication. Still later, a pair of gears having a 3:1 or 4:1 ratio were used to obtain more power. These were known as brake boosters, and a typical design is shown in Fig. 7.

During the period of 1900-1915, when the electric street railway flourished in innumerable cities in our country, a vertical wheel brake for trolley cars appeared on the scene. One popular model that was installed on tens of thousands of street cars is shown in Fig. 8. It was floor mounted and the hand wheel contained a ratchet and pawl feature for "pumping" the brake should it be necessary. A holding pawl and ratchet were located on the hand wheel shaft and operated by a foot treadle. The chain drum was equipped with a pocket to accept each link of the chain and this resulted in a constant force and unlimited travel hand brake.

Steam road freight cars generally used horizontal wheel brakes of the Plate A variety up until the 1920's. In the mid 1920's, geared hand brakes for freight cars were developed (Figs. 9 and 10). Vertical wheel types offered the brakeman the mechanical advantage of the club and in addition permitted him to set the brake from a safe operating position. One such geared hand brake is shown in Fig. 11. It is equipped with a 24 in. diameter hand wheel.

By the early 30's, so many geared hand brakes were being offered to the railroads that train and

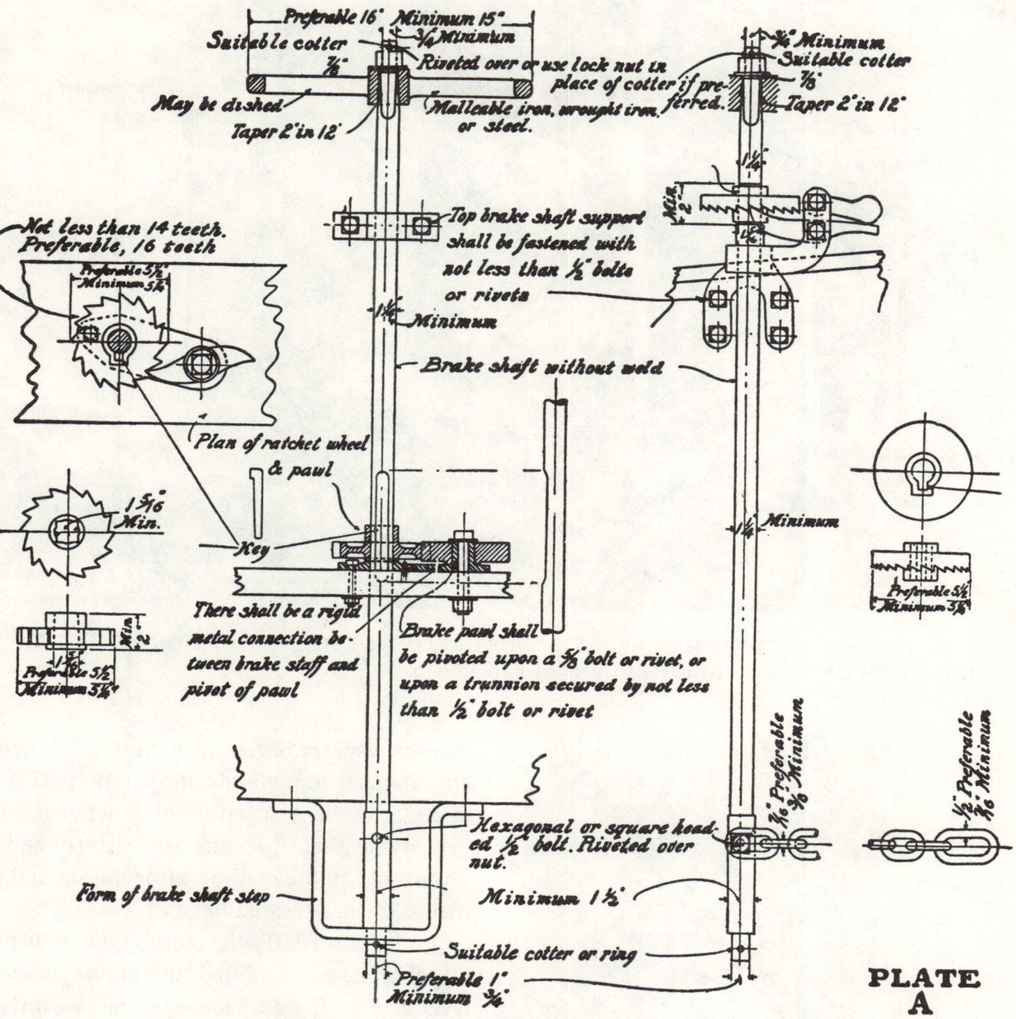

[*Any efficient arrangement of ratchet-wheel and pawl may be used.*]

FIG. 4 STANDARD MCB ARRANGEMENT OF VERTICAL STAFF RATCHET AND PAWL STOP HAND BRAKE OF EARLY 1900's VINTAGE.

FIG. 5 TYPICAL VERTICAL STAFF HAND BRAKE, SHOWING LOSS OF MULTIPLICATION BY "DOUBLE WRAPPING" OF CHAIN – NEW FREIGHT CAR BUILT IN 1920'S

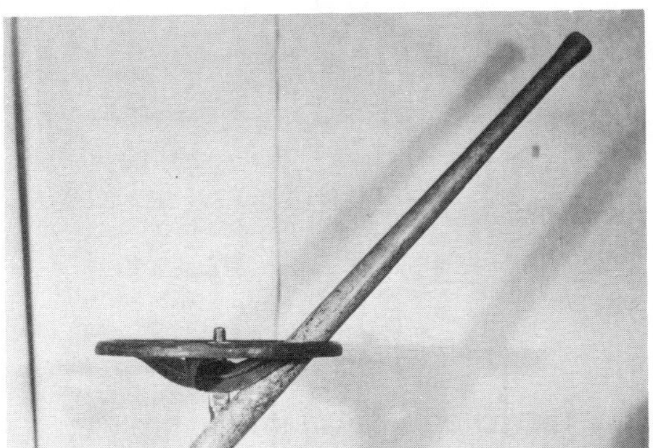

FIG. 6 USE OF A STOUT HICKORY OR ASH BRAKE CLUB TO INCREASE MECHANICAL ADVANTAGE WITH TYPICAL 16 IN. HAND BRAKE WHEEL. THE BRAKE CLUB WAS THE SWITCHMAN'S AND BRAKEMAN'S "GOOD FRIEND" THROUGH THE DECADES FROM THE 1850'S UNTIL THE WIDESPREAD USE OF MODERN GEARED HAND BRAKES BEGINNING IN THE 1930's.

yard crews were becoming confused by their varied appearance and methods of operation. This resulted, in 1935, in the Association of American Railroads developing a set of specifications to try and get some uniformity in the method of operation and performance of these geared hand brakes.

In the late 1930's the clutch type, nonspinning wheel brake was invented. It had no quick release lever and the hand wheels were cast with the words "On-Off" and directional arrows to indicate to the brakeman how it operated. This innovation was a major "land-mark" in the development of the geared hand brake. (Fig. 12).

In 1942 the Association of American Railroads published specifications for certification of geared hand brakes which established the minimum average forces produced. These specifications are found in Section E of the Manual of Standards and Recommended Practices, published by the Association of American Railroads. The specifications for geared hand brakes were revised in 1950, 1956 and

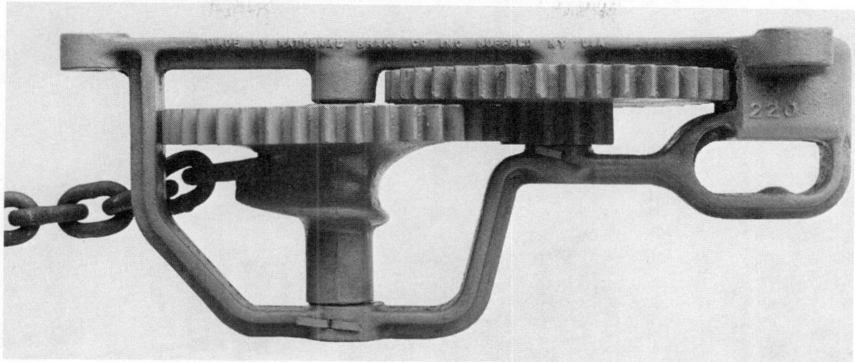

FIG. 7 BRAKE BOOSTER – EARLY MEANS OF INCREASING MECHANICAL ADVANTAGES OF HORIZONTAL WHEEL STAFF-TYPE BRACES

FIG. 8 VERTICAL WHEEL ENCLOSED MECHANISM BRAKE FOR STREET RAILWAY, SUBWAY AND INTERURBAN CARS

in 1966, and have recently been revised to include, in addition to the so called "Power Test," a life cycle test.

For interchangeability on freight cars, every brake must provide minimum force and travel requirements. In addition to these basics, there are specifics for the diameter and dimension of the hand wheel or length of lever, the overall depth of the brake unit, the size, number and spacing of mounting holes and the length, size or strength of the brake chain. Additionally, there are standards for bell cranks and sheave wheels to be used with the brakes.

A.A.R. certified brakes are listed in the "Field Manual of the Interchange Rules," and Rule 13 covers hand brakes. The latest issue lists 64 active A.A.R. certified brakes which include 46 vertical wheel brakes, including high power brakes, ten horizontal wheels, and eight lever types (Figs. 13-15).

In European practice the screw type hand brake was very popular, but in this country the geared hand brake winding chain has been the standard. Mechanical type hand brakes have generally been able to supply the required braking forces with all conventional freight cars. However, as high capacity cars with multi-truck arrangements have become increasingly common, it has become more difficult to apply a mechanical brake.

The Federal Railroad Administration suggests that a hand brake must apply a force to at least half the axles on the car. Since cars with up to 36 axles are now being built, mechanical hand brake applications have become extremely complicated. For these unusual cars the thought of a hydraulic hand brake becomes intriguing.

Several different types of hydraulic hand brakes have been tried over the years, but with no particu-

FIG. 9 LEVER TYPE RATCHETING HAND BRAKE ON PERE MARQUETTE 50 TON BOX CAR BUILT DECEMBER 1922 – NOTE EARLY USE OF BELL CRANKS.

FIG. 10 HAND BRAKE BOOSTER GEAR ON TYPICAL HORIZONTAL WHEEL BRAKE OF LATE 1920'S.

FIG. 11 EARLY VERTICAL WHEEL GEARED HAND BRAKE ON SILICA GEL ICELESS REFER BUILT OCTOBER 1929 BY PULLMAN

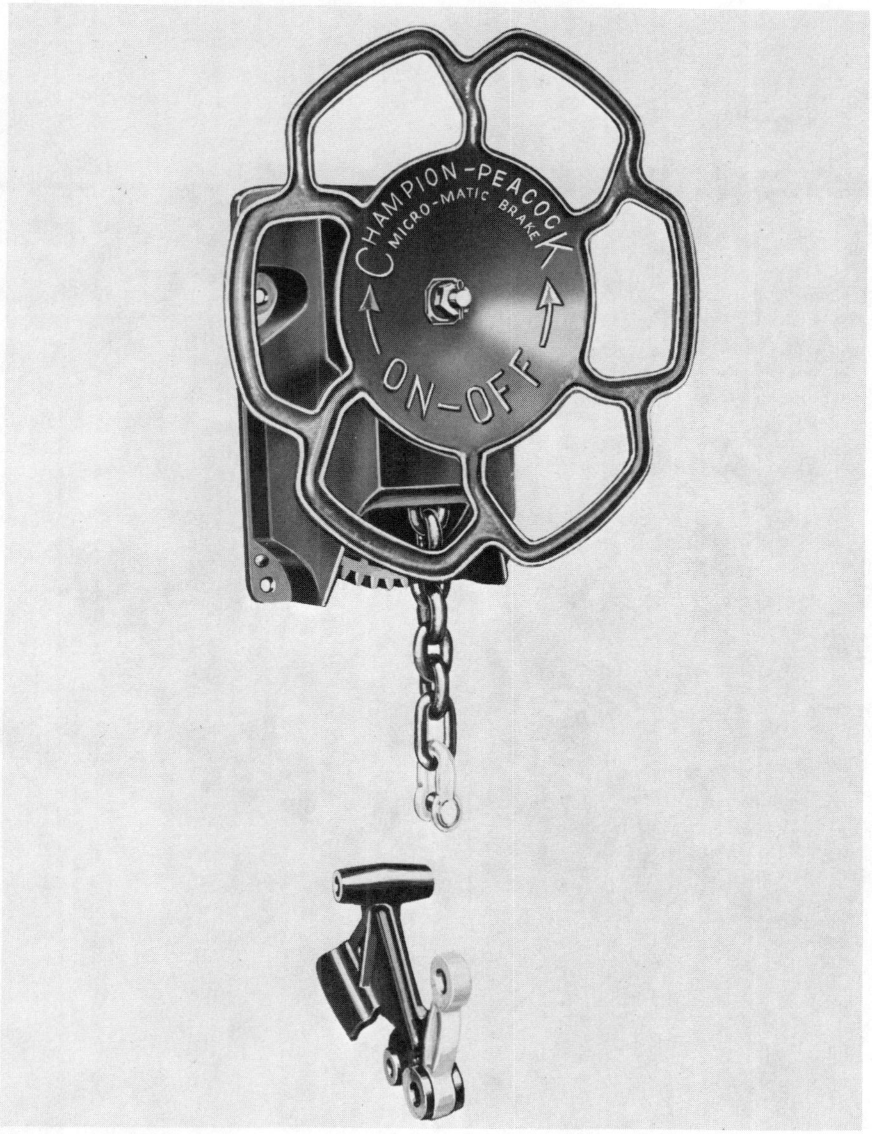

FIG. 12 ORIGINAL TYPE OF "NONSPINNING WHEEL" HAND BRAKE. NOTE ABSENCE OF QUICK RELEASE LEVER.

lar marketing success. One of the requirements for a hydraulic hand brake is that it keeps the brake applied for an indefinite length of time. This requirement mandates that some type of a mechanical locking arrangement be employed with the hydraulic actuators or cylinders. A multitude of patents have been issued covering various means of holding hydraulic cylinders extended.

Although hydraulic brakes do work and do eliminate much complicated rigging, their use has been limited because of their high cost.

As long as federal regulations require that all cars be equipped with a manually operated brake, it would appear that hand brakes will be around for a long time. However, the design departments of the various hand brake manufacturers are constantly at work developing more efficient, more economical and longer lived products.

FIG. 13 LEVEL TYPE GEARED HAND BRAKE DEVELOPED FOR "DROP END" GONDOLA CARS.

FIG. 14 VERTICAL WHEEL GEARED HAND BRAKE ON DROP END GONDOLA CAR, 1941.

FIG. 15 MODERN VERTICAL WHEEL HAND BRAKE OF "HIGHER POWER" TYPE ON 100-TON CAPACITY COVERED HOPPER CAR. NOTE EXTERNAL SHOCK RETARDER.

17 History of the Railway Freight Car Truck in the United States

CARL E. TACK
and
ROBERT B. LOVE

INTRODUCTION

The American railway freight car truck was and is a product of its environment. Demands for freight service introduced it and demands of freight service have progressed it. The real history stories are of the blood, sweat and tears, and the men who planned and labored to make the trucks for one of the world's great freight systems. This is a brief history of what they accomplished.

Eras stand out and provide the chapter titles in this history of the freight car truck in the U.S.:

In the Beginning
Arch Bar
Cast Steel
Research on Steel and Strength
Research on Suspensions

IN THE BEGINNING

The first wagons were just that, modeled after road wagons. Wheel and axle assemblies supported

FIG. 1 EARLY FOUR-WHEEL AMERICAN BOX CAR (FROM EARLY CAR BUILDERS' DICTIONARY).

FIG. 2 B&O TANK-HOPPER CAR

car bodies in the period from 1825 to 1850. They probably looked like this comparatively modern version from the 1884 Car Builders' Dictionary (Fig. 1).

Since early in the history of freight car and freight car truck development in the United States, the emphasis has been on the use of the four-wheel truck as opposed to the four-wheel "wagon" more common in Europe. This is undoubtedly because of our desire to carry heavier loads fast and far. Figure 2 shows an early B&O tank car with a crude four-wheel truck constructed of timbers.

Even prior to 1870, longer, higher capacity cars required four wheel trucks with a pivot center. Principal structural portions of the trucks consisted of heavy wooden beams, later trussed for strength (Fig. 3).

ARCH BAR

About 1865, the next notable truck development took place. It resulted in the introduction of the so-called diamond truck (Fig. 4), forerunner in name to the arch bar truck.

On the earliest types, the spring plank and truck bolster were heavy wooden beams, and the bolster was reinforced with wrought iron truss rods or plates. Refinements involved substitution of rolled steel beams in place of wooden bolsters, and combinations of angles and channels for spring planks (Fig. 5).

Column castings, originally cast iron, were changed to malleable iron and later to cast steel. Many variations of the basic design were introduced, such as the Washburn side frame shown in Fig. 6.

Bolster designs used oak timber, later reinforced with wrought iron. Figure 7 shows an early design of fabricated truss bolster.

Arch bar trucks served the era through the development of cast steel. Constantly increasing severe service took its toll on the 35 or so relatively loosely held together parts (Fig. 8).

In addition to the maintenance nightmare, increasing failures and dangerous accidents brought legislation from the ARA (American Railroad Association) in 1930 that prohibited arch bar trucks from service effective January 1, 1936, which was later extended to January 1, 1938.

About 1890, the Fox pressed steel truck was introduced and attracted a great deal of attention, but not great popularity (Fig. 9).

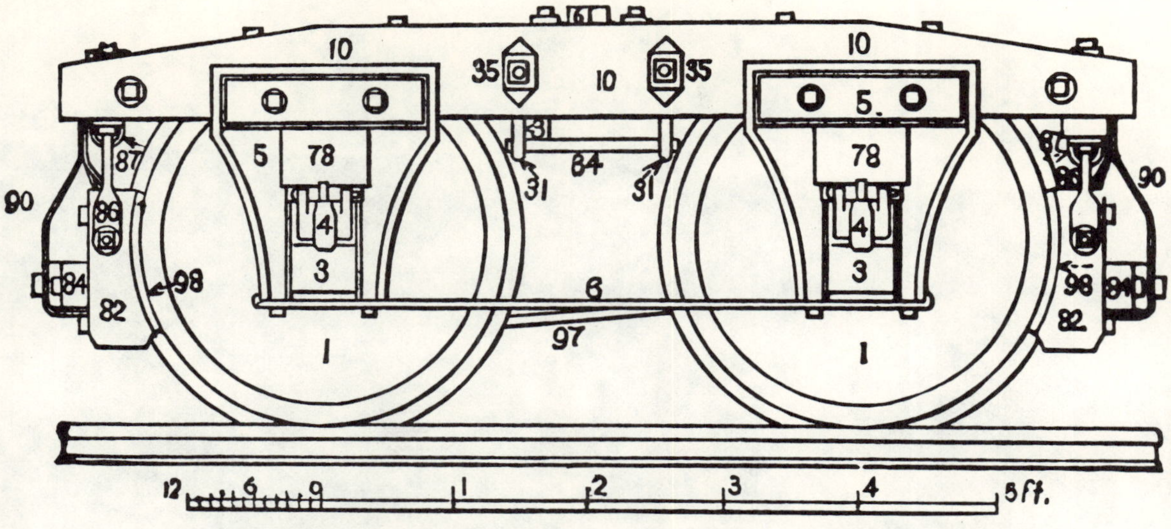

FIG. 3 WOODEN FREIGHT-CAR TRUCK, NEW YORK & HARLEM RAILROAD. SIDE VIEW. 31 IN. WHEELS, 4 FT. WHEEL BASE, RIGID FRAME COIL SPRING ABOVE EACH JOURNAL BOX.

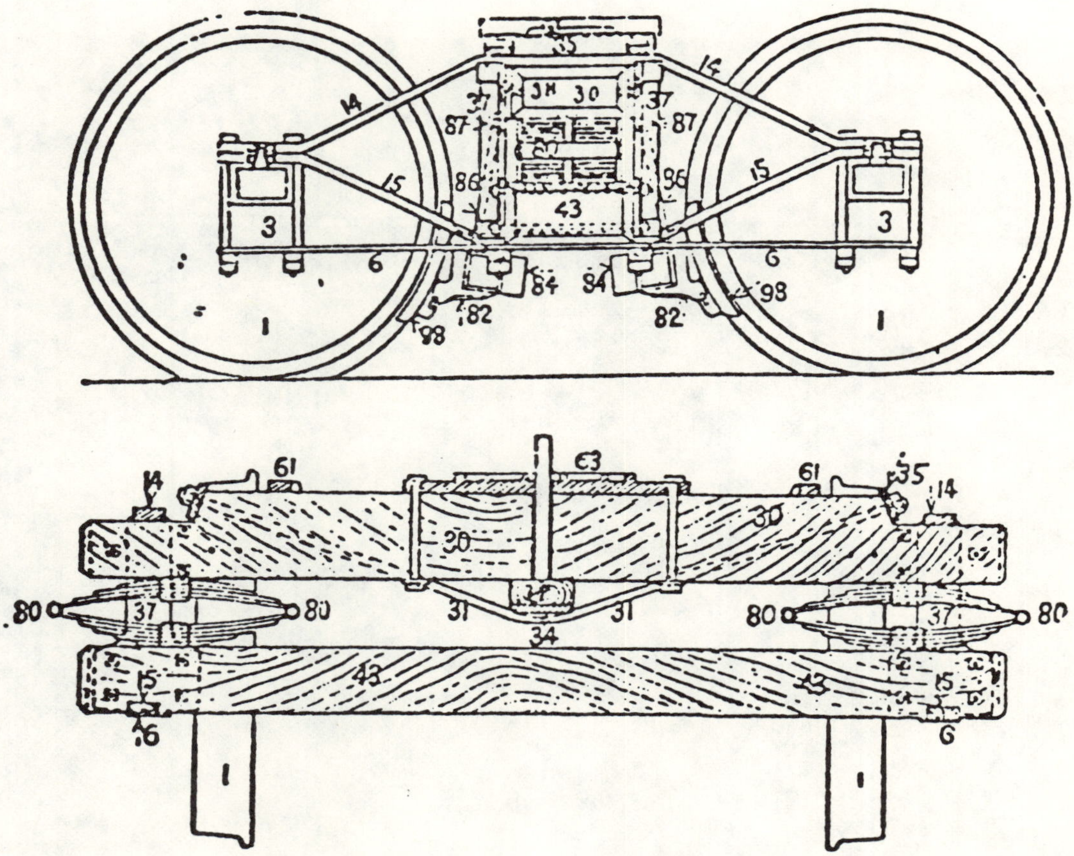

FIG. 4 DIAMOND-TRUCK, PENNSYLVANIA RAILROAD.

FIG. 5 ARCH-BAR TRUCK - 1902

FIG. 6 WASHBURN SIDE FRAME

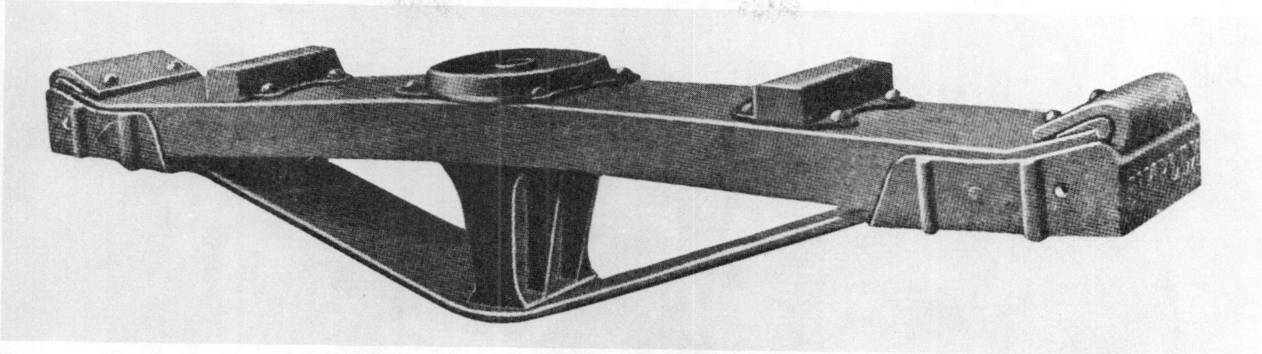

FIG. 7 FABRICATED TRUSS BOLSTER

FIG. 8 ARCH BAR TRUCK SIDE-PARTS

Although short lived, pressed steel-riveted arch bar and other designs also came on the scene. These were generally rigid types with side frames and transoms riveted rigidly together, all components being pressed and riveted shapes. Simplicity was appealing, but its short life is attributed to the impossibility of keeping rivets tight, rapid corrosion, and lack of adequate strength.

THE ERA OF CAST STEEL
(Or the Transition from Dry Rot to Metal Fatigue)

By 1895, the art of casting steel had progressed to the point where some foundries were proposing its use for truck bolsters and for designs of column castings in which the two columns, together with a spring seat and including the brake hanger brackets, were being cast in one piece.

FIG. 9 FOX PRESSED STEEL TRUCK

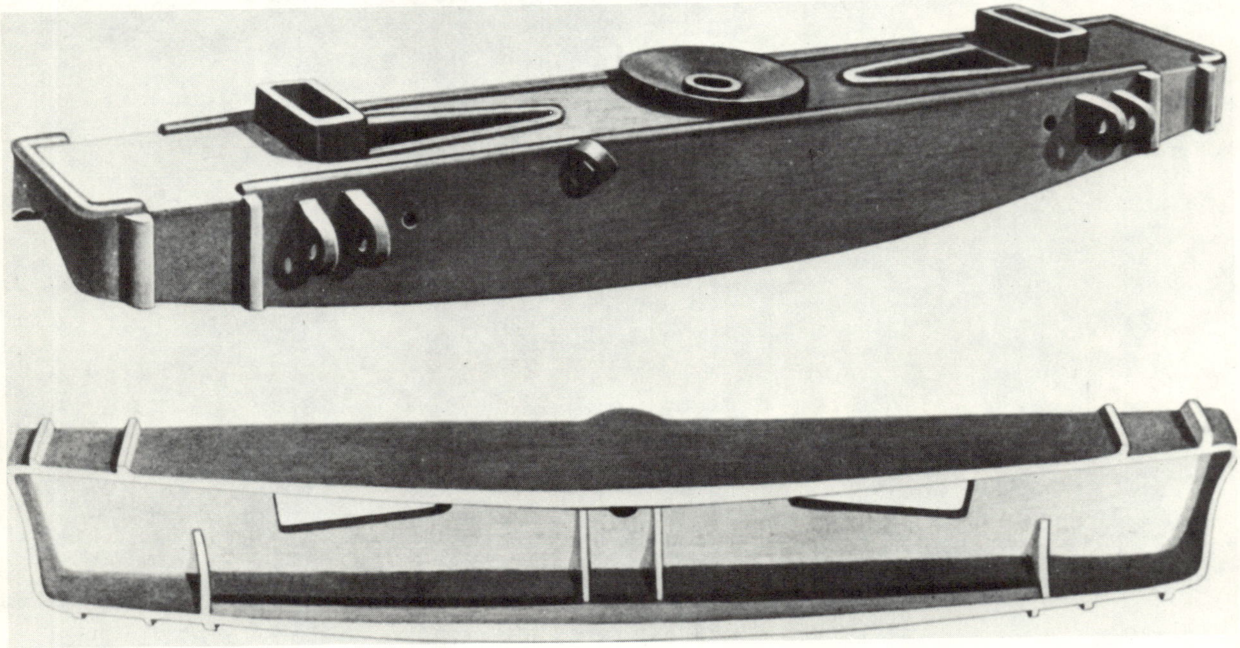

FIG. 10 "U" SHAPE BOLSTER PRIOR TO 1904

The first designs utilized were a "U" section. The top and sides of the bolsters were cast solid with large bulbs along the bottom inside edge of the side members (Figs. 10 and 11). Several drawings are reproduced in this paper as a matter of interest to show shapes rather than to be dimensionally readable.

This design was a compromise between good foundry practice and pressed steel or structural steel bolster design. The bolster spring seat was generally an oak block set into the open bottom side of the end.

The appeal of the one piece truck bolster soon attracted competition between the manufacturers of cast steel and pressed steel or structural truck bolsters. Pressed steel designs forced the early casting designers to unwise extremes in weight reductions or the adoption of sections easy to produce

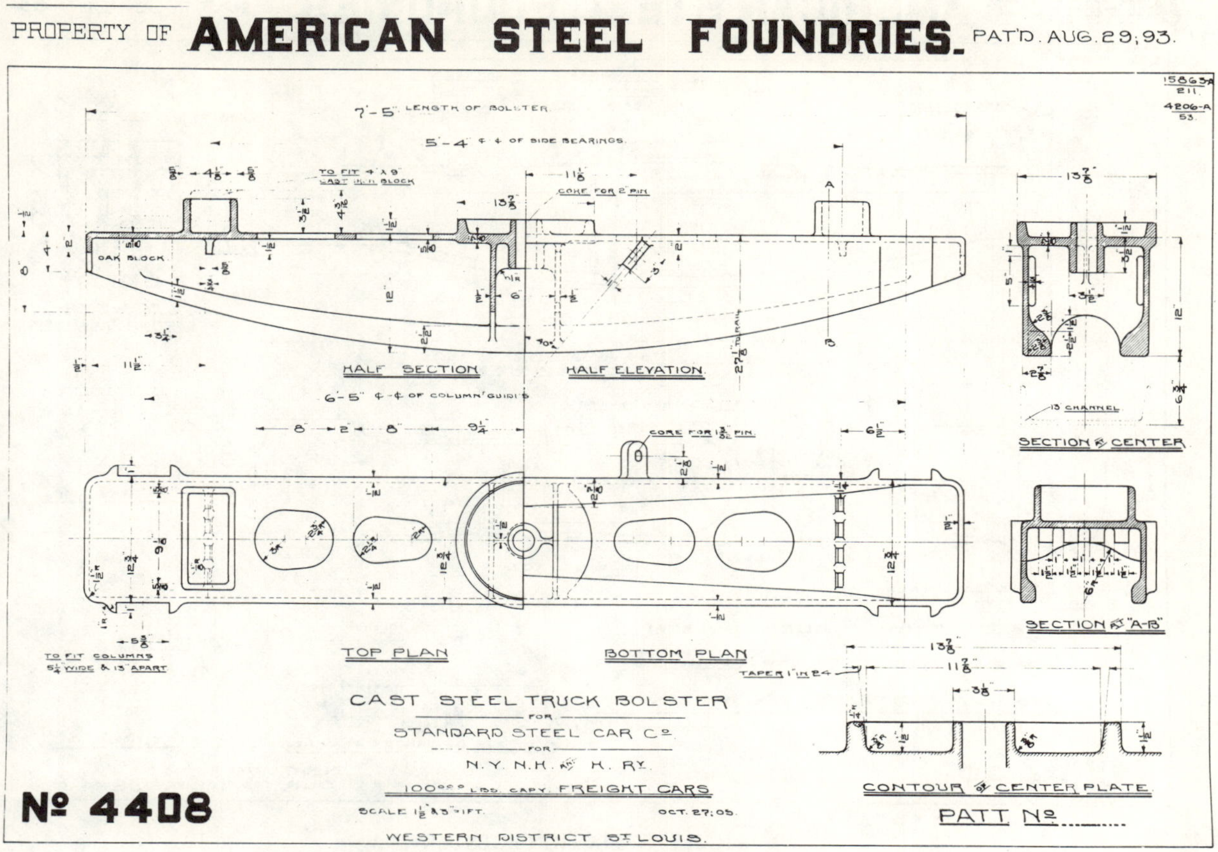

FIG. 11 "U" SECTION BOLSTER

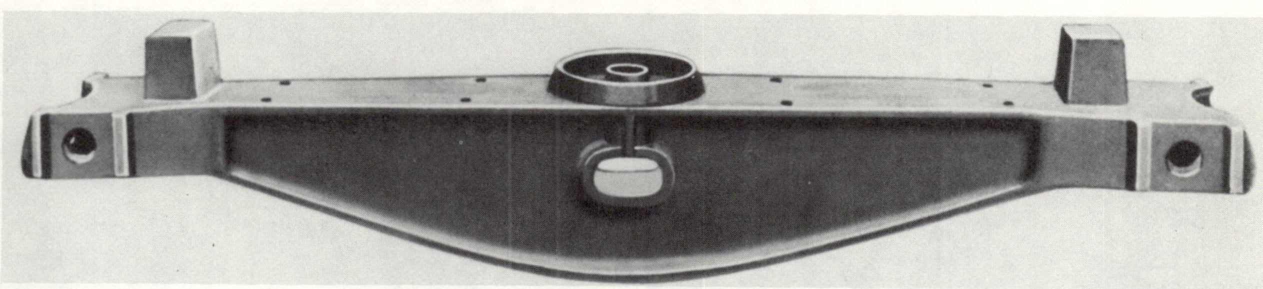

FIG. 12 CAST STEEL "I" SECTION TRUCK BOLSTER

in order to accomplish a price advantage. The early "I" section truck bolster (Figs. 12 and 13) and the subsequent "T" section (Figs. 14 and 15) were the result of this and, although widely used, none of these early designs was entirely satisfactory.

It was not until the box sections, such as illustrated in Figs. 16 and 17, became generally used that the performance of cast steel in truck bolsters improved by more completely utilizing the properties inherent in the material.

About 1900, experiments with the idea of making cast steel side frames were begun, and in 1903 the first commercially available cast steel frame, of the type shown in Fig. 18, was produced by the Bettendorf Company.

Its design, because of the integral boxes and "I" section structure, presented fairly serious manufacturing problems, so about this same time (1904), ASF (American Steel Foundries) produced the Andrews type of side frame which had separable journal boxes (Fig. 19). This design was popular both with the railroads for the separable journal boxes and with the manufacturers for ease of manufacture.

In addition, the structural section was changed from "I" to "L" or angle sections, and to channel

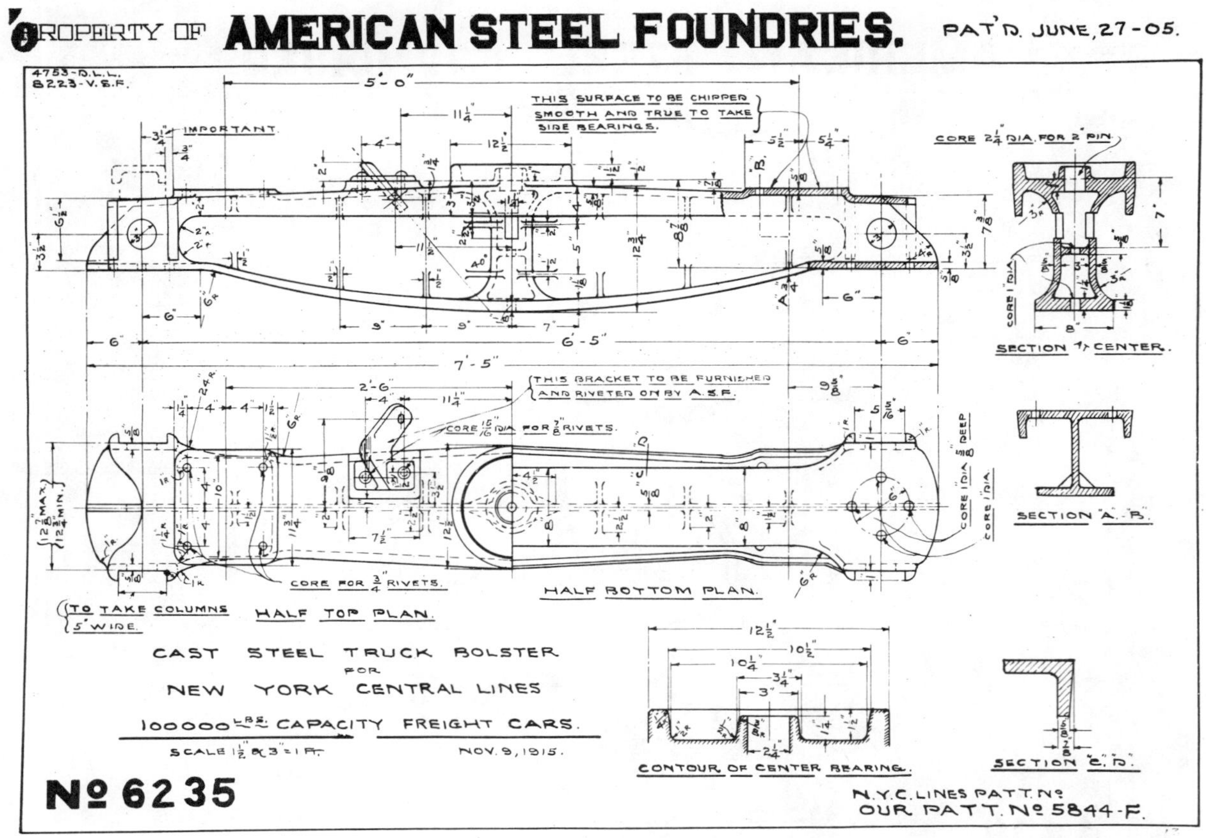

FIG. 13 "I" SECTION BOLSTER

FIG. 14 CAST STEEL "T" SECTION TRUCK BOLSTER

sections or combinations as shown in Fig. 20. Figure 21 shows some representative drawings of that era.

THE ERA OF RESEARCH

The era of research on stresses expanded from only empirical methods to more sophisticated use of research tools. Static loads were generally known in the early stages of cast steel development, but there was little knowledge of the magnitude and direction of the dynamic or augment forces. Also, early specifications were formulated using the requirements for malleable iron as a guide.

The period 1910-1917 saw appreciable improvement in cast steel side frames and bolsters, resulting in part from a rapid advance in metallurgical and foundry art, research by certain of the truck manufacturers and several railroads on freight car truck performance and requirements, and development of new testing devices.

In 1910, Prof. Endsley of Purdue was engaged by a steel foundry to study dynamic forces, and as a starter, a "scenic" railway was constructed and

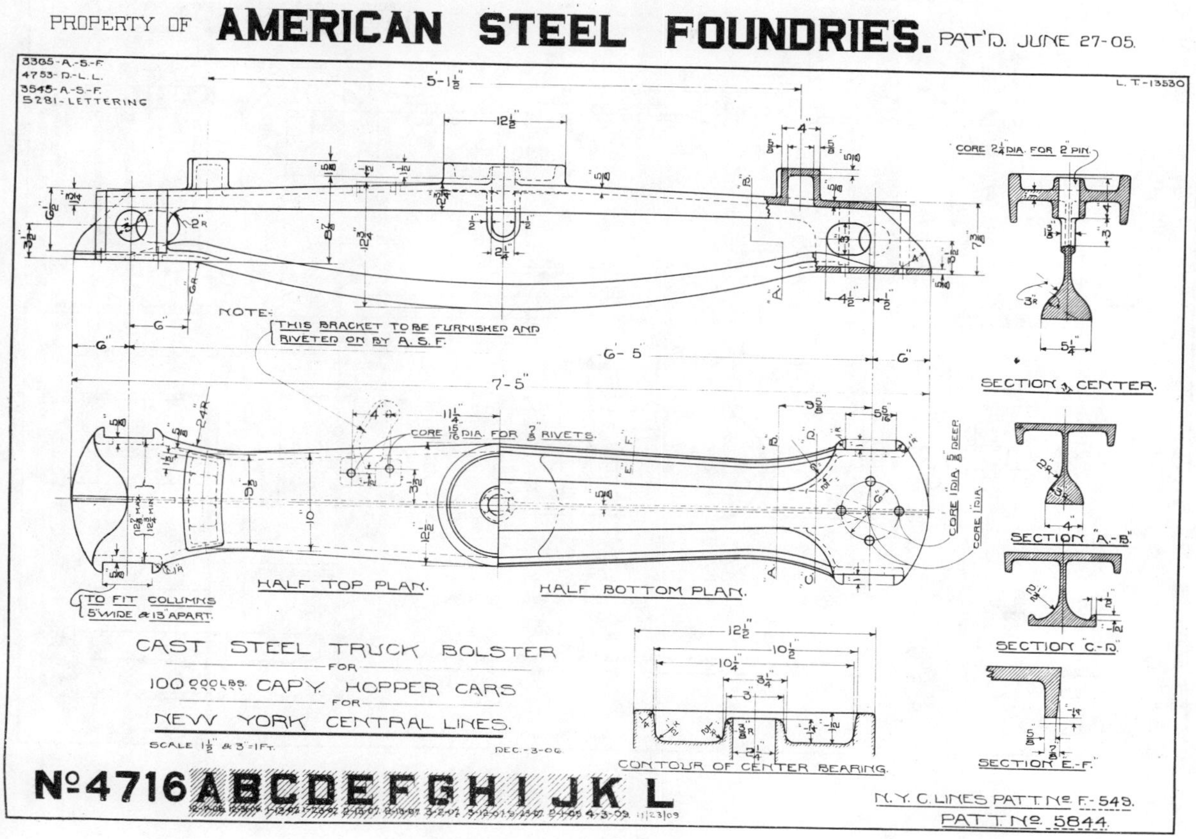

FIG. 15 "T" SECTION BOLSTER

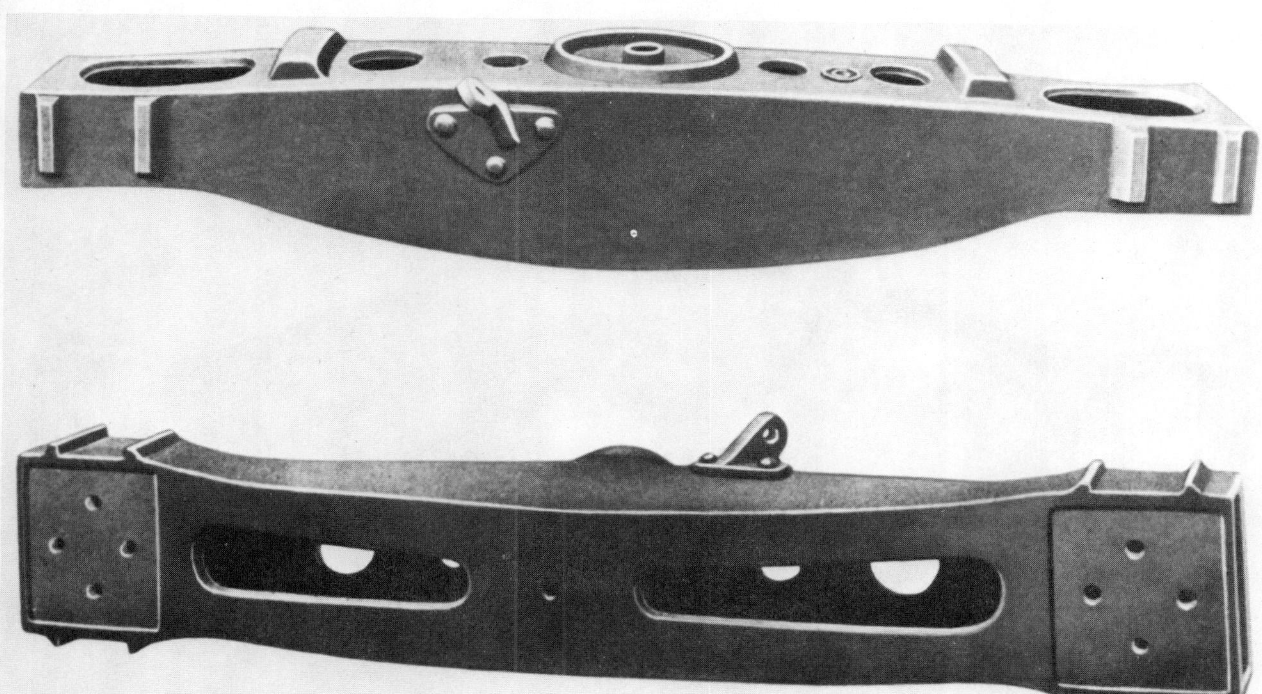

FIG. 16 "BOX" BOLSTER ABOUT 1900

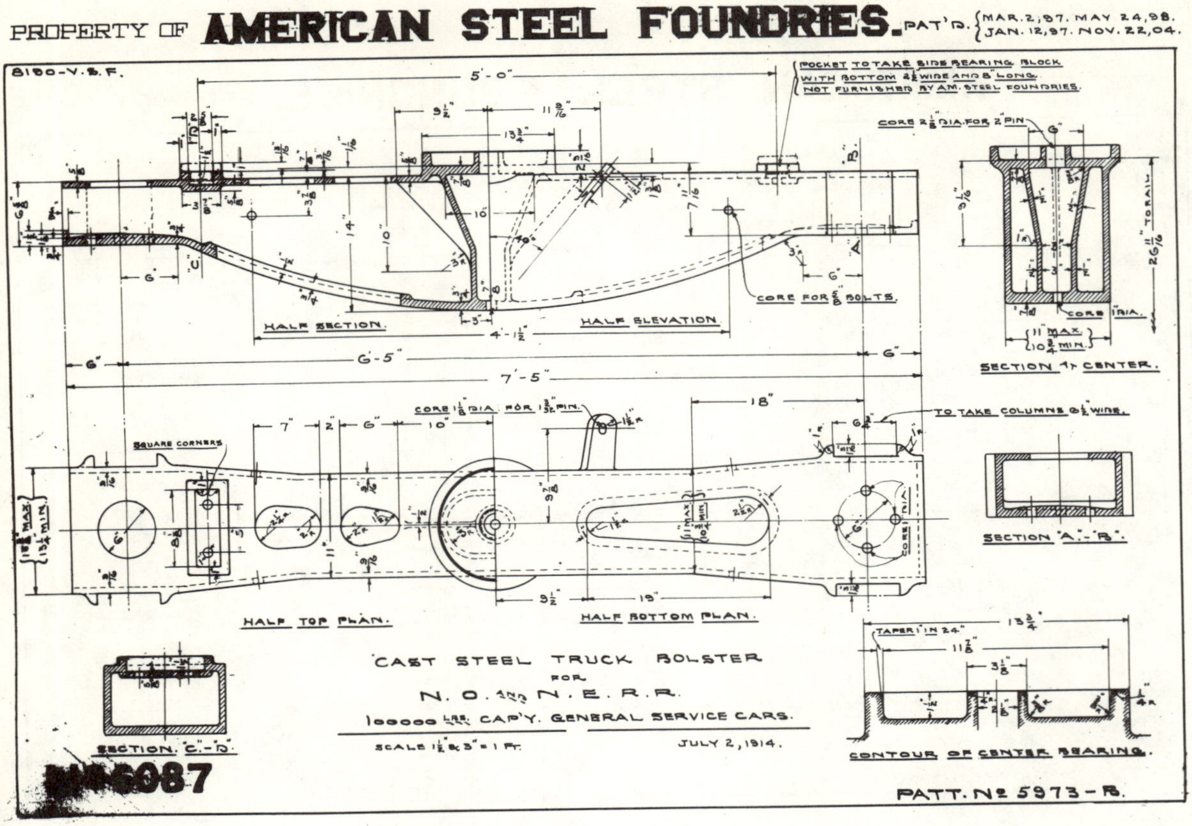

FIG. 17 BOX SECTION BOLSTER

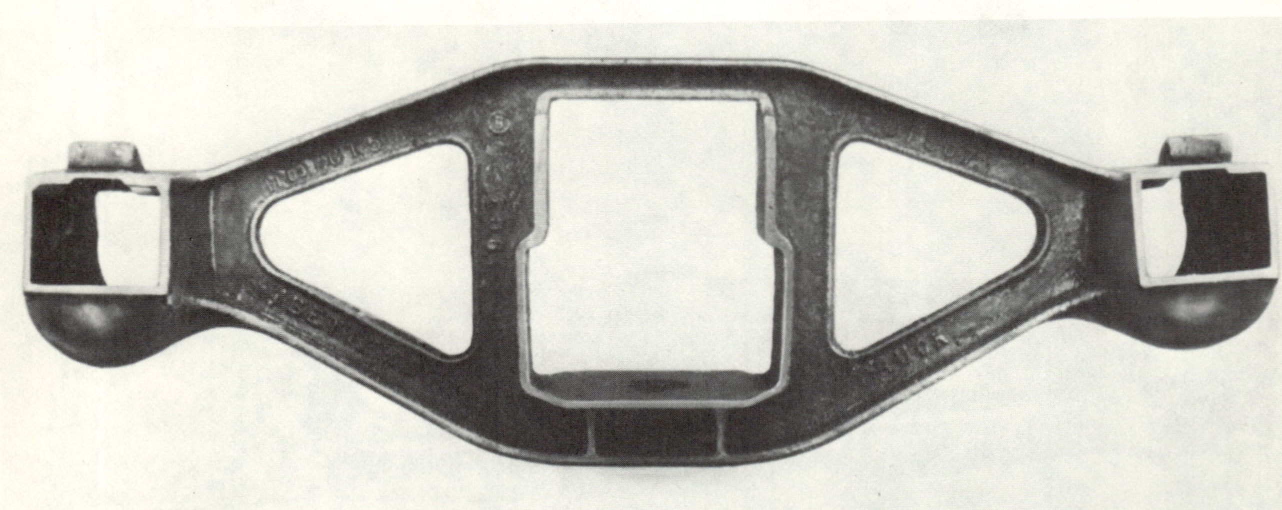

FIG. 18 BETTENDORF "T" SECTION IBSF 1909

FIG. 19 "T" SECTION ANDREWS SIDE FRAME 1912

FIG. 20 "L" SECTION WITH FULL "U" TOP MEMBER ANDREWS TYPE, 1913

observations made of the performance of various designs of four wheel trucks as they coasted downhill and entered curves and tangent track.

In 1914 the trend was to increased freight car capacity. Changes in car design to accommodate greater loads made it imperative to determine the effect of such loads on truck parts during actual service. So in that year, special truck castings and testing apparatus were developed for what is generally recognized as the first of all test cars – a Pennsylvania standard H21 hopper, equipped with specially instrumented four wheel trucks to measure vertical, lateral, twist, and "out-of-square" forces. This car was coupled to regular Pennsylvania freight trains and made repeated trips between Pittsburgh and Alliance, Ohio.

After careful analysis, the results of these tests indicated the need for fatigue testing equipment that would duplicate the actual conditions of truck operation. Thus began one of the most significant contributions to modern side frame integrity and railroad safety.

American Steel Foundries engineers designed a side frame fatigue test machine which could apply the service type loads and put it into operation in 1921 in a shed near their Granite City foundry (Fig. 22). It was upgraded in the 1960's for 150-ton capacity truck frames and is still in operation for ASF and official AAR tests.

In 1923, the Symington Company (now Dresser Industries) also designed and built a side frame fatigue test machine at their Depew, New York facility. It also serves as an official AAR tester today (Fig. 23).

Recognition that the coordination of the research programs needed attention brought about improved cooperation between truck manufacturers and railroads.

Sometime during the fall of 1915, the Subcommittee on Designs and Specifications of the ARA Committee on Standard Freight Car Equipment invited these four leading manufacturers of cast steel side frames to cooperate in designing a side frame for proposed standard box cars: American

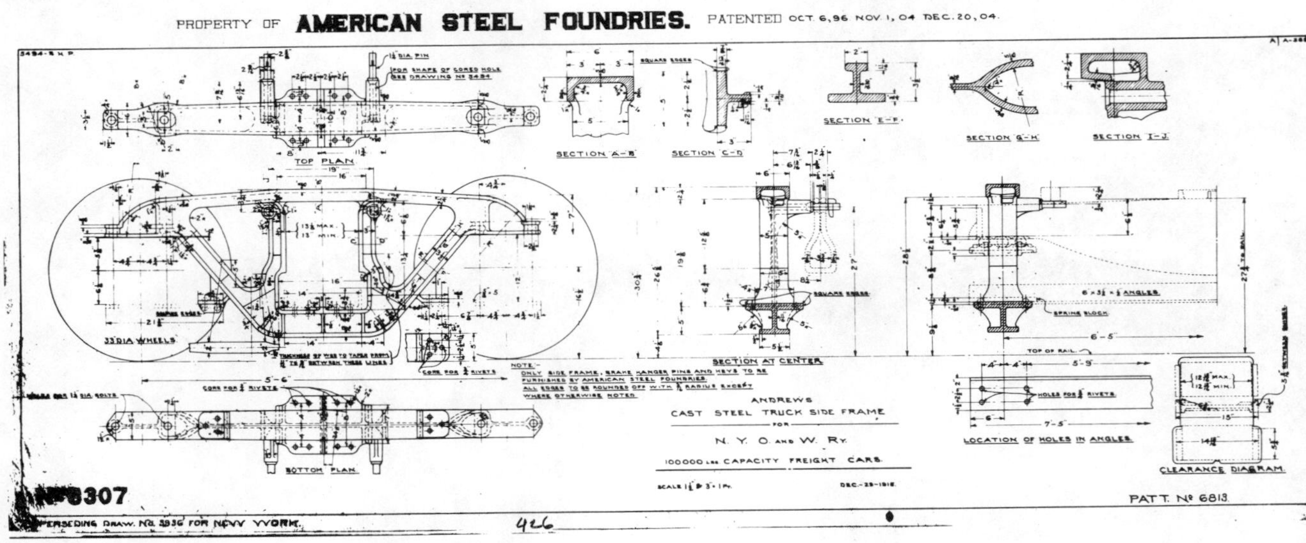

FIG. 21A "T" SECTION ANDREWS TYPE SIDE FRAME

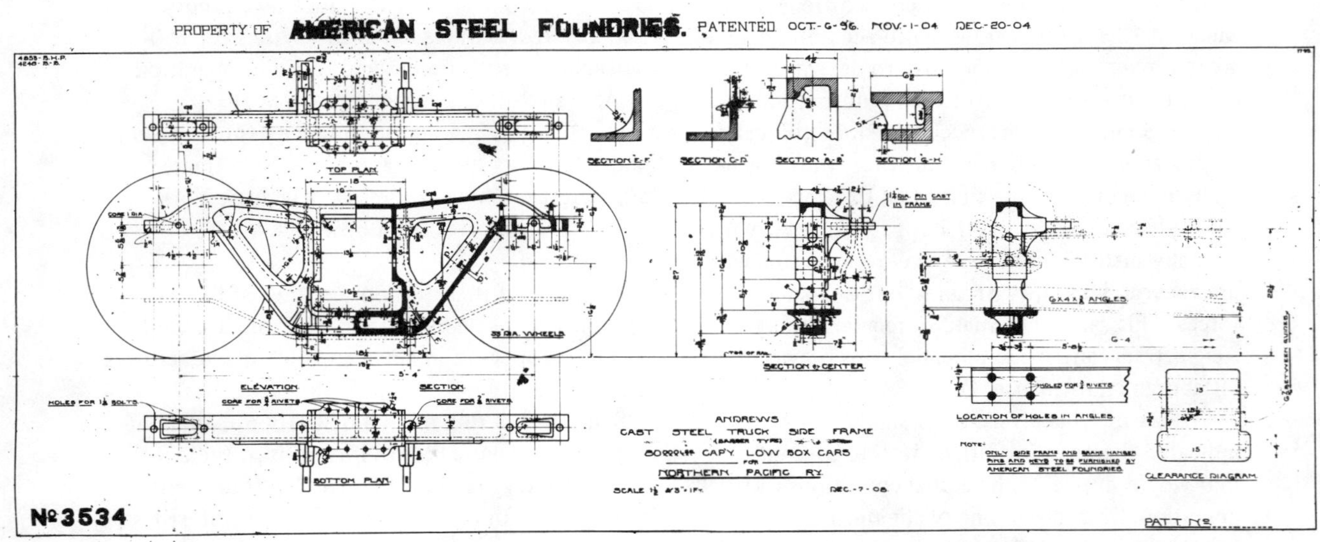

FIG. 21B "L" SECTION ANDREWS TYPE SIDE FRAME

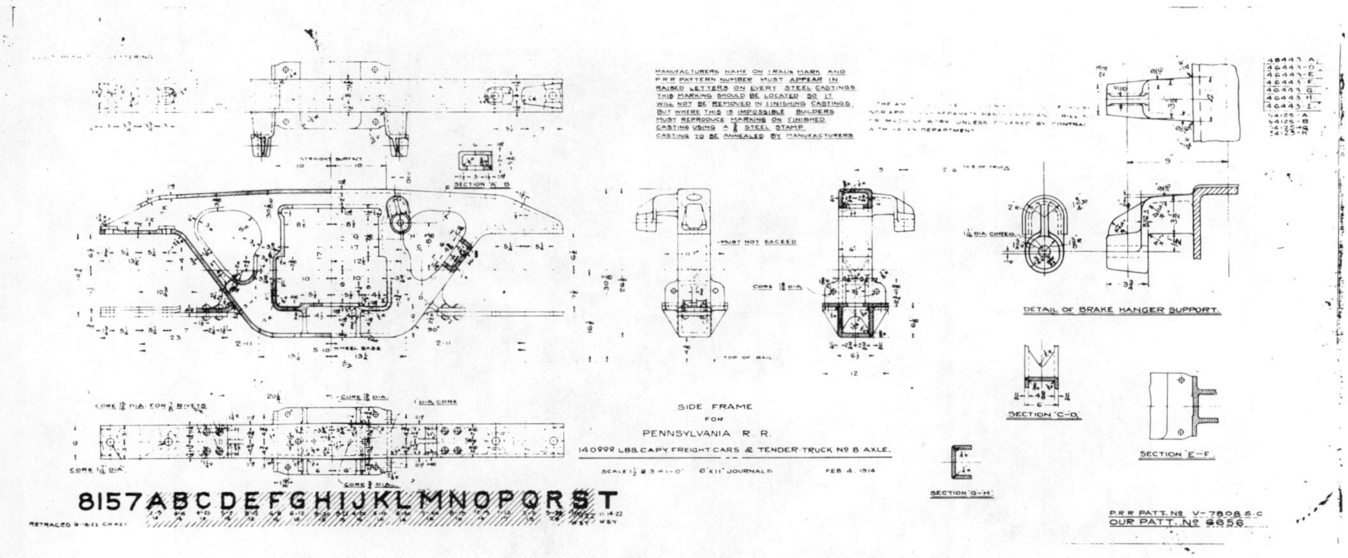

FIG. 21C "U" SECTION ANDREWS TYPE SIDE FRAME

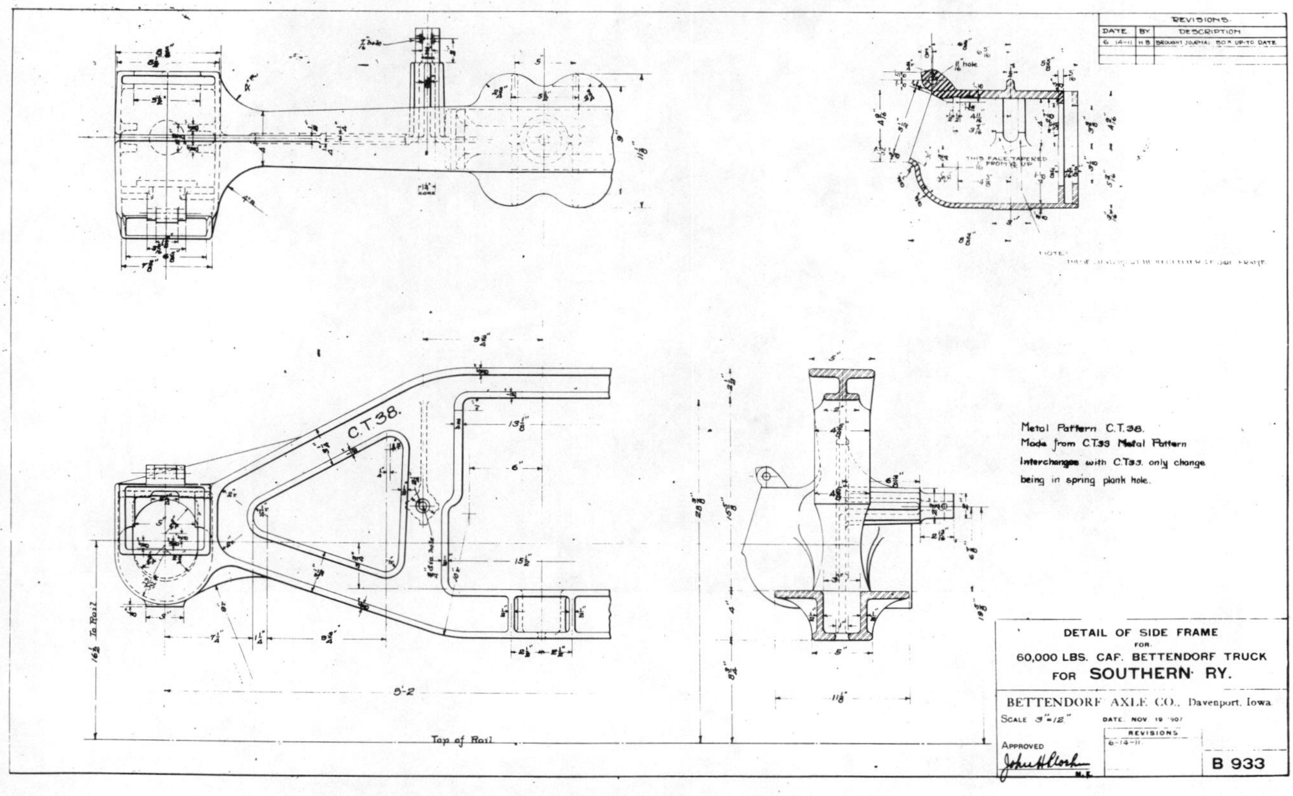

FIG. 21D "I" SECTION BETTENDORF TYPE SIDE FRAME

FIG. 22 ASF SIDE FRAME FATIGUE TEST MACHINE

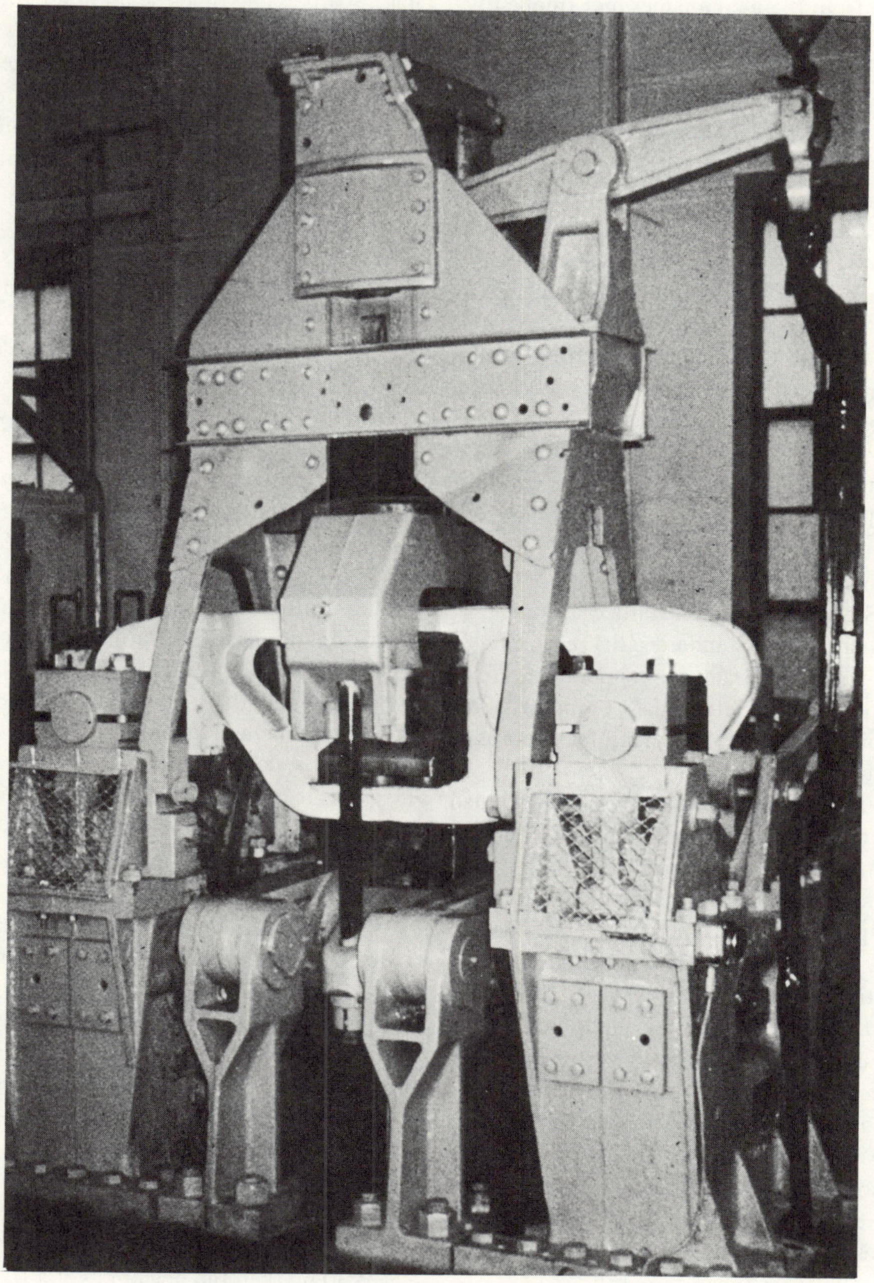

FIG. 23 DRESSER FATIGUE TEST MACHINE

Steel Foundries ("ASF") of Chicago, The Buckeye Steel Castings Company ("Buckeye") of Columbus, Ohio, Gould Coupler Company ("Gould") of Depew, New York, and Scullin Steel Company ("Scullin") of St. Louis. Accordingly, during 1916 and 1917 a committee consisting of the chief mechanical engineers from these companies met frequently with the ARA subcommittee to coordinate the development of railroad design specifications with the ability of truck manufacturers to meet those specifications.

When the United States government took over the American railroads during the First World War in 1917, it gave added impetus to standardized side frame and bolster design. The government committee of the United States Railroad Administration ("USRA"), and the appropriate ARA committees and the TMEC (Truck Manufacturers' Engineers' Committee) worked together on matters of design and production of truck specialties. As part of its wartime operations, the USRA ordered 100,000 freight cars and, for these cars, specified cast steel side frames and bolsters of a standardized design (Fig. 24).

These showed marked improvement and for years the USRA designs were regarded as something in the nature of a standard. This railroad-government-truck manufacturer cooperative effort was enlarged when the Bettendorf Company ("Bettendorf") joined the other four manufacturers as a member of the TMEC.

These wartime joint efforts (1918-1920) yielded a rich body of experience from which both railroads and truck manufacturers could draw during the years to come. Consequently, the TMEC and the ARA, through its various committees, continued to work jointly in progressing standardized designs.

Most of the railroads purchasing cast steel side frames and truck bolsters for new equipment were drafting their own specifications covering the general design and strength requirements of these devices, and basing their specifications largely upon their own peculiar conditions. There were naturally wide variations in requirements and for some time it was apparent that uniformity of specification would be desirable from the standpoint of both the producer and the user. Therefore, in the fall of 1920 the ARA Car Construction Committee appointed a subcommittee on freight car trucks to cooperate with the TMEC to develop and recommend a standard for side frames and bolsters, and also for complete trucks. The proposal was completed in 1925. Hence, the 1926 ARA acceptance test specifications were adopted as standard. Needless to say, some design modifications were necessary to meet these specifications.

The severity of the test requirements for side frames was increased in 1929 by reducing the deflection and set requirements while maintaining load. At the same time the minimum breaking load was increased from 9 times to 12 times axle capacity. The transverse requirements also were stiffened as the deflection and set were reduced.

The design test requirements for truck bolsters were revised in 1931. Side frame and bolster specifications were again revised in 1937 and 1938, but no change was made in the test requirements.

The first and tentative requirements for side frames and bolsters made of high tensile steel came in 1936. This project, more than any other, occupied the attention and energy of the TMEC during a 10-year period.

In 1935 (after 14 years of research with dynamic testing), proposals by TMEC were accepted to include dynamic testing of side frames in the specifications. AAR Circular DV-872, dated 5-25-36, is the author's last reference until AAR Circular DV-1168, dated 1948, relates action taken in 1947 to establish official approval procedures and identification numbers. The fatigue machines were kept very busy. Specifications were revised in 1947 to increase slightly the ultimate load requirement for side frames and include dynamic tests, and to reduce materially the deflection requirements for high tensile castings. Through this research era the strength of castings was greatly increased, and this was accomplished without a corresponding weight penalty.

Credit for this progress in freight car trucks rightfully belongs not only to the engineer and designer, but likewise to the metallurgist and foundryman who have made very substantial improvements in steel and quality of castings. The close association and cooperative efforts of these men have made possible the present strong, safe and dependable cast steel side frames and bolsters.

THE SUSPENSION ERA

1930 marked the beginning of a trend for faster freight train speeds and higher capacities. The need for improved ride quality was recognized along with continuing emphasis on weight reduction. Spring plankless truck ideas were generated in this decade, and softer suspensions and snubbers were also among the designers' goals.

Suspension tests conducted on the New York Central Railroad at Ashtabula in June 1929 demonstrated a need for damping in truck suspensions, and also resulted in a spring development later adopted as the AAR-1936 spring.

Standard Car Truck Company introduced friction snubbing in the freight car truck in 1935. This Barber truck provided improved ride quality over the unsnubbed truck.

In the latter part of 1938, at the request of the Vice President of Operations and Maintenance for the AAR (Association of American Railroads) (ARA became AAR in 1936), the General Committee considered the investigation of trucks for high speed freight service. The purpose of the investigation was to aid in the development of freight car trucks suitable for high speeds by making an extensive study of the problems involved. The data secured would then be available for use in designing and building freight car trucks suitable for high speed.

The AAR subcommittee met on January 11, 1939 to formulate plans for procedure. It was decided that the scope of the investigation would be:

1. To determine whether the conventional freight truck:
   (a) is safe to operate at high speeds, i.e., 80 mph or over; or
   (b) has satisfactory riding qualities at 80 mph or over.
2. To determine riding qualities and performance of trucks designed specifically for the above high speed service.

Official testing began June 6, 1939 after a series of preliminary runs, and included the following trucks (all 5½ in. x 10 in. journal size):

1. AAR Truck – the conventional design of spring plank truck equipped with 1936 plain coil springs. These trucks were secured from the stock of the Pennsylvania Railroad Company.
2. ASF Truck – furnished by American Steel Foundries.
3. Barber Truck – furnished by Standard Car Truck Company.
4. Buckeye Truck – furnished by the Buckeye Steel Castings Company.
5. Full Cushion Truck – furnished by the Allied Railway Equipment Company.
6. National Truck – furnished by the National Malleable & Steel Castings Company.
7. Ohio Truck – furnished by Ohio Steel Foundries.
8. PRR No. 1 Truck – furnished by the Pennsylvania Railroad Company.
9. Railway Truck – furnished by the Railway Truck Corporation.
10. Scullin Truck – furnished by Scullin Steel Company.
11. Simplex Truck – furnished by the Bettendorf Company.
12. Symington Truck – furnished by Symington-Gould Corporation.

Results reported in the AAR 1940 report entitled "Tests of Trucks for High Speed Freight Service" indicated there was much work to do ("Back to the Drawing Board").

ASF began its test train operation in 1940 to accumulate a broader base of data for truck performance with varying road and lading conditions. In 14 months, 27,396 miles of test routes were covered on the Milwaukee, New York Central and the Missouri Pacific railroads. Component strength and significant truck performance improvements resulted from their effort, and the spring plankless truck with built-in snubbing and long travel springs was introduced.

There was very little research published through the later World War II years but the AAR again ran a high speed test program during 1948-1950. Tests were conducted on the Illinois Central Railroad by the AAR using the ASF 5 car test train. Costs were prorated between the participants with the AAR covering about 25 percent. Sixteen participants or suppliers tested various devices during the three year program, consisting of 15 unit snubbers, 11 package snubbers, three modified conventional trucks and 34 high speed trucks. Other related research programs included:

Investigation of solid type journal bearing performance.

Roller bearing research program (Timken Roller Bearing Co.).

Lateral action of freight car trucks.

Accelerometer performance and side frame stress.

Studies of vibration in freight cars (U.S. Signal Corps, C&O R.R.).

Obviously, the researchers were busy. These tests revealed the dramatic advancement in freight car truck suspensions in that decade. The high speed trucks with long travel springs and built-in snubbing outperformed the short spring, unit and package snubber trucks by better than a 10 to 1 lading damage index reduction. Twenty-eight truck designs were deemed satisfactory for speeds up to

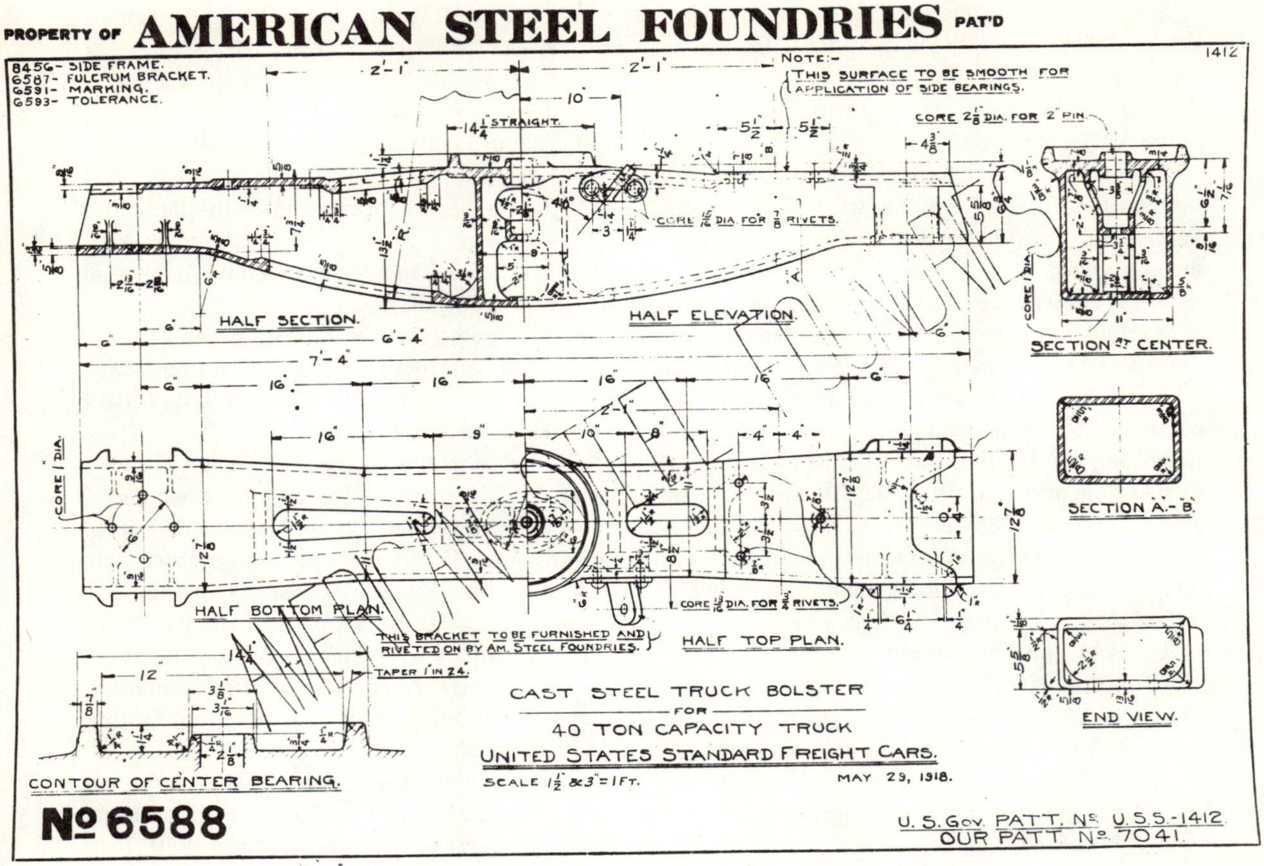

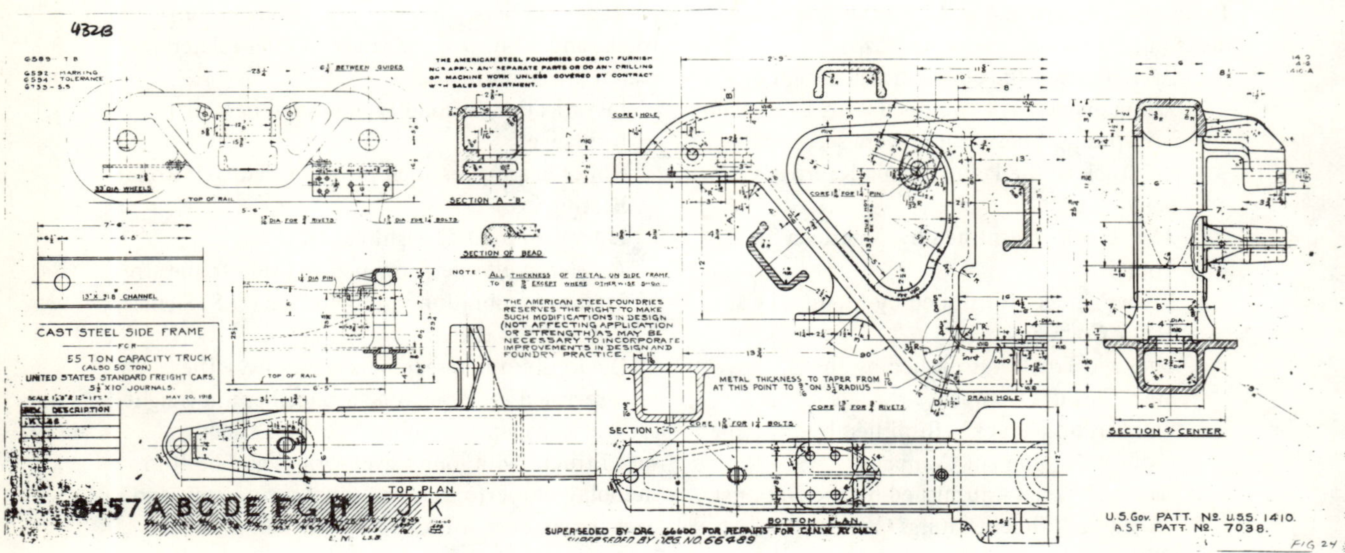

FIG. 24 USRA STANDARDIZED SIDE FRAME AND BOLSTER

90 mph. Space does not permit a complete listing here, but these are all contained in the AAR reports.

Although used in small numbers as far back as the 1930's, 70-ton trucks became popular in the mid-1940's and 100-ton trucks in the early 1960's. D-2 springs were officially adopted by the AAR in 1936, and D-3, D-4 and D-5 springs in 1947. D-7 was adopted in 1975.

The popularity of 100-ton equipment and roller bearings in the 1960's exposed at least two performance problems that have been researched heavily. Rock and roll invited auxiliary snubbing devices for control, and truck and car hunting likewise invited various control devices. Of interest, the car rocking problem was studied in 1925 and a thorough analysis presented by B. R. Leffler to the AREA in 1926.

The 1970's produced funded and cooperative research programs like the Truck Design Optimization Project (TDOP) and Track Train Dynamics (TTD); FAST (Facility for Accelerated Service Testing) at the Federal Railroad Administration High Speed Ground Transportation Test Center at Pueblo, Colorado (later changed to Transportation Test Center); mathematical models of trucks, validation tests and computers for the analysis of truck, car and train performance. Hydraulic actuating fatigue test equipment for railroad components began in 1972 at ASF, 1976 at Dresser and the AAR. The Southern Railway, and very likely other railroads also, have their own.

Many truck and suspension configurations have been introduced to provide improved performance. Car rocking, truck hunting, wear, ride, etc., have been addressed with these configurations. A complete survey of designs was published by the TDOP Project and is available from the National Technical Information Service. The researchers continue their efforts to improve the freight car truck, as evidenced by the many publications including those of the Rail Transportation Division.

Recognizably, a high volume of activities and many names involved in the American history of the freight truck have been omitted, but not intentionally. Please accept our apology.

REFERENCES

[1] *1884 Car Builders' Dictionary.*

[2] *1888 Car Builders' Dictionary.*

[3] *Car Builder Encyclopedias.*

[4] Ernst, F. B., "Development of Freight Car Trucks," report to the Federal Coordinator's Mechanical Advisory Committee.

[5] "Report of Mechanical Advisory Committee to Federal Coordinator of Transportation on Freight Cars," Section IX, 1935.

[6] Cottrell, Robert Boyd, ASME Paper 44-A-41.

[7] ASF Report and Microfilm Files.

[8] "Tests of Trucks for High Speed Freight Service," AAR Report, April 1940.

[9] AAR Reports—Cooperative Freight Truck and Snubber Research Program and Related Freight Car Research Programs; 1st Progress Report, September 1, 1949; 2nd Progress Report, July 1, 1950; 3rd Progress Report, December 15, 1950; Summary Report-3800, June 30, 1951.

[10] "Track Train Dynamics Progress Reports".

[11] "Truck Design Optimization Project Truck Survey".

18 The History and Development of Standard Railroad Car Couplers

D. J. ALBANESE

INTRODUCTION

In the earliest days of railroading in America, beginning with the inaugural service of the famous "Dewitt Clinton" in 1831, the only means available to join cars together was the link and pin coupling shown in Fig. 1. This crude, hazardous arrangement had to be manipulated by hand and required trainmen to go between the cars. Despite the resulting heavy toll of serious injuries and death, it was not until 1863 that the first automatic coupler of any real value came into being. This device, shown in Fig. 2, and known as the Miller Hook, was greatly improved during the next several years and was heralded as the greatest life-saving invention of the age. It soon received wide usage on passenger cars both in this country and abroad.

In 1877, the first vertical plane knuckle type coupler was invented by Major Eli Janney and in the years that followed, many coupler designs of this type came into being. By 1889, there were about 80,000 freight cars in use and on them were 39 varieties of couplers, some of which were capable of coupling with each other, but the majority would not intercouple.

This situation obviously created a major problem for the railroads of that period and was undoubtedly the greatest single factor responsible for the organization of the Master Car Builders' Association. Due to their initial efforts, which included extensive tests, the number of coupler designs was reduced to 46 and these subsequently to twelve.

The first truly automatic knuckle type coupler, known as the "Tower" coupler, was invented in 1892. This coupler would not only couple and lock on impact, as would many other designs than in use, but it could also be unlocked and the knuckle thrown fully open by means of a lever at the side of the car.

Many railroads were slow to adopt the use of automatic couplers and other safety devices with the result that the Safety Appliance Act was passed by Congress in 1893, which made it mandatory, on and after January 1, 1898, that all cars used in interstate commerce must be equipped with couplers capable of coupling automatically by impact and uncoupled without the need of trainmen going between the cars.

This action served to stimulate coupler development and as a result, many new designs came into existence, all incorporating the pivoted knuckle principle and capable of coupling with each other. Among the important improvements which became Standard were the lockset feature, which was made mandatory in all couplers applied to new cars after January 1, 1904, and the automatic knuckle throw feature, which was adopted in 1910.

The advent of many new knuckle-type coupler designs and the increasing interchange of rolling stock created a very serious coupler maintenance problem. In order to keep cars moving, it was necessary for the railroads to carry tremendous stocks of repair parts. In fact, one manufacturer advertised 119 different designs of knuckles for use in repairing existing couplers.

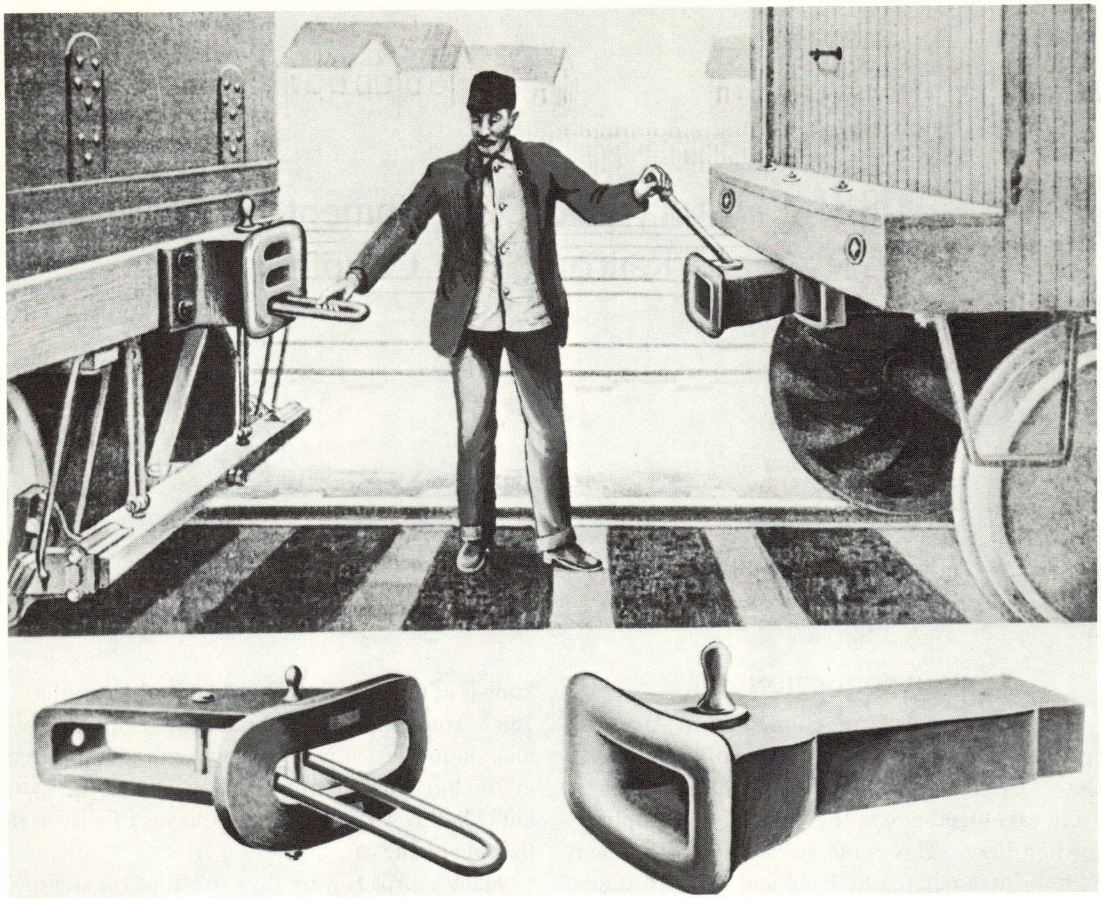

FIG. 1. LINK AND PIN COUPLER – 1831 TO 1863

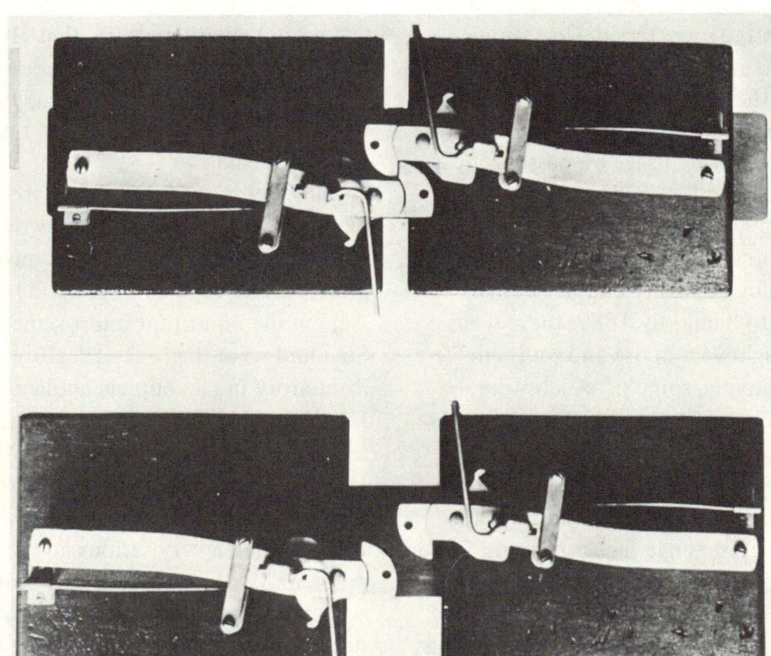

FIG. 2. MILLER HOOK – 1863

By 1911, the situation had become so confused that the Master Car Builders' Association invited all of the coupler manufacturers to participate in the development of a single coupler design as a standard for all railroads. This invitation received enthusiastic response and after extensive laboratory and service tests, all but two of the many designs submitted were eliminated.

THE STANDARD TYPE D COUPLER

The two surviving designs, identified as the Experimental Standard Type C and Type D couplers, were carefully studied by the M.C.B. Coupler Committee with the result that the Type D coupler, illustrated in Fig. 3, was finally chosen and recommended for adoption as Standard. This recommendation was submitted to a special letter ballot in 1916 and through favorable action the Type D coupler became the first Standard coupler.

In order that the fullest benefits could accrue to the railroads in having a single Standard coupler, it was necessary for all of the manufacturers to establish and maintain consistently uniform practices. To this end, complete construction drawings of the coupler and parts were made for use by all. Through cooperative efforts, over 50 gages were designed to insure complete interchangeability of parts and satisfactory functioning of complete couplers irrespective of the source.

Each manufacturer was provided with a master coupler and parts to control the accuracy of all new gages and for checking the wear of used gages. Standard manufacturing specifications were also prepared and adopted by the association for the further benefit of all concerned.

It was recognized early that service problems would arise through more extensive use of the Standard coupler and to meet this contingency a committee was formed, consisting of an engineering representative of each manufacturer. This committee, known as the Mechanical Committee of the Standard Coupler Manufactuers, met as often as necessary to consider not only coupler service problems, but any design improvements suggested by the railroads or the manufactuers themselves. After thorough study and investigation, appropriate recommendations were made to the MCB Coupler Committee for final action.

Another function of the Mechanical Committee was to periodically check the quality of each manufacturer's product. For this purpose, a complete coupler, representing current production, was obtained from each manufacturer. Each coupler and parts was completely gaged, the functional features checked and a miscellaneous interchange made with

FIG. 3. TYPE D – FIRST MCB STANDARD RAILROAD CAR COUPLER – 1916

no two essential parts of any one manufacturer in any assembly. The fit of these parts and the functioning of each coupler so assembled was required to be entirely satisfactory. Finally, all of the representative couplers were subjected to physical tests to verify the strength. It is worthy of mention that the strength of the Type D coupler was about twice that of the preceding nonstandard couplers and weighed about one-third more. During the years following the adoption of the Standard D coupler, numerous design improvements and adjustments were made, all accomplished without affecting interchangeability.

THE STANDARD TYPE E COUPLER

The steady increase in train tonnage and motive power made it apparent that greater strength and, if possible, increased efficiency, should be provided in the Standard coupler to keep pace with this trend. It was recognized by the Mechanical Committee that many desirable improvements could be made, but not without affecting interchangeability of parts. Since it was believed that the advantages would far outweigh the disadvantages, a new coupler was designed based on the years of service experience with the Type D coupler.

This new coupler, designated as the Type E and illustrated in Fig. 4, was recommended by the A.A.R. Committee on Couplers and Draft Gears (formerly the M.C.B. Coupler Committee) for letter ballot approval of the Association in 1930. By this ballot, the Type E coupler was approved as Recommended Practice to replace the Type D coupler and the following year, by similar action, was advanced to Standard, effective March 1, 1932.

The advantages of the Standard E coupler, as compared with the Type D, are many, the more important of which are: increased strength of the coupler body, particularly in the guardarm, the front face, the knuckle side wall and the shank; a stronger knuckle and lock; improved anticreep and increased operating efficiency.

With the exception of the knuckle pivot pin, none of the D and E coupler parts are interchangeable. However, the Standard E coupler knuckle can be used for repairs to D couplers and provides some benefit when used. Another advantage in the Standard E coupler is that it may be fitted for either top or rotary bottom operation by application of the desired type of locklift assembly.

In 1933, the contour of the Standard E coupler was improved which further strengthened the front face, reduced contour free slack and increased contour life before the condemning limit is reached.

Several years later, the railroads became interested in weight reduction, without sacrifice of strength, which was made possible by employing high tensile steel. Accordingly, the coupler manufacturers and the A.A.R. Coupler and Draft Gear Committee again cooperated in the development of a reduced weight E coupler including coupler

FIG. 4. TYPE E – RAILROAD CAR COUPLER – AAR STANDARD – 1932

yokes. The strength of the Grade B steel coupler was maintained in the high tensile steel design with a weight reduction of from 20 to 25 percent in the complete coupler.

Subsequent to adoption of the Standard E coupler, efforts were continued to incorporate improvements to meet the ever increasing demands of service. In order to provide higher yield strength in the knuckle, a high tensile steel design was adopted as Standard for all Type E couplers in March 1947. The lock also was made in the same material to correspond with the knuckle and increase wear life. In 1954, an articulated type of locklift assembly, shown in Fig. 5, was adopted as Standard. This arrangement provided decidedly improved anticreep protection, particularly when vertical and longitudinal shocks occur simultaneously in service.

In 1969, high tensile steel regular weight coupler bodies were approved as standard to increase service life of the coupler. Continuing the effort of making improvements to keep up with changing service environment, in 1973 the material for yokes was upgraded from Grade B steel to high tensile steel and the coupler shank was strengthened. As service demands continued to increase, in 1976 Grade E material also was adapted as standard for the coupler system and yoke.

Occurrence of tank car explosions during train derailments necessitated designing an E coupler with top and bottom shelves, Fig. 6, for tank cars. These shelves minimize accidental vertical coupler separation from happening during a derailment situation. The coupler was approved as the tank car standard in 1977.

Both the Type D and Type E couplers were used not only in freight service but on passenger cars as well. For the latter applications, longer and different style shanks were provided and the couplers were fitted with passenger type lockshift assemblies to cooperate with the coupler operating mechanism used.

In the Type E coupler, about 3/4 in. of free contour slack is normally present when two such couplers are mated. This amount of slack was intention-

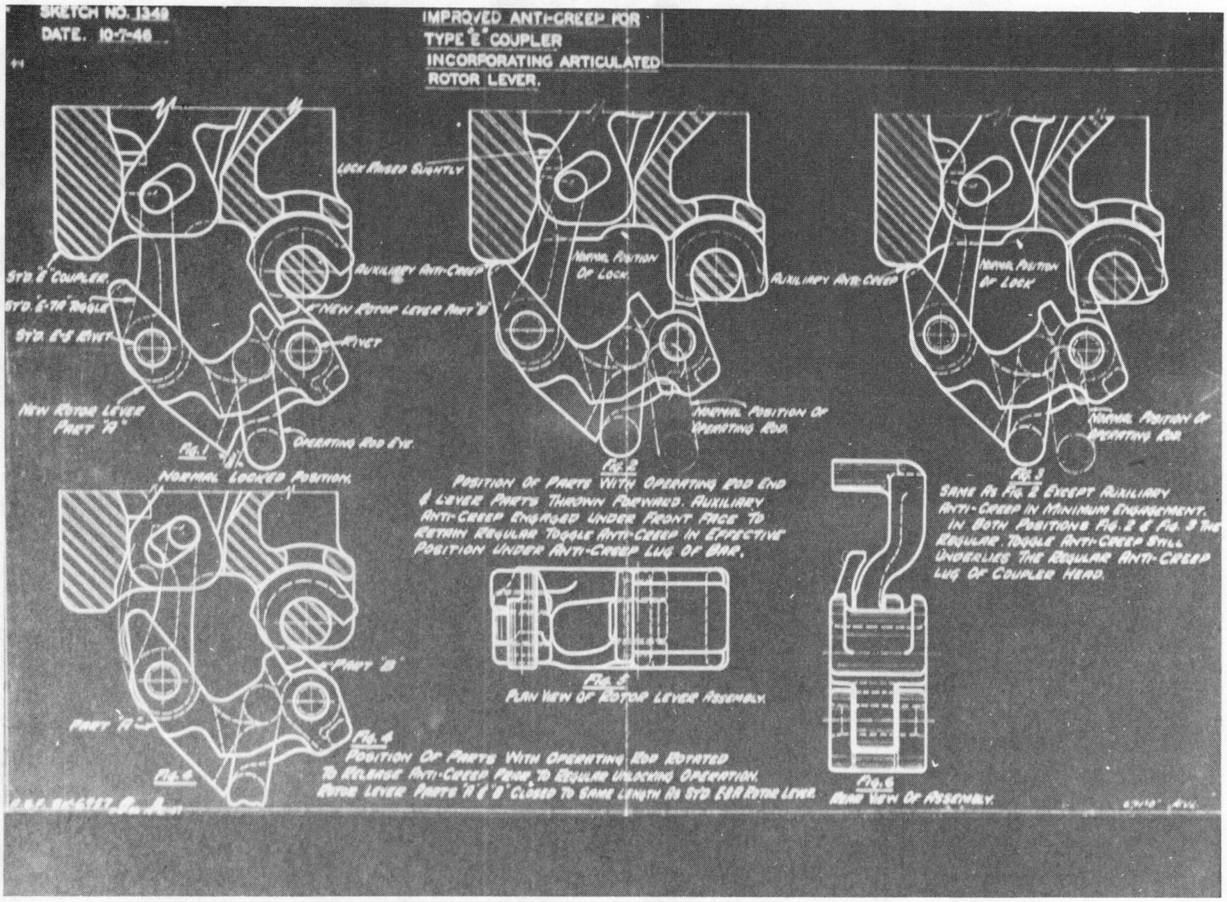

FIG. 5. TYPE E RAILROAD CAR COUPLER – ARTICULATED LOCKLIFT ASSEMBLY – AAR STANDARD – 1954

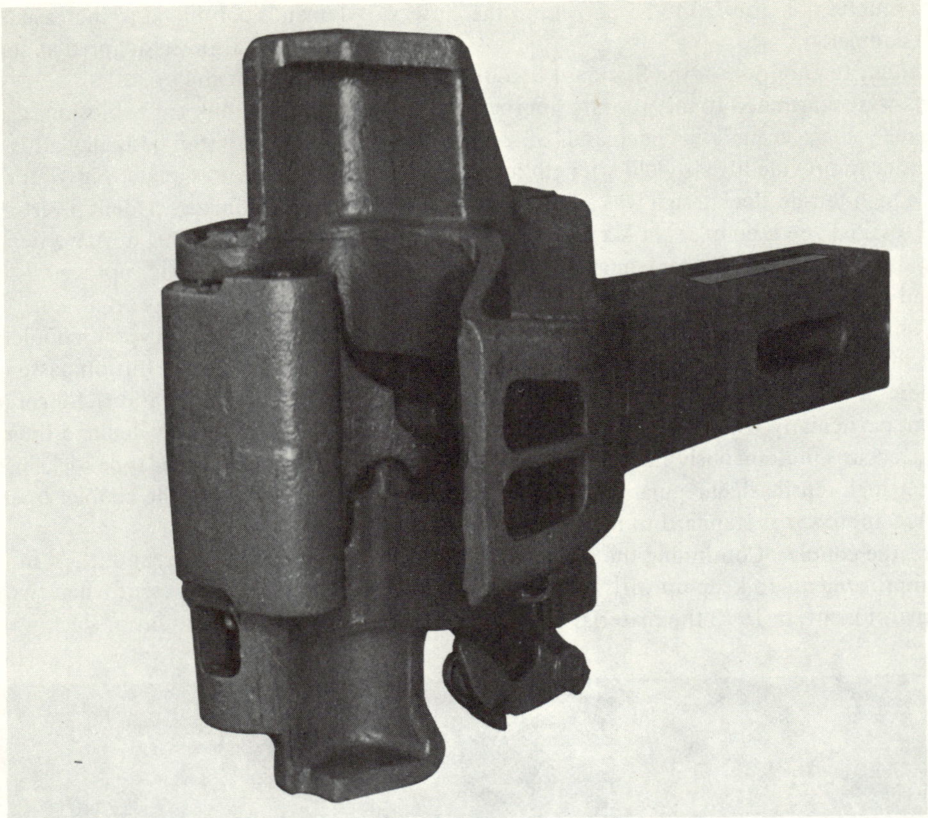

FIG. 6. TYPE E COUPLER WITH TOP AND BOTTOM SHELVES – AAR STANDARD – 1977

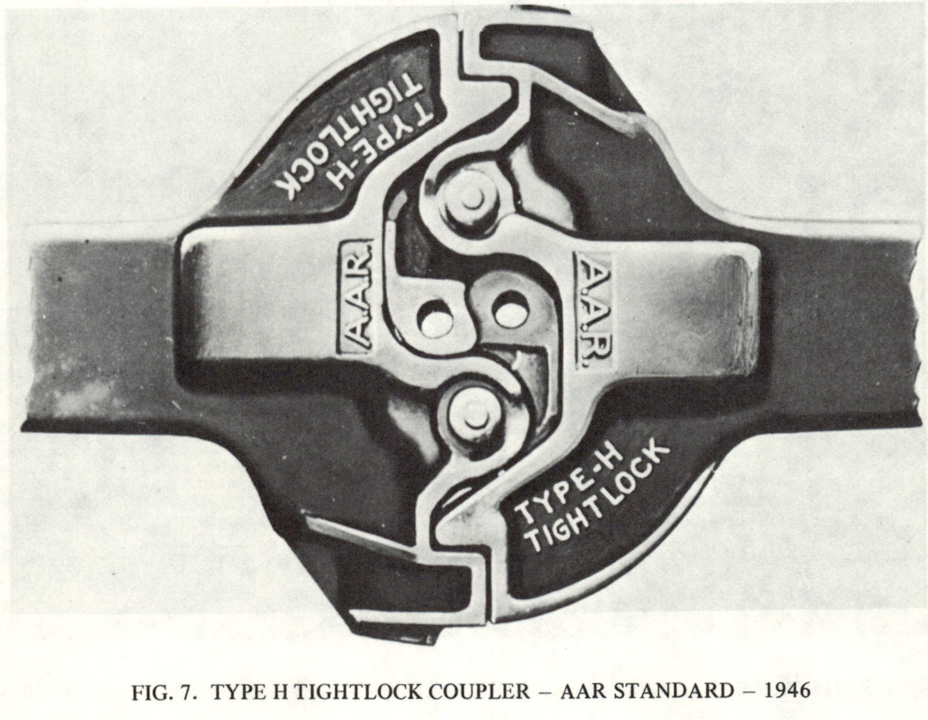

FIG. 7. TYPE H TIGHTLOCK COUPLER – AAR STANDARD – 1946

ally provided to permit negotiation of classification yard humps and 30 degree curves (193 ft minimum radius).

The rigid coupler carrier and standard 11½ in. lateral opening in the striking casting do not provide sufficient coupler angularity and the additional amount required, both vertically and laterally, is obtained by the 3/4 in. clearance in the coupler contours. This free slack is also an aid in starting heavy trains by permitting the individual cars to be started serially. However, with the advent of Diesel power, this procedure is no longer necessary.

THE A.A.R. STANDARD H TIGHTLOCK COUPLER

The necessity for free contour slack in freight couplers does not exist in passenger service because ample carrier clearance can be provided for lateral coupler movement, the curves to be negotiated are less sharp, and no humping is involved. In fact, the elimination of free slack is very desirable from the standpoint of passenger comfort and noise reduction. There was also the ever-present danger of train separation by vertical slip-over of coupler knuckles and cars overturning in event of derailment.

The obvious need for greater safety led to the development of a coupler design which eliminated free contour slack and incorporated interlocking features to prevent vertical separation and overturning of cars. In 1936, the first couplers of this design were applied experimentally in passenger service. This coupler, first identified only as the Tightlock coupler, underwent several years of development directed by the cooperative efforts of the railroads and the standard coupler manufacturers.

In 1939, the Association adopted Specifications for the Construction of New Passenger Equipment Cars which established certain basic requirements involving couplers. The cars were required to resist a static compression end load of 800,000 lb without permanent deformation in any member of the car structure. This meant that the couplers between two cars would be loaded as a column and must meet the same requirement. It was further specified that the coupler and its supporting members must resist vertical and lateral loads of 100,000 lb without permanent deformation. Interlocked tightlock couplers meeting these requirements were considered satisfactory compliance with the anti-telescoping provisions of the specifications.

These increased strength requirements and many other progressive refinements were considered in the development of a substantially improved tightlock coupler, identified as the Type H Tightlock coupler, two heads of which are shown locked together in Fig. 7. This coupler, including the yoke and the radial connection casting, which joins the coupler and yoke, are all made of high tensile steel. The Type H Tightlock coupler and attachments, including an improved design of coupler operating mechanism, designated as Type No. 6, were approved by the Association in 1944 as Tentative Standard for passenger equipment and advanced to Standard in 1946.

One of the outstanding design features in the Type H coupler is the anticreep arrangement. When the coupler is fully locked, the top of the locklift toggle fully underlies the shelf provided on the tail of the knuckle. Slight upward movement of the lock brings these two surfaces into engagement and no further upward lock movement is possible during train operation.

The elimination of free slack in the Type H coupler required extremely close fitting which necessitated machine finishing of various contacting surfaces of the head and parts. Figure 8 shows the machined surface of the front face of the coupler head and both faces of the lock. Figure 9 shows the machined surfaces on the knuckle, namely the buffing and pulling faces and both sides of the knuckle tail. The adjacent locking faces of the lock and knuckle tail are provided with a vertical taper which is necessary to insure free dropping of the lock to fully locked position when coupling and to eliminate contour slack when the lock is seated.

The Type H Tightlock coupler is provided with wings on either side of the head which, when coupled with another tightlock coupler, provide an efficient interlock of the coupler heads. Two such couplers in position for mating are shown in Fig. 10. This interlocking feature prevents cars from telescoping and prevents overturning in event of derailment. Another function of these wings is to assist in aligning the couplers when angled relative to each other either vertically or laterally, or both. They also make satisfactory coupling possible when the height of adjacent couplers varies as much as 3 in., which position is illustrated in Fig. 11. The interlocking feature provides a further benefit in that the vertical movement between mated couplers is restricted, thereby appreciably reducing front face and knuckle nose wear. Since contour slack is eliminated when tightlock couplers are mated, there can be no vertical angling of the couplers relative to each other. For this reason, it is necessary to support the coupler on spring carriers to compensate for vertical track curvatures and any appreciable difference in the coupler height of adjacent cars. In order to comply with the passenger car specifications, spring carriers must be capable of supporting a load of 100,000 lb with the coupler in any position laterally. This is the equivalent of a derailment condition where one car, by means of the couplers, must be capable of supporting the weight of one end of the adjacent car. It was indicated by experience that a number of coupler carrier designs in use did not meet this requirement and the A.A.R. has since adopted specifications for coupler carriers which now govern the fundamental requirements.

The necessary vertical and lateral angling at the shank end of the tightlock coupler is provided by means of a so-called radial connection casting which joins the coupler shank and yoke head. The coupler shank is joined to the radial connection casting by a 3 in. diameter vertical pin, in quadruple shear, which provides for lateral angling. In turn, the radial connection casting is joined to the yoke by a 4 in. diameter horizontal pin, in double shear, which provides for vertical angling. Figure 12 shows these

FIG. 8. TYPE H COUPLER SHOWING MACHINED SURFACES OF HEAD

FIG. 9. TYPE H COUPLER KNUCKLE SHOWING MACHINED SURFACES

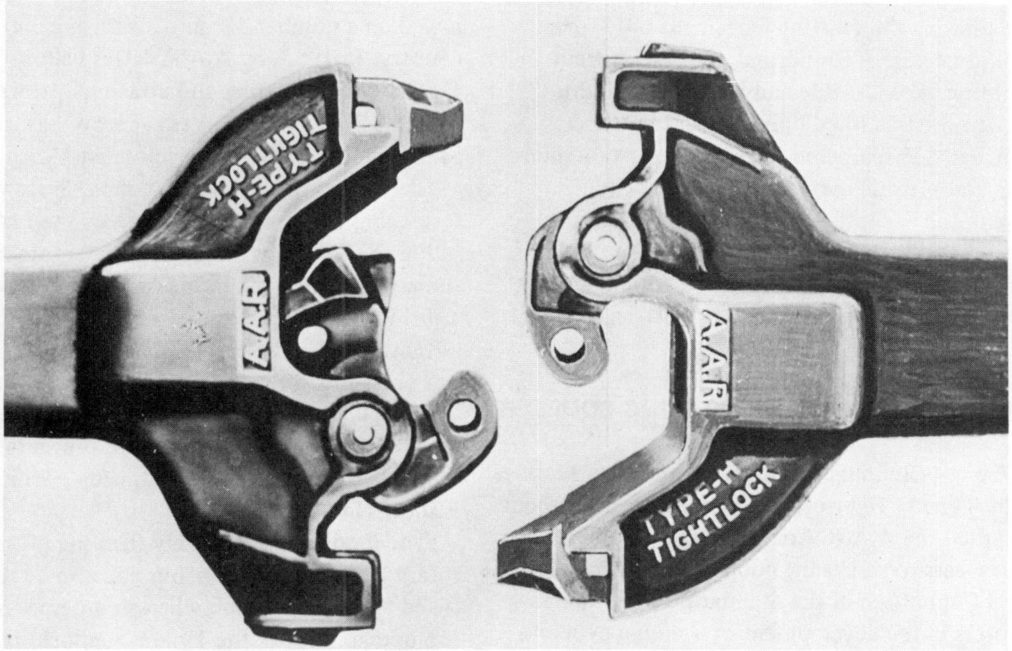

FIG. 10. TYPE H COUPLERS IN MATING POSITION

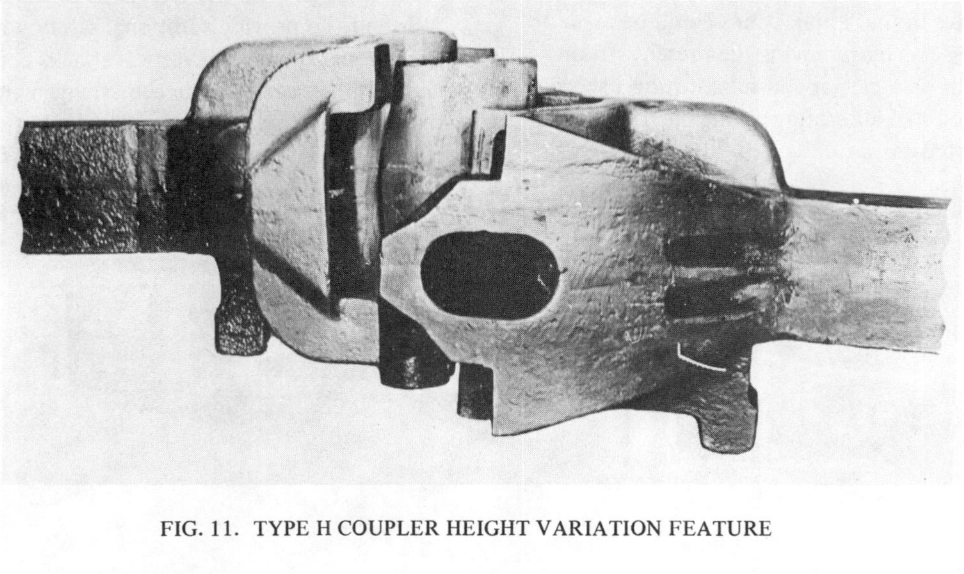

FIG. 11. TYPE H COUPLER HEIGHT VARIATION FEATURE

parts in their relative positions prior to assembly.

It will thus be seen that this arrangement provides in effect a universal joint, but at the same time limiting the twisting movement only to an amount necessary for normal car sway in train operation. The knuckle and locks for all tightlock couplers are furnished only in high tensile steel, quenched and tempered to reduce wear to a minimum and provide maximum safety.

With the exception of the knuckle pivot pin, none of the Type H coupler parts can be used in any other A.A.R. coupler; nor can parts from any other type coupler be used in Type H Tightlock couplers.

THE A.A.R. TYPE F INTERLOCKING COUPLER

The outstanding performance and safety features of the Type H Tightlock coupler in passenger service prompted the A.A.R. Committee on Couplers and Draft Gears to solicit the cooperation of the Mechanical Committee of the Standard Coupler Manufacturers in the development of a similar type of coupler for freight service. Among the stipulated fundamental requirements for this coupler design were the elimination of all machine finishing, the lock to have parallel faces with no vertical taper, and that a minimum of free contour slack should be provided.

Another desirable objective was to design the coupler and attachments for application not only to new cars but to existing cars as well with the least possible modification. As the design work progressed, it was found possible to achieve this objective in that applications could be made to existing cars by removing the present striking casting with rigid carrier and substituting a specially designed striking casting having the necessary spring supported carrier.

The first trial service applications of this new development, identified as the Type F Interlocking coupler, were made early in 1948, and others followed on a number of railroads throughout the country. In 1953, by A.A.R. letter ballot, the Type F Interlocking coupler and attachments, which included a suitable yoke, were approved as Alternate Standard for freight service, effective March 1, 1954.

The design of the Type F coupler head with interlocking wings and safety support shelf is shown in Fig. 13. When two Type F couplers are mated, these wings interlock in the same manner as the tightblock passenger couplers and provide the same advantages. A new feature in the Type F coupler not present in any other coupler is the safety shelf, which serves to support a conventional type coupler in event of a pull-out and thus prevents it from falling to the track. A demonstration of this feature is shown in Fig. 14.

The F coupler is normally furnished fitted for rotary bottom operation, but can also be furnished fitted for top operation when so ordered. In the top operated form, the Type E top locklift assembly is used together with a different design lock than that used for rotary bottom operation. The top operated F coupler provided anticreep protection in the same manner as in the top operated Type E coupler. However, in the rotary bottom operated F coupler, which is predominantly used, double anticreep protection is provided.

The principle anticreep functions in much the same manner as in the Type H Tightlock coupler, namely, by engagement of the top surface of the locklift toggle with the underside of the knuckle tail shelf. To provide additional safety when combined longitudinal and vertical shocks occur simultaneously, a second anticreep arrangement is utilized. Under these conditions, the locklift lever moves diagonally forward and upward. When in

FIG. 12. TYPE H COUPLER RADIAL CONNECTION CASTING AND JOINING PARTS

this position, rotation of the locklift assembly is prevented and the lock therefore cannot be raised.

All of the desired objectives have been achieved in the Type F coupler design with no machine finishing. The free contour slack between two mated F couplers has been reduced to 3/8 in. or less than half of that provided in Type E couplers. This is the least amount of slack possible with assurance of positive dropping of the lock when a coupling is made. With this reduced slack it is necessary to utilize a spring supported coupler carrier to compensate and thus obtain the total vertical coupler angularity required to negotiate classification yard humps.

The Type F coupler and attachments, including the striking casting with spring supported carrier, completely assembled in a car application, is shown in Fig. 15. Here it can be seen that a vertical pin provides the connection between the coupler shank and yoke. The butt end of the coupler is spherical and bears against a correspondingly shaped front follower when buffing forces are applied. Just forward of the spherical surface of the coupler butt are laterally extended shoulders which engage the follower when the coupler is moved laterally 2 in. either side of the center measured on the coupling line. Further lateral movement will compress the draft gear and this force serves to return the coupler to a partially centered position when released. These shoulders also serve to maintain longitudinal alignment of coupled diesel units in a heavy pusher service and thus reduce the lateral forces which may otherwise cause derailment.

The Type F coupler body, knuckle and lock were only designed in high tensile steel, normalized and tempered, and the yoke and front follower in Grade B steel.

As service requirements increased, higher strength materials were used. In 1973, high tensile steel was adapted as standard for the yoke and in 1976, Grade E material was also approved as standard for the coupler systems and yoke.

The shank design was modified from a pin bearing block system to a solid butt arrangement and adapted as standard in 1976. A top shelf was added to the coupler and approved as a tank car standard in 1978, Figure 16. The use of the top shelf minimizes accidental vertical coupler separation during a derailment, when a noninterlocking coupler is coupled with the F coupler.

The A.A.R. Standard E coupler and the A.A.R. Alternate Standard F Interlocking coupler for freight service and the A.A.R. Standard H Tightlock coupler for passenger service represent the result of over 60 years of concentrated study and engineering cooperation between competitive manufacturers and their customers, the railroads of America.

FIG. 13. TYPE F INTERLOCKING COUPLER – ALTERNATE STANDARD FOR FREIGHT SERVICE – 1954

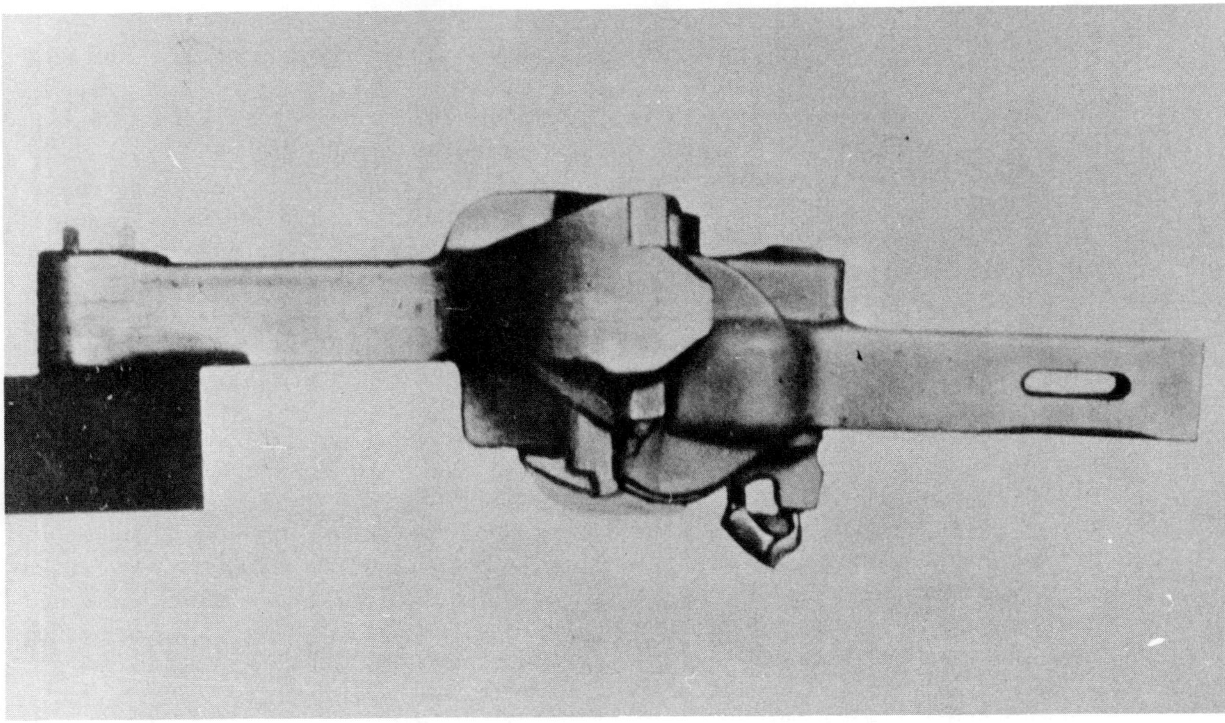

FIG. 14. TYPE F COUPLER – FUNCTION OF SAFETY SUPPORT SHELF

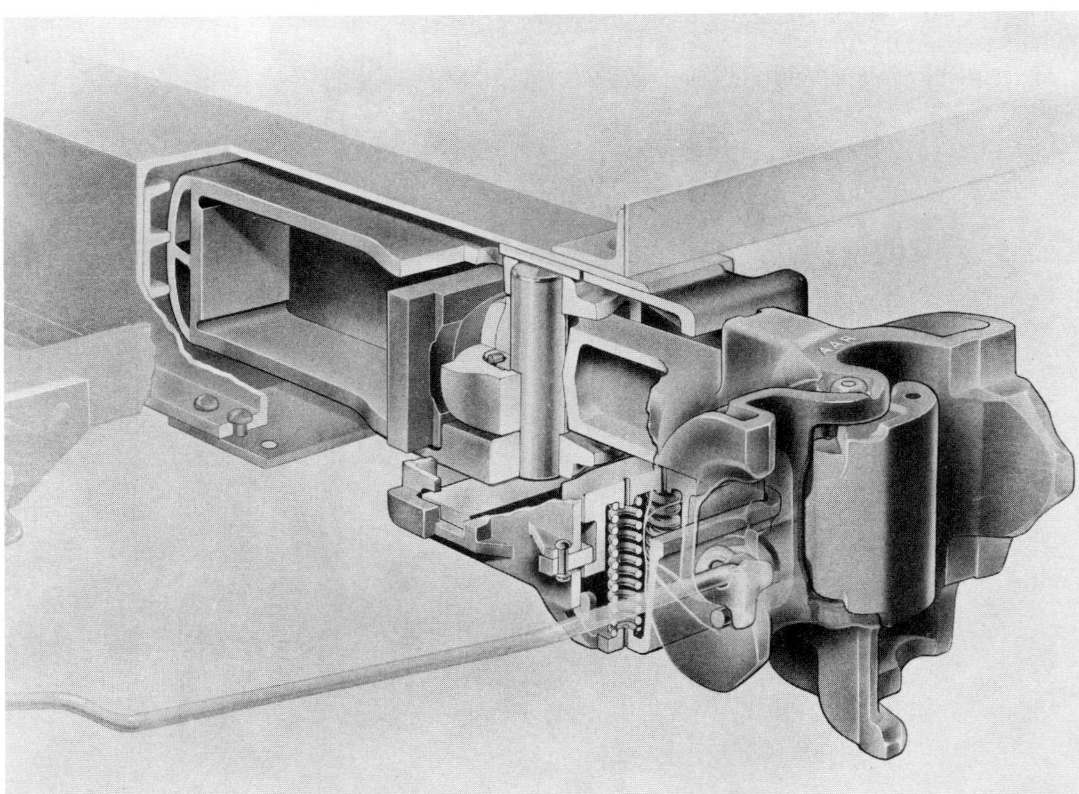

FIG. 15. TYPE F COUPLER – ASSEMBLED IN A FREIGHT CAR

FIG. 16. TYPE F COUPLER WITH TOP SHELF – AAR STANDARD – 1978

19 An Abbreviated History of Draft Gears for American Freight Cars

WILLIAM D. WALLACE

One of the first lithographs mentioning the earlier connection of railroad cars was the Mohawk and Hudson Railroad train of 1831. The locomotive was the American-built Dewitt Clinton and the cars of this train, on its first trip, were connected by short and heavy chains of three links each. Another earlier event was recorded as follows:

The train started.

It did not start all at once, but in sections. The tender was attached to the engine by a chain nearly 3 ft long made of three large links of wrought-iron; the first passenger coach was attached to the tender in a like manner; the second passenger coach was similarly connected to the first passenger coach; and so on down the train to the very last car. With the first forward jump of the engine the captain was nearly snapped backward out of his seat, but he seized a roof-support and luckily held on. The passengers in the first coach were unprepared for what was about to happen, and those sitting in the front and middle sections were projected backward and piled in a heap against the passengers at the rear. Those in the second car followed suit, and as the slack length of each chain was gathered up by the abrupt forward jerk, the succeeding group of passengers was overturned. After the whole train was straightened out, it rolled on in good order as far as smoothness of movement was concerned.

This event readily explains why one of the first subjects to be taken up for development by the Master Car Builders Association was that of suitable draft attachments for freight cars. The MCB is the direct descendent of the Mechanical Division of the Association of American Railroads.

The earlier freight cars in use in America were very small compared with today's modern cars and were constructed almost wholly of wood. Cars were attached together in a train by means of hooks and chains. There was a certain amount of slack in the fastenings between cars which provided for serial starting. Buffing action between cars was taken care of by buffing blocks attached to the ends of the sills.

These crude draft attachments served fairly well for light cars and small trains, but it soon became apparent that some form of yielding resistance must be placed in the draft apparatus to permit handling of longer and heavier trains. The link and pin drawbar with a coil spring for a draft gear became the accepted form of railway vehicle draft apparatus. The drawbar was housed between two draft sills which were bolted below the ends of the center floor sills of the car. The coil spring took care of pulling forces and, in later arrangements, assisted the buffing blocks in taking care of compression forces.

The constantly increasing size of cars and locomotives imposed more and more severe duties on the draft apparatus.

At the ninth M.C.B. convention held in 1875, the Committee on Draw Bars and Buffers reported that something would have to be done to better the situation in regard to draft attachments. They stated the following requirements which should be met by an adequate draw bar:

1. It should be a self-coupler.
2. The arrangement for buffing should be very strong and placed on the line of most effective resistance.
3. It must be able to be used with present equipment.

The air brake initially came into use on American railroads during the 1870's, and its introduction immediately created a more difficult situation for the draft rigging. The speed of handling trains was increased as a result of the adoption of the air brake and more use was made of brakes while trains were in motion. This caused more slack action, which gave the draft gears more work to handle.

In 1878 the M.C.B. Association appointed a committee to recommend a form and dimensions for a standard draw bar and draw spring, and the best method of bringing about uniformity in their length and construction. In 1879 it made recommendations for various dimensions of the draw bar and attachments. This was a link and pin bar using a draw and buffer spring of 13,000 lb capacity, 8 in. long.

The demand for an automatic coupler had become very insistent, and a committee had been appointed by the M.C.B. to study the problem of the numerous designs which were proposed by manufacturers and inventors. In 1882, the executive committee of the M.C.B. took over the task of securing a satisfactory automatic coupler. Tests were made in 1885, and twelve couplers out of fifty were selected for a service test. In 1887 it was recommended and adopted that the Janney type of coupler be chosen as the standard because it gave the smallest amount of slack in long trains. The bad effect of slack in long trains was demonstrated in the Burlington air brake tests that were made a short time before; it is thus seen that the problem of train slack and its relation to draft attachments has been recognized for at least one hundred years plus.

In 1884 the capacity of the standard M.C.B. draw spring was increased from 13,000 lb to 18,000 lb. In 1885 an M.C.B. committee was appointed to report on a standard draw gear for the nonautomatic coupler; but the report, which was ready in 1886, was withheld by the executive committee. In 1887 a report was issued which recommended the use of a gear having two springs instead of one, but the adoption of this as a standard was defeated by letter ballot.

By 1893 the M.C.B. automatic coupler had become pretty well developed, and in that year a committee recommended the adoption of various standards in the matter of attaching this coupler to freight cars and also an increase in size of the draw spring to 22,000 lb capacity, this spring to be 6¼ in. in diameter, 8 in. long and have 2⅛ in. free motion.

With the perfection of the automatic coupler, the general use of air brakes, and the introduction of steel in the construction of cars in the late 1890's, the service demanded of the draft gear was immeasurably increased. The automatic coupler permitted the use of impact coupling, which subjects cars to much greater impact forces. Air brakes increased the speed of operating trains and the slack action in the trains as longer trains proceeded over undulating terrain. Steel permitted the construction of larger cars and also removed some of the resiliency which was inherent in wooden construction and which acted to cushion energy in impacts. Increased size of locomotives made possible the operation of 4000 to 5000 ft trains. This, in conjunction with the quicker stops made possible by air brakes, caused trouble from the recoil of the plain spring draft gears in use after the train stopped and brakes were released. The situation at this time might be described by saying that general improvements in rolling stock had proceeded faster than draft gear development on the general interchange fleet. The crying need was for a gear which had more capacity and less recoil, and it was inevitable that the friction draft gear should appear in the picture.

The first patent for a device in the energy absorbing friction draft gear class was No. 391,997 for a "Buffing Apparatus" issued on October 30, 1888, to George Westinghouse, Jr. It is related that Mr. Westinghouse first conceived the need for a friction draft gear when the recoil of the buffer springs caused a most inopportune parting of an English train on which he was conducting some demonstration trials of his newly developed air brake. Some ten years previously he set to work to build a device which would eliminate a recurrence of this unfortunate incident, the primary purpose in his mind being to provide a mechanism which would absorb energy and not give it back in recoil as the plain spring did. The severe slack action experienced and recorded on the 50-car trains of the M.C.B. Burlington Brake trails convinced Westinghouse a practical friction draft gear was essential for freight cars.

Apparently one of the first commercial installations of friction draft gears in America was an installation of Westinghouse friction draft gears on fifty Pennsylvania Railroad coke cars in the year 1896. In the spring of 1897 an impact demonstration with some of these cars was made for the benefit of the M.C.B. committee on improved buffers. This demonstration was supplemented by a shop test of some of the gears in which a 16,000 lb weight was used. This test definitely proved that the Westinghouse friction draft gear provided better cushioning protection for cars in impact.

In 1903 the committee reported that so many new designs of draft gear were coming out that it was impossible to secure reliable data on the service performance of any of them. In 1904 they reported that they were unable to specify a standard draft gear, but they did recommend the adoption of various standards for draft gear attachments.

In the period from 1902 to 1908, the friction draft gear seems to have become fairly well established in superseding the spring gear.

The Southern Pacific Railroad in 1908 made a series of train tests of draft gears in conjunction with 80 car train tests of the type "K" new triple valve. These tests were of friction draft gears in comparison with spring draft gears and showed that from a train-handling standpoint the friction gears were superior on account of the absence of recoil. Following the recognition of the possibilities of the friction draft gear, development was so rapid and the need for an improved product so great that little attention was paid to some of the finer problems involved. Several of the manufacturers had provided themselves with laboratories and testing facilities for testing friction gears, but much of the work done in these laboratories consisted of sales demonstrations in which the scientific side of the problem was slighted for the spectacular. Out of this chaotic situation order was bound to be restored, and the Inspection and Test Section undertook the task. It was assisted in its work by the Coupler and Draft Gear Committee of the Mechanical Division of the American Railway Association.

The program plan included laboratory tests, service tests, and road tests; but unfortunately only the laboratory tests were completed. These consisted of static tests, 9000 lb drop hammer tests, and tests in the car impact plant. Even though the program of tests was not completed, much good was accomplished because weaknesses which existed in certain methods of testing were brought out and standard methods of conducting laboratory tests were developed. The car impact tests cleared up much of the misunderstanding of just what a draft gear does in cushioning two cars when they collide.

With the cessation of activities of the World War I United States Railroad Administration (1917-1920), the Coupler and Draft Gear Committee of the A.R.A. continued its efforts to improve the draft gear specification and qualification situation. A 1919 proposal for a new pendulum type of laboratory testing machine design did not find much favor. The U.S.R.A. tests had shown that the drop hammer gave a good laboratory test of the buffing characteristics of a draft gear, and the A.R.A. committee, with the cooperation of those manufacturers who had built drop hammers, worked out the design for a hammer to be equipped with both a 9000 lb and a 27,000 lb tup (hammer weight) arranged so that they could be easily interchanged. This machine was erected in a special building at Purdue University in 1927 by the American Railway Association, where tests of the "new" AB type long train freight brake were being conducted in the Air Brake Laboratory.

By 1925 so many changes had been in draft gears that the data secured in the U.S.R.A. tests were no longer suitable for gaging their performance. In that year four railroad companies, the New York Central, the Santa Fe, The Nickel Plate, and the Chicago, Milwaukee and St. Paul jointly undertook to make a series of comparative laboratory tests of the gears then on the market with the object of securing information to guide draft gear purchases. These railroads made tests in the laboratories of several different draft gear manufacturers. For the first time a laboratory endurance test was included in a program of comparative tests. There was some question on the part of manufacturers and users of draft gears about the possibilities of measuring service life by a laboratory wear test, but the successful completion of this test program served to remove much of the doubt. Even though there was no method of converting performance in the laboratory endurance test into terms of service life, it could not be denied that all gears had been subjected to the same treatment and that their relative performance had been measured thereby. The making of a laboratory endurance test of draft gears proved to require considerable time and care, but the end attained justified the means because it was vitally necessary to know how long a draft gear would retain its protective value after it was placed on a car.

The American Railway Association constructed its draft gear testing laboratory at Purdue University for the purpose of developing draft gear specifications, the first step in the program being the making of a complete series of tests of modern friction draft gears. Each manufacturer was invited to enter one type of freight gear in the tests, which were started in July, 1927, ten manufacturers accepting the invitation.

Ten gears of each type were secured on the open market, all of them being first subjected to a capacity test and then half of them being given a sturdiness test and the other half an endurance test. This was the first series of draft gear tests made under a 27,000 lb tup; in order to be able to make some comparison with previous test results, two gears of each type were given the capacity test with the 9000 lb tup. The heavier 27,000 lb tup proved to be much superior for testing modern draft gears, and many refinements of measuring draft gear performance under the drop hammer were made possible by its use.

Following completion of this series of tests early in 1929, a report was made at the 1929 convention of the American Railway Association which included recommended specifications for freight-service draft gears. The endurance test had proved to be a vital part of any laboratory determination of the service worth of a draft gear. However, it was impossible to see how an endurance test could be included in any routine acceptance specification test because of the length of time required to make a satisfactory endurance test. Thus, it was proposed to have each manufacturer of draft gears submit his gear for an approval test at the Purdue laboratory of the A.A.R. and, if the gears passed this test, to issue a certificate of approval.

The specifications were finally adopted by the Association in 1931, and later an interchange rule was made requiring certified gears on all cars built new on or after January 1, 1934. Forty-five years later, all cars of the then 1,930,000 North American interchange fleet were equipped with A.A.R. certified energy absorbing draft gears.

The draft gear specifications adopted by the American Railway Association in 1931 contain requirements for capacity, endurance, sturdiness, uniformity of action, freedom from sticking, and dimensions, including travel. The fundamentals of construction of standard draft rigging of today are not materially different from what they were in 1870. These include a draw bar on each end of the car, connected to the car body through an elastic medium, which is the draft gear. The spacing of center sills has been standardized at 12⅞ in. since 1905, and the length of the draft gear pocket on practically all standard modern cars is 24⅝ in.

It is a far cry from the old spring draft gear with its low capacity and high recoil to the modern freight car draft gear with characteristics which provided satisfactory shock protection for car bodies and lading. Modern draft gears of today employ heat-treated alloy steels in substantial amounts for materials of construction, and at present it is not apparent where much improvement can be looked for through the use of better steels.

With today's 6½ in. combined maximum draft and buff travel and assuming a maximum buff travel of 3¼ in. and a reaction force not exceeding 500,000 lb, which is considered a safe value for modern car construction, it is theoretically possible to secure in a square wave work curve of 135,000 ft-lb as a limit or 100 percent efficiency. Such a cushioning performance is entirely too stiff, so a more realistic cushioning figure of 115,000 ft-lb or 85 percent efficiency as a limit should be strived for in 3¼ in. maximum travel.

Cushioning capacity alone is not enough; a draft gear must also have smooth action and proper release action, and it should have a reasonable life in order to afford lasting car protection.

As already mentioned, the A.A.R. Coupler and Draft Gear Committee tested draft gears for the first time in 1927. This was done to standardize the size and performance of draft gears and eliminate those devices which did not meet the standards agreed upon. Finally, in 1931 the first draft gear specification was made mandatory and these specifications were revised to the M-901 specification in 1949 which further refined requirements. Essentially, this called for draft gears to have a capacity of 18,000 ft-lb, and to meet a rigid endurance and sturdiness test.

Up until this time, with few exceptions, friction draft gears dominated the scene. Shortly after the finalization of the M-901 specification in 1949, along came a tremendous flurry of activity pertaining to all rubber cushion type draft gears. These earliest all rubber freight gears were modifications of the then used and accepted all rubber passenger gears. It wasn't long before some draft gear design engineers were questioning the wisdom of certifying these all rubber draft gears with certain waivers of basically important requirements under a specification designed primarily to test friction draft gears, so it was inevitable that a special specification was drawn up and adopted in 1953 to cover these all rubber gear types and it was entitled Spec. M-901-A.

As mentioned, there was tremendous activity by all draft gear manufacturers which resulted in a total of 13 24⅝ in. pocket all-rubber gears which were certified within 15 years.

In the early 1950's the A.A.R. Coupler and Draft Gear Committee, consisting of railway mechanical men, confirmed the opinion of some railroad mechanical men that contemporary gears were possibly inadequate to protect cars and lading from damage under then existing "high speeds" and heavier trains. In 1956 the committee adopted Specification M-901C which covered a new draft gear pocket size of 36 in. and a new draft gear travel of 4½ in. The committee recognized that high capacity alone is meaningless unless it is obtained at a reasonable reaction force, and for the first time the reaction force level (sill pressure) was made a part of this specification. The M-901C specification for a 36-in. draft gear pocket required a gear having a capacity of 75,000 ft-lb at a reaction force level of 500,000 lb.

The draft gear manufacturers tried to certify draft gears under this high capacity specification without success. They petitioned the A.A.R. Committee and Coupler and Draft Gear's to lower the capacity requirement from 75,000 to 60,000 ft-lb. This capacity reduction was eventually granted and remains the same.

The year 1959 was an active one for the A.A.R. Coupler & Draft Gear Committee in that they adopted Specification M-901-F which was entitled: "Specifications -M-901F- Requirements of Instrumentation to be used in Measuring and Recording Draft Gear Reaction Forces. Then came the adoption of Spec. M-901E in 1959. Specification M-901E required 36,000 ft-lb of capacity at a reaction force level of 500,000 lb.

At the time M-901-C and M-901-E specifications were written, none of the draft gears being sold could comply with these two new specifications. The draft gear manufacturers, however, quickly made every effort to design and build new draft gears to meet the new specifications.

This activity also caused the upgrading of the old Spec. M-901 and the Committee in 1959 also adopted Spec. M-901-D, which called for a minimum capacity of 18,000 ft-lb at a reaction force of 500,000 lb.

From the year 1959 on there was considerable debate in the industry concerning an upgrading of standard and alternate standard draft gears and their testing and ultimate certification.

In keeping with the desire of the A.A.R. Committee on Couplers and Draft Gears to make changes in the best interest of the industry as a whole and in consultation with the Draft Gear Manufacturers' Engineering Committee, it was not surprising that in 1964 Spec. M-901G was adopted. This specification was unique in that it was the first time standard pocket draft gears were to be rated by freight car impact testing. To pass this test, a standard 24⅝ in. pocket draft gear had to have a minimum impact rating velocity of 5 mph at a reaction force of 500,000 lb under certain test requirements. It should be noted that this impact rating requirement did not eliminate the very necessary drop hammer endurance and sturdiness test required by all the prior specifications.

Two manufacturers quickly certified draft gears under Spec. M-901-G and lobbied for a similar specification pertaining to the A.A.R. Alternate Standard Pocket of 36 in. The Committee quickly obliged, and in 1966 adopted Spec. M-901-H which required a minimum impact rating velocity of 6 mph.

In the evolution of the draft gear, manufacturers have built their own laboratories, developing various means of testing to determine qualities of capacity, endurance, sturdiness, recoil, and reaction, the prime requisites of an effective friction draft gear. Making full use of new developments in materials and methods, exploring every avenue of technology, the manufacture of modern draft gears is a science of its own.

Probably the last few years have seen a more intensive effort to develop new and more effective draft gears than any other period in railroad history. At the present time, a number of new developments are in progress using combinations of rubber and friction, hydraulics and friction, and hydraulics alone. All of these developments have contributed to safety and efficiency of operation as well as to large savings in car repairs and damage claims.

A great deal of progress has been made in the past in the design and development of improved draft gears, and this progress will no doubt make giant strides in the future. Certainly railroading can point with pride to the fact that because George Westinghouse gave the industry the friction draft gear and the air brake, fountainheads of its growth and development with safety, the Hall of Fame became his home.

Few devices have been the object of so much inventive effort as the freight car draft gear, and the number of U.S. patents on these devices is legion (over 21,000 patents up to December 1, 1978). Most of these developments have occurred since ASME was founded in 1880.

REFERENCES

[1] Gray, W. E., and Messersmith, C. W., *Development of Draft Gears for American Freight Cars,* 1936.

[2] Dunbar, S., *History of Travel in America,* 1937.

[3] Wallace, W. D., "Evaluation of Railway Draft Gears," A.S.M.E. Railroad Conference, Chicago, Illinois, April 25-26, 1957 – Paper No. 57-RR-8.

[4] Cardwell, Lloyd, "Finding Facts . . . About Draft Gears," *Progressive Railroading,* May-June, 1963.

358 *Draw-gear.*

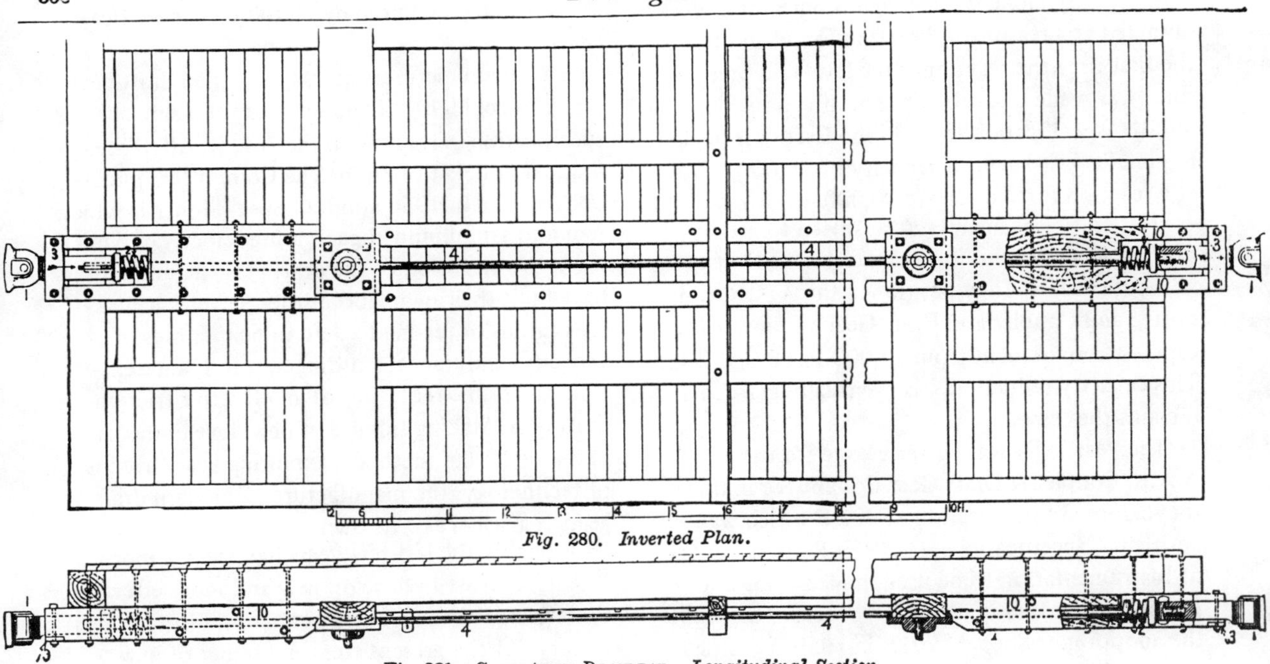

Fig. 280. Inverted Plan.

Fig. 281. CONTINUOUS DRAW-BAR. *Longitudinal Section.*
1. *Draw-bar.* 2. *Draw-spring.* 3. *Draw-bar Carry-iron.* 4. *Draw-rod.* 10. *Draw-timber.*

FIG. 1 CONTINUOUS DRAW-BAR WITH SIMPLE DRAFT SPRING ADJACENT TO EACH COUPLER – 1870's-80's

For list of names of the parts designated by the numbers in the engravings, see page 348.

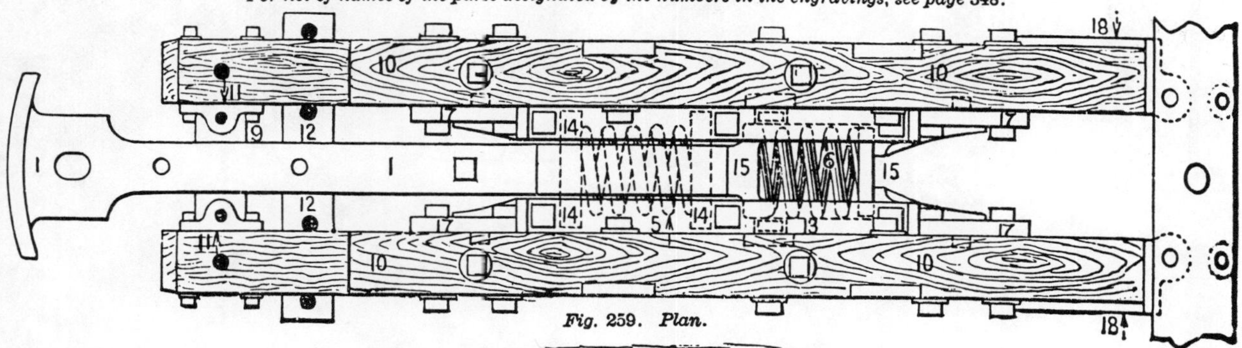

Fig. 259. Plan.

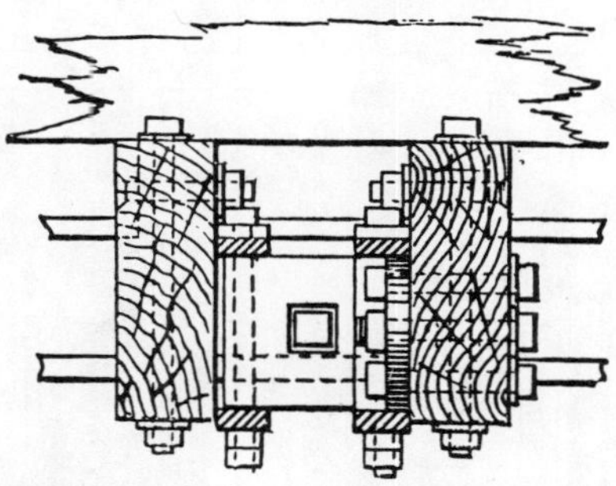

Fig. 260. Side View, with Draw bar Removed. DRAW-GEAR FOR FREIGHT-CARS, NEW YORK CENTRAL RAILROAD.

FIG. 2 DRAW-BAR WITH DRAFT AND BUFF SPRINGS

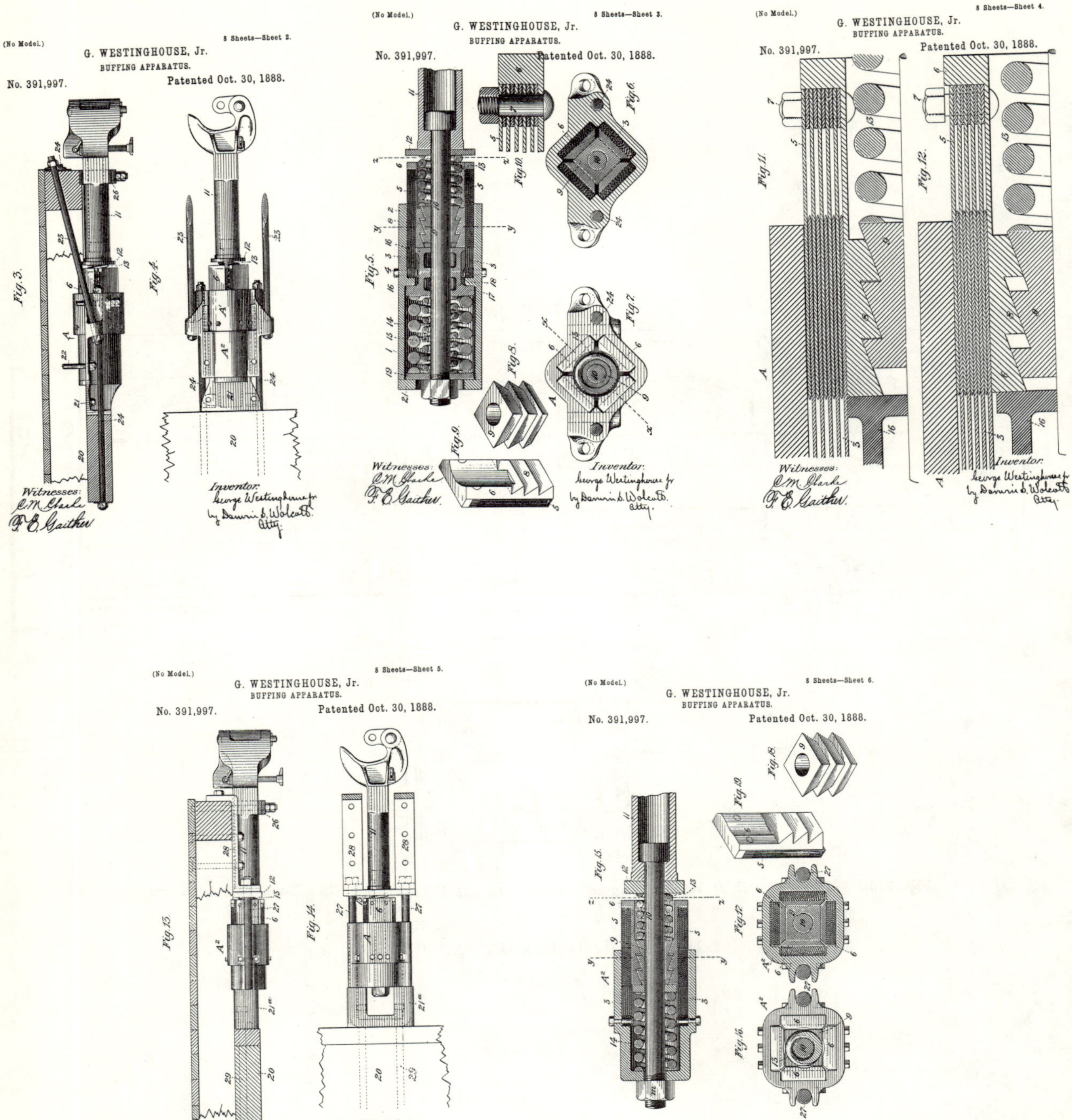

FIG. 3 GEORGE WESTINGHOUSE's ORIGINAL PATENT

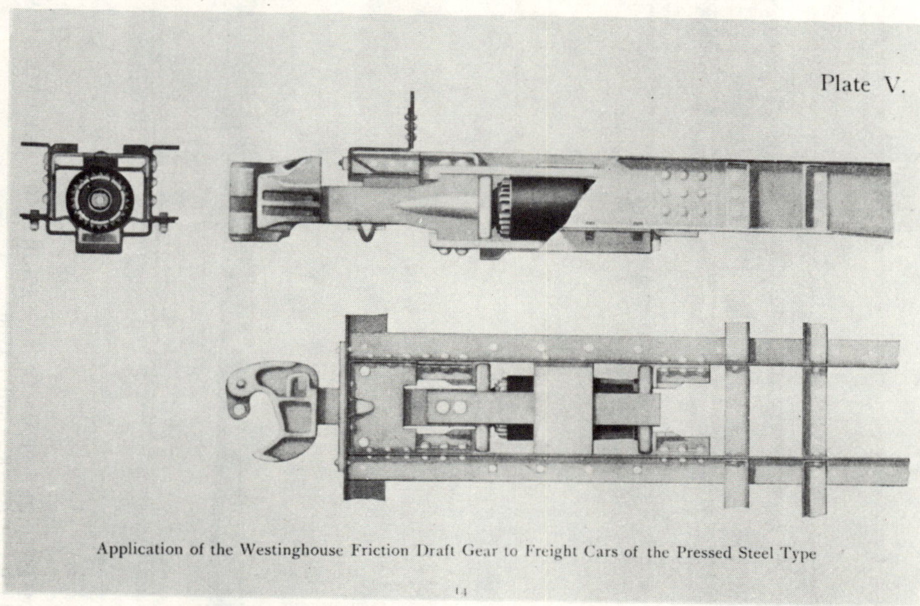

Application of the Westinghouse Friction Draft Gear to Freight Cars of the Pressed Steel Type

FIG. 4 APPLICATION OF WESTINGHOUSE FRICTION DRAFT GEAR TO STEEL FREIGHT CARS – EARLY 1900's

FIG. 5 FIRST COMPLETE ORE TRAIN ON BESSEMER AND LAKE ERIE RR WITH FRICTION DRAFT GEAR, ABOUT 1900

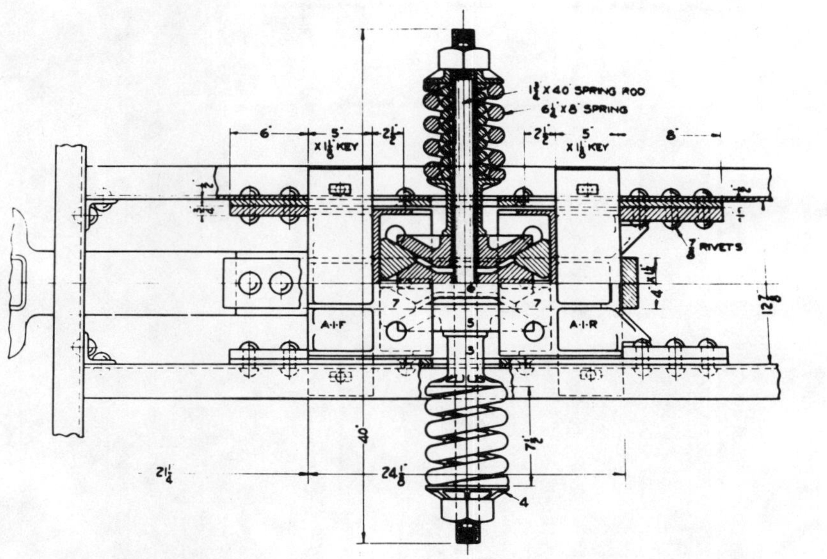

FIG. 6 CARDWELL FRICTION DRAFT GEAR, EARLY 1900's

FIG. 7 MONON ROUTE 40 TON BOX CAR EQUIPPED WITH CARDWELL GEAR, 1914

FIG. 8 PURDUE UNIVERSITY DRAFT GEAR LABORATORY MID 1920's

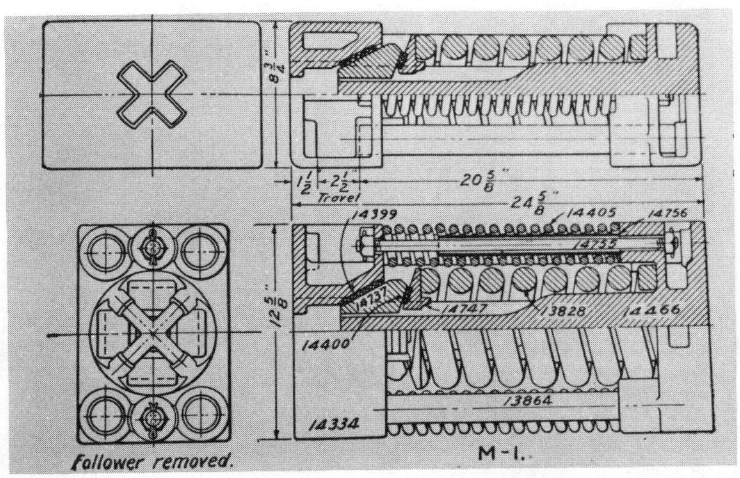

FIG. 9 NATIONAL M-1 TRIPLE SPRING FRICTION GEAR

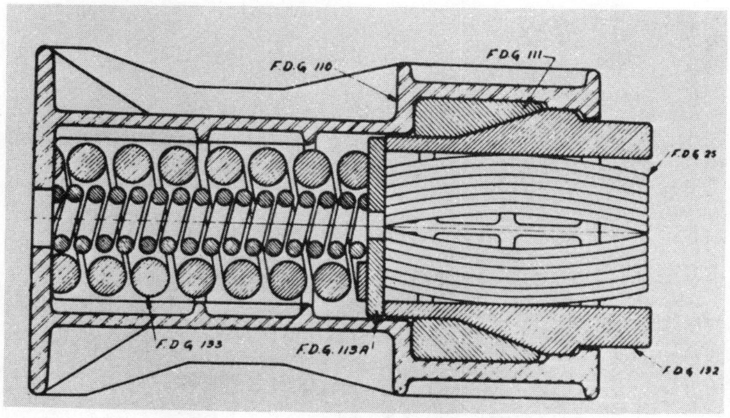

FIG. 10 GOULD TYPE 110 DRAFT GEAR

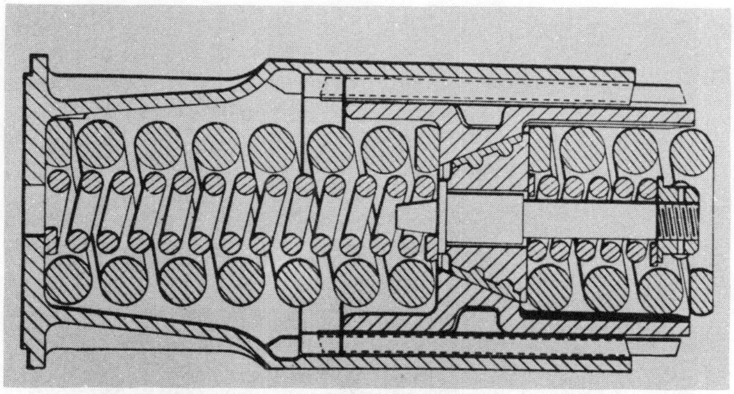

FIG. 11 WESTINGHOUSE D-3 DRAFT GEAR
1910 – 1920 PERIOD

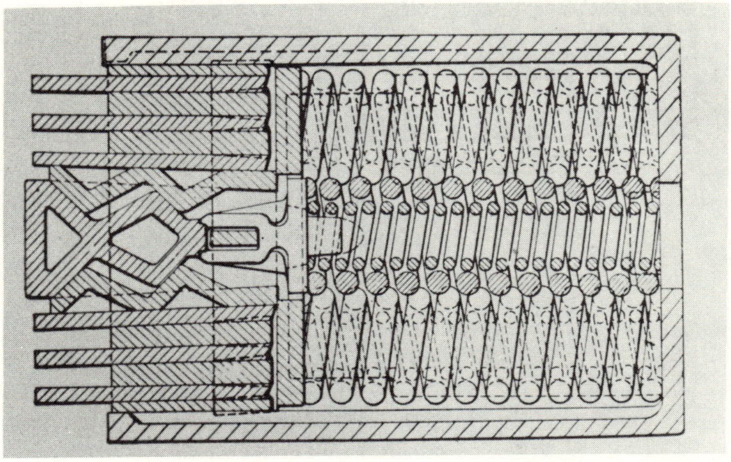

FIG. 12 WESTINGHOUSE NA-1 DRAFT GEAR, LATE 1920 PERIOD

FIG. 13 MINER A-19-X DRAFT GEAR

FIG. 14 MINER H-2 DRAFT GEAR

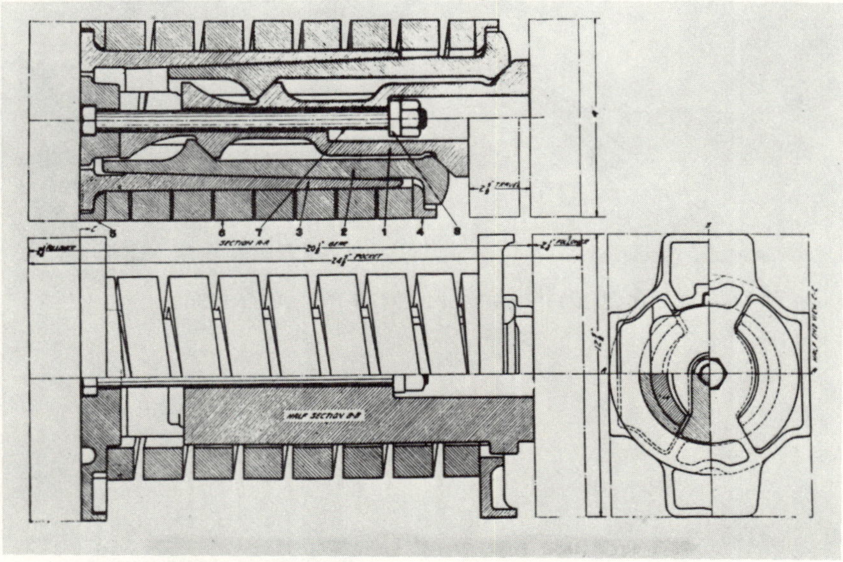

FIG. 15 CARDWELL SPRING BARREL GEAR
TYPE F-3

FIG. 16 SOUTHERN RY FREIGHT IN EARLY 1950's

FIG. 17 R-500 RUBBER FRICTION DRAFT GEAR

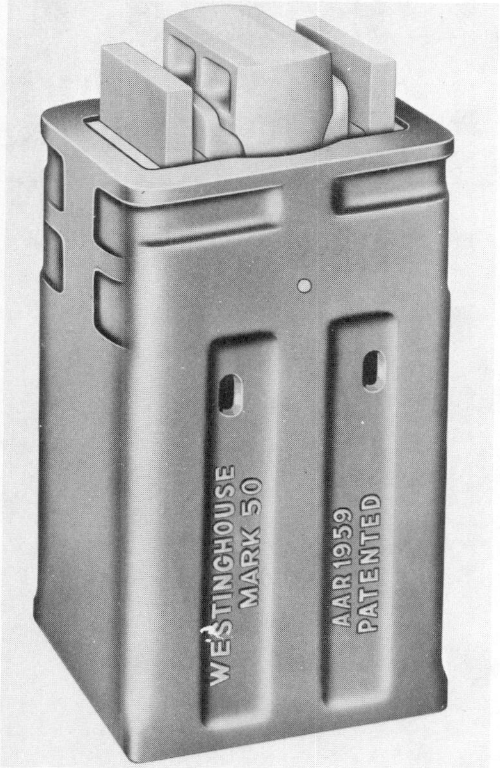

FIG. 18 MARK-50 FRICTION GEAR

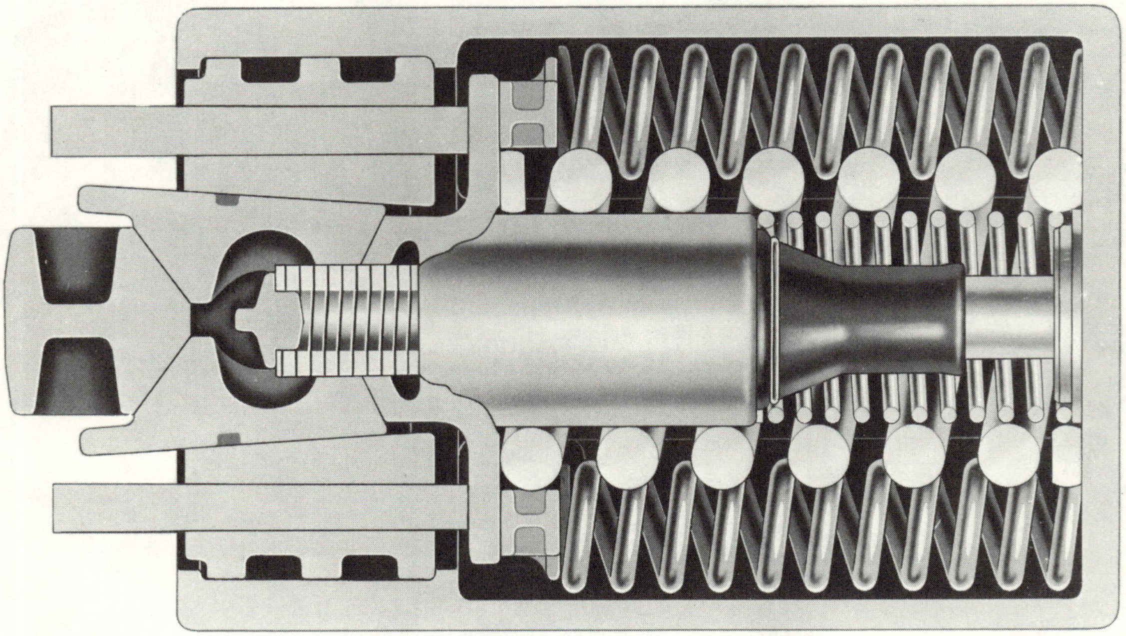

FIG. 19 H-60 FRICTION-HYDRAULIC GEAR
FREE OR RELEASE POSITION

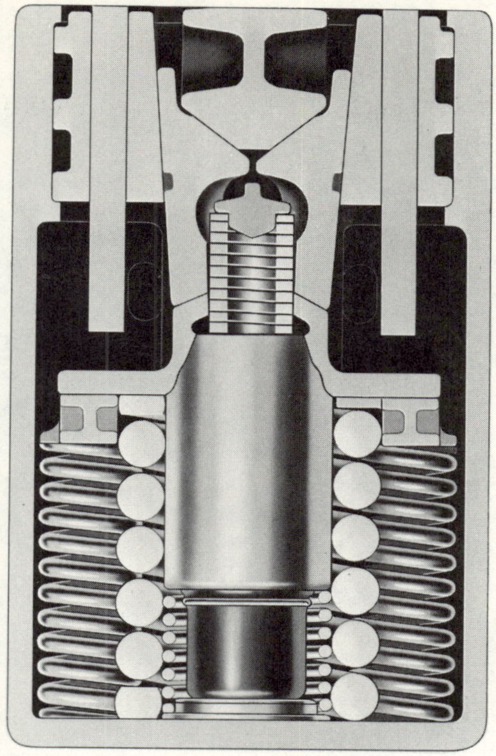

FIG. 20 H-60 FRICTION-HYDRAULIC GEAR
COMPRESSED POSITION

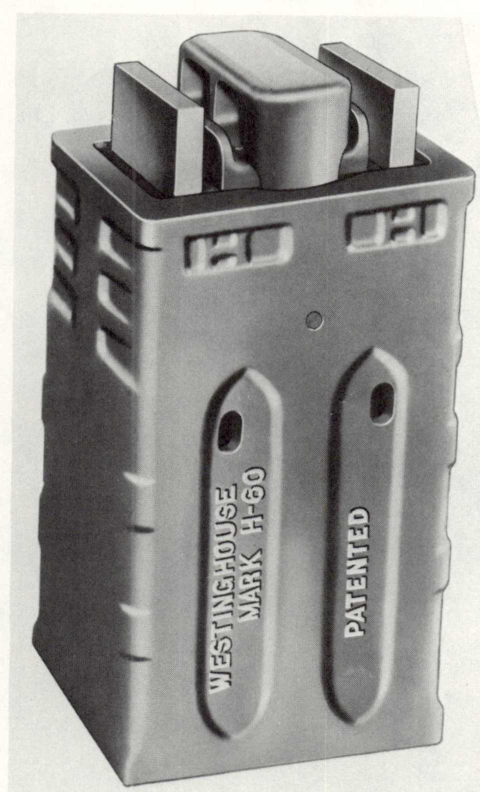

FIG. 21 MARK H-60

20 100 Years of Progress in Railway Mechanical Engineering-Special Freight Car Cushioning

V. TERREY HAWTHORNE

Special freight car cushioning was developed to reduce lading and car damage caused by overspeed impacting in the classification yards and severe longitudinal train action. By increasing the travel and space available, the designer was able to apply equipment to the freight car, which had significantly increased energy absorbing capabilities. Thus, the purpose of a special cushioning system is to absorb more of the energy received by the coupler than a draft gear cushioning system.

Two types of special cushioning systems have evolved into acceptance in the modern freight car; e. g., the sliding center sill system (Fig. 1), and the end-of-car system (Fig. 2). The sliding center sill is a continuous beam through the length of the car with a coupler mounted at each end of this beam. A draft gear cushioning or a buff only cushioning system is mounted behind the coupler to absorb energy when the car is in a "squeeze" mode between other cars in the impact because the sliding sill system does not offer adequate protection to the car structure under this condition. In the case of end-of-car, one unit is mounted at each end of the car with an interfacting portion to mate with the coupler. As a general rule, the sliding center sill system is used in shorter freight cars and the end-of-car system is used in long freight cars where the continuous beam is impractical.

The predecessor of modern sliding center sill cushioning systems was the Duryea cushioned underframe (Fig. 3), which was first built in 1927. By the end of 1929, there were approximately 8600 carsets of the Duryea system in service. By the end of 1937, this had increased to approximately 24,000 carsets. This increased to approximately 48,000 carsets before the modern concept of hydraulic cushioning systems was adopted which provided greater protection than the 7 mph limit for the Duryea arrangement. A modern version of the Duryea concept is still popular in cabooses developed by Waugh Equipment Company and now manufactured by Dresser, T. E. D.

Using an improved version of the Duryea concept of sliding center sill in the early 1950's, the Southern Pacific Railroad, in conjunction with Stanford Research Institute, developed a higher capacity cushioning system, which became known as HYDRA-CUSHION (Fig. 1). This design consisted of friction plates mounted within the sliding center sill and these plates were loaded by a vertically mounted hydraulic unit attached to the fixed underframe. In 1964, Messrs. W. E. Thomford of the Southern Pacific and W. K. MacCurdy of Stanford were awarded the Franklin Institute Design and Development Award for their work. Waugh Equipment Company and Evans Products, Inc. developed the HYDRA-CUSHION jointly under license from the Southern Pacific. Dresser Industries, Inc. now owns HYDRA-CUSHION completely since taking Waugh Equipment Company in a merger.

In the mid-to-late 1950's, several companies began to develop hydraulic units to fit within the sliding center sill. Thus, the development of hydraulic sliding center sill cushioning was well underway. The Santa Fe Railroad developed a hydraulic unit, which had an auxiliary reservoir and orifice

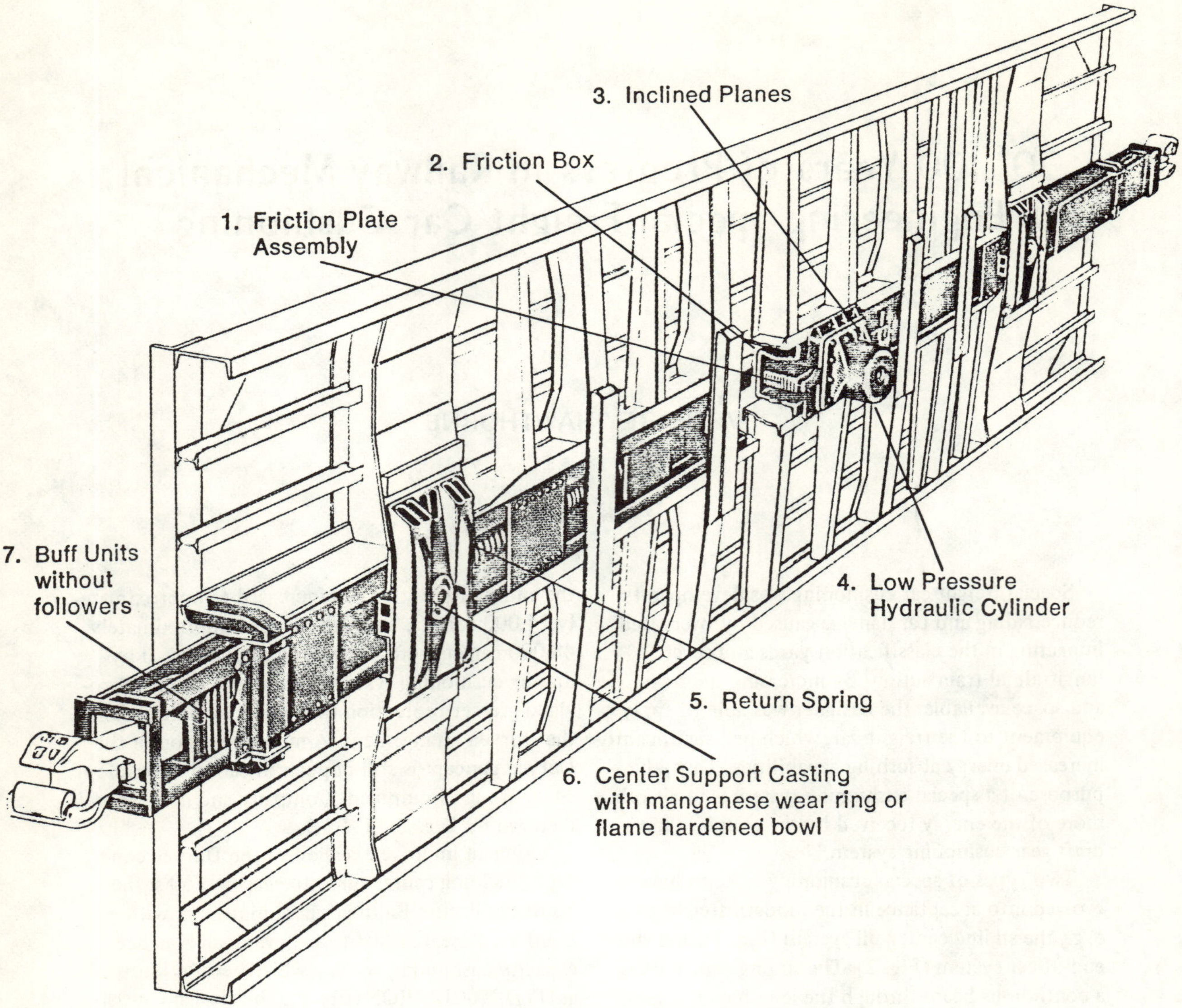

FIG. 1 SLIDING CENTER SILL CUSHION SYSTEM-HYDRA-CUSHION

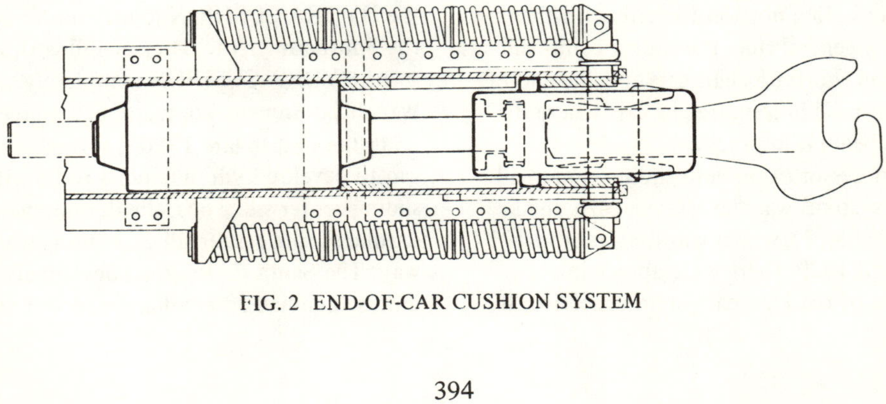

FIG. 2 END-OF-CAR CUSHION SYSTEM

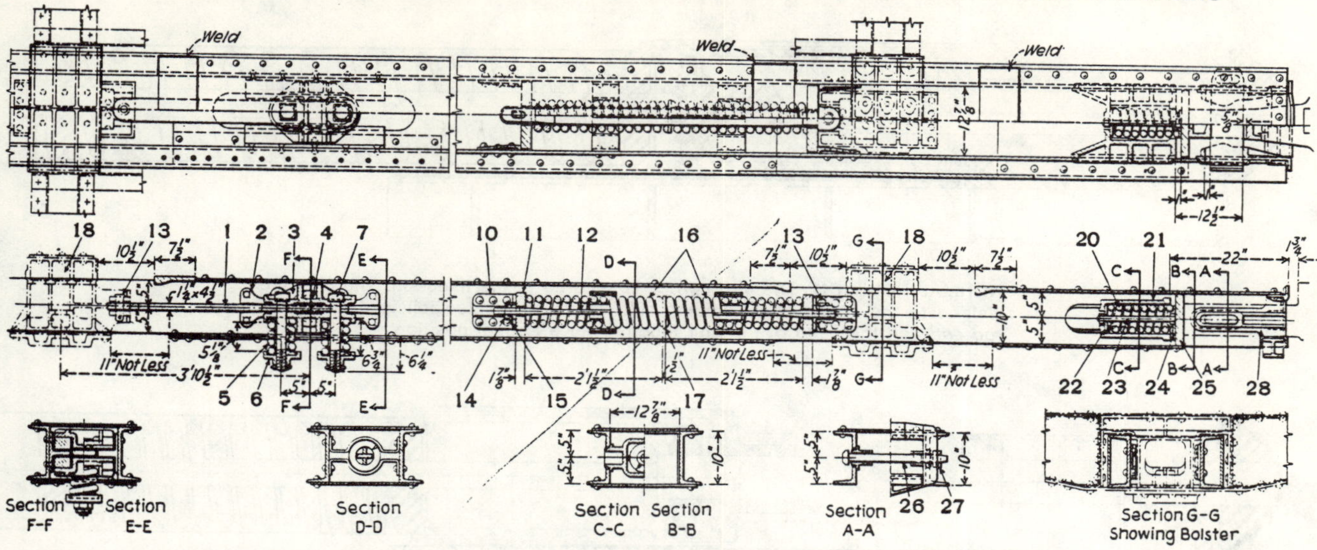

Fig. 573—Duryea Coupler Cushion and Snubbing Gear Applied to Caboose Car with Duryea Cushioned Underframe.

Snubbing Gear

1. *Snubbing Gear Link*
2. *Snubbing Gear Cam*
3. *Snubbing Gear Housing*
4. *Snubbing Gear Shoe*
5. *Snubbing Gear Spring*
6. *Snubbing Gear Spring Seat*
7. *Snubbing Gear Spring Bolt*

Cushion Gear

10. *Cushion Gear Follower Lug*
11. *Cushion Gear Follower Plate*
12. *Cushion Gear Link*
13. *Link Pins*
14. *Cushion Gear Link Key*
15. *Cushion Gear Link Key Lock*
16. *Cushion Gear Spring*
17. *Cushion Gear Spring Spacer*
18. *Body Bolster Center Brace*

Coupler Gear

20. *Coupler Gear Spring*
21. *Coupler Gear Spring Housing*
22. *Coupler Gear Spring Washer*
23. *Coupler Gear Spring Bolt*
24. *Coupler Gear Spring Cap*
25. *Coupler Gear Follower Plate*
26. *Draft Key*
27. *Draft Key Retainer*
28. *Coupler Carrier Casting*

FIG. 3 DURYEA SLIDING CENTER SILL SYSTEM

holes drilled in an internal high pressure cylinder for metering the fluid. This design eventually became the Keystone Railway Equipment Company "Double Acting" unit. Keystone Railway Equipment Company later developed the tension single-acting unit to simplify the application to the car (Fig. 4). Buckeye Steel Castings developed double-acting units with a unique metering concept, which consisted of a uniformly varying cylinder diameter. Pullman Standard developed a design which was single-acting and employed a self-contained return spring with a rubber boot accumulator mounted within the spring. Aircraft landing gear technology was transferred in the use of the Bendix landing gear by American Car and Foundry in their development of the FREIGHT-SAVER hydraulic unit (Fig. 5). This unit introduced a pneumatic recentering system in lieu of a heavy return spring.

Generally, the sliding center sill units are applied to the car by two different methods. The first is to use piston stops mounted to the fixed underframe at the end of the unit with mating stops mounted in the sliding center sill. Another approach is the use of horizontal keys (Fig. 6), which are held by the fixed underframe and have mating slots and stops in the sliding sill.

Paralleling the development of the sliding center sill units, William T. Blake developed an end-of-car device concept in his home workshop during the mid 1950's. This concept (Fig. 7) consisted of a large hydraulic unit with orifice hole metering and means for supporting the coupler. Eventually, Wellex Jet Services, Inc. hired Mr. Blake to continue his development. This work continued and eventually Hallibruton Company purchased Wellex and established the Freight Master Division for the purpose of producing railroad end-of-car devices. Since the end-of-car concept has a very practical use in the longer freight car designs, several other companies, which were in the draft gear or sliding center sill cushion field, developed end-of-car devices. The ACF, Keystone and Pullman Standard end-of-car designs used technology, which was developed for the sliding center sill units. Miner Enterprises used their draft gear technology to assist them in the design of an end-of-car device.

In the early 1970's, the Cushion Unit Manufacturer's Engineering Committee was formed by the AAR to establish a universal pocket for end-of-car and sliding sill devices. This effort resulted in the universal pocket arrangement published by the AAR in 1975.

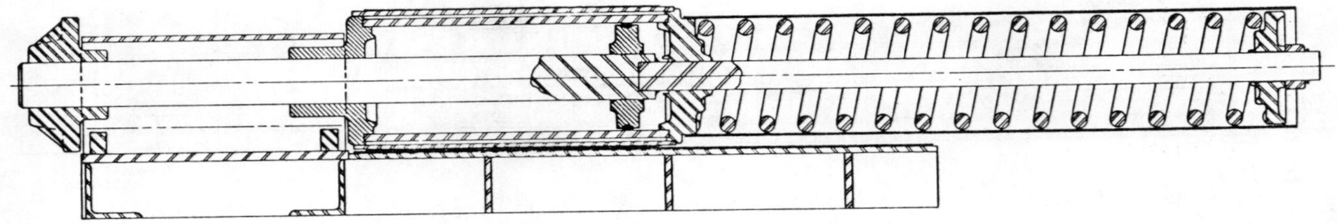

NORMAL POSITION: The *unit carrier* is secured to the fixed sill. The *lugs* are welded to the inside of the sliding sill. Return spring tension is drawing the *rod end* and the *head* of the unit together for a normal position.

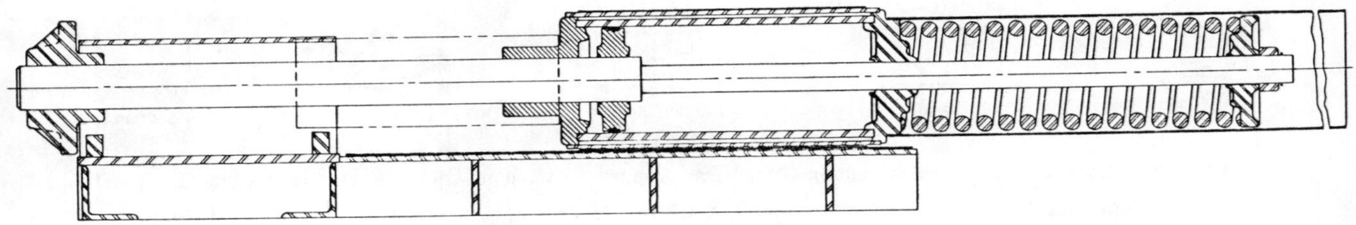

CUSHIONING TRAVEL OF 20″—An impact directed into the sliding sill from the left is forcing the head of the unit to withdraw from the fixed rod end. This impact is absorbed by a capacity of 20″ of cushioned travel distance. After impact, the return spring returns the hydraulic unit to its normal position.

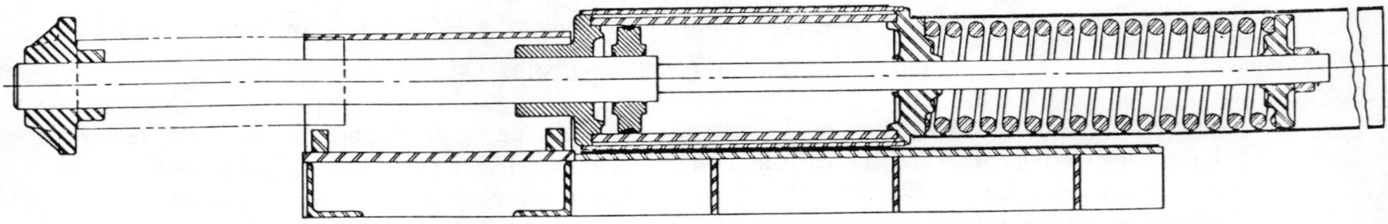

CUSHIONING TRAVEL OF 20″ FROM OPPOSITE DIRECTION—In this case, the impact is directed at the opposite coupler. The lugs on the sliding sill are now forcing the rod end to withdraw from the unit. The piston is traveling in the same direction within the cylinder to provide the 20″ capacity of cushioned travel.

FIG. 4 KEYSTONE SINGLE ACTING UNIT

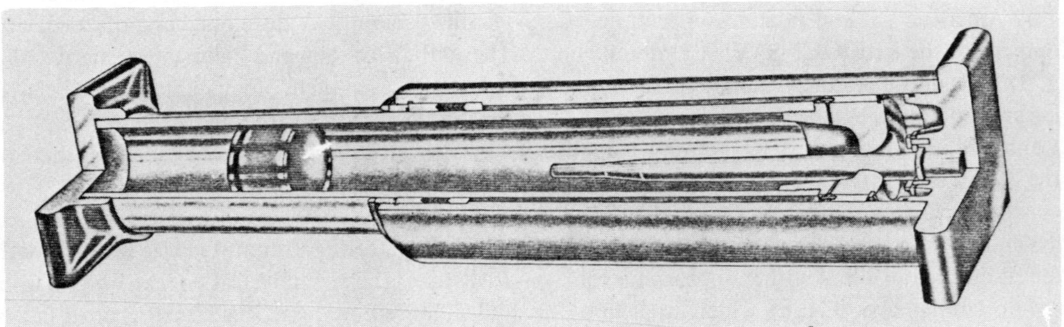

FIG. 5 ACF FREIGHT-SAVER UNIT

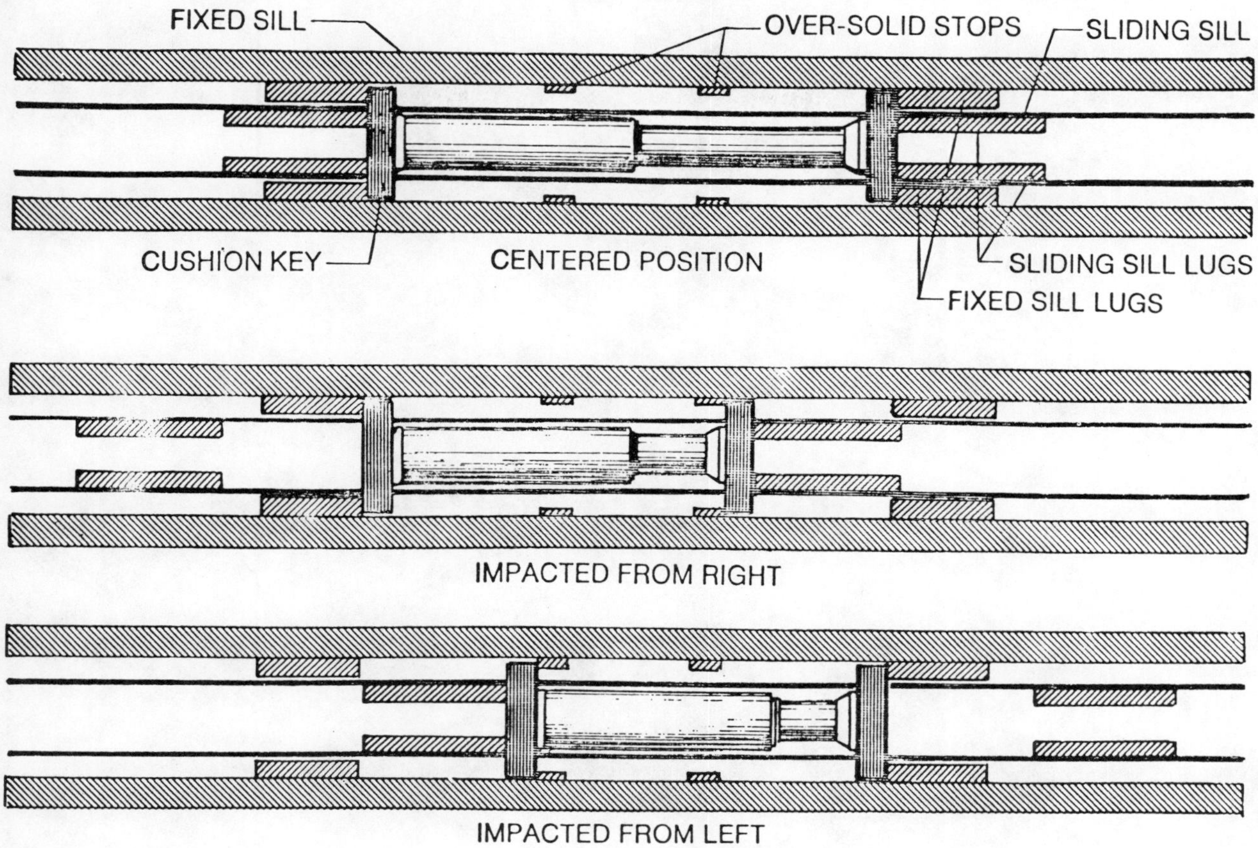

FIG. 6 KEY APPLICATION OF SLIDING CENTER SILL UNIT

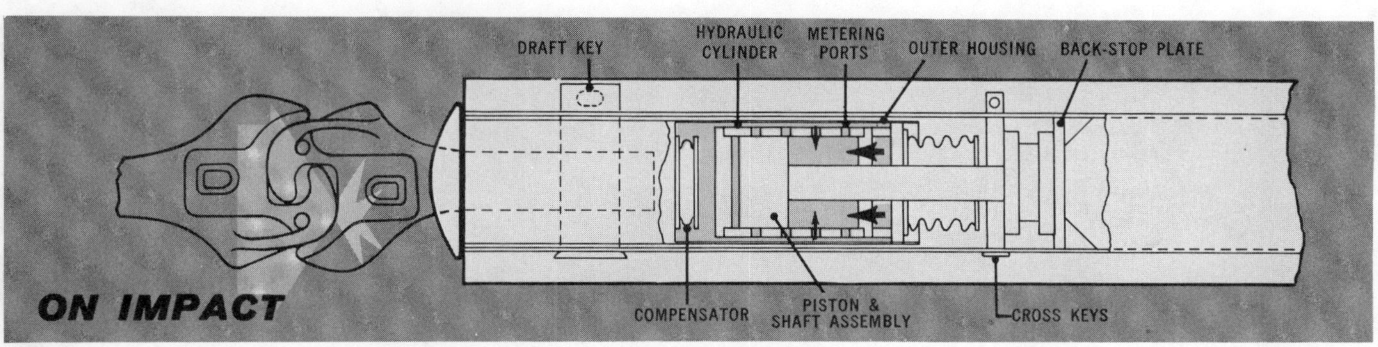

FIG. 7 EARLY CONCEPT OF FREIGHT-MASTER UNIT

21 Development of Brake Shoes: Part I—Cast Metal

GRANT ARRASMITH

For the first two decades of steam railroads, wood was universally used for brake shoes. Hard wood, properly arranged in the brake block, was efficient and durable when the speed and weight of the train were limited and the brake pressure required was low and of short duration. However, when the friction was greatly increased there was danger of setting fire to the brake shoe and also the cars.

Stephen Morse, in his patent of 1853, explained the danger of the heat setting fire to the wooden block, and emphasized the advantages of an iron brake shoe. He even gave his brake shoe a large thin web at the back with openings for air circulation to cool the shoe.

F. W. Sargent stated in the Railroad Gazette, May 5, 1905, that in the search for a satisfactory

Transition from wood to a combination of wood and metal brake shoes

FIG. 1

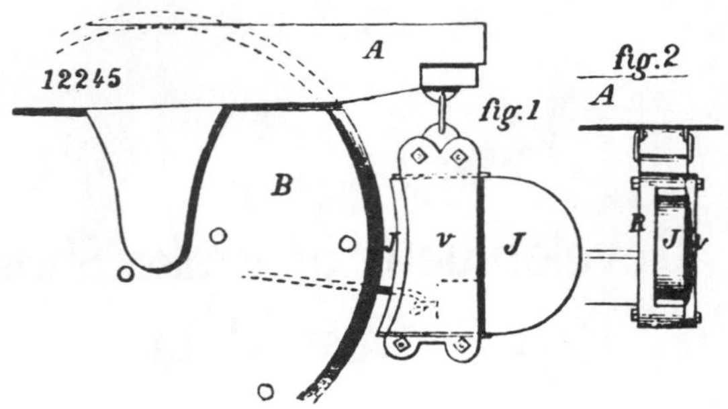

FIG. 2

Common cast iron and Congdon brake shoes, 1870 and 1876

FIG. 3

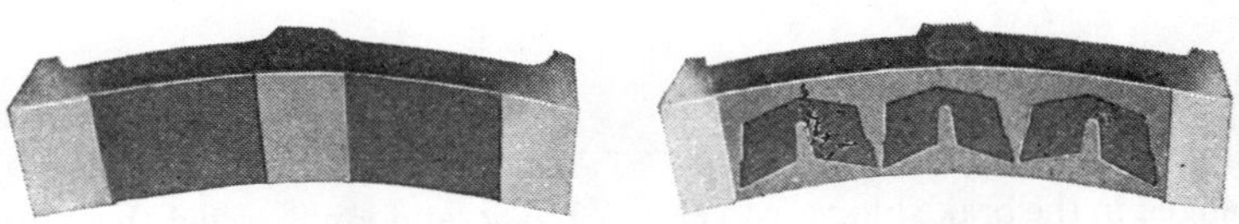

Lappin and Corning brake shoes, 1884 and 1896

FIG. 4

incombustible brake shoe one road even tried stone, but it was found that this did not make a satisfactory braking material.

An interesting patent for a brake block was that issued to Lucius Page of Cavendish, Vermont, in 1855. This allowed a shoe to be readjusted as it wore.

A new type of brake shoe was patented by James Christie, of Philadelphia, in 1865. This was a new departure and a basic idea in the construction of brake shoes. It covered a separate head and shoe which exposed most of the back of the shoe and left an open space for air circulation between the shoe and the head. This type was adopted by the Master Car Builders' Association at an early date and is a common standard at the present time.

At first the objection to the common cast iron brake shoe was its rapid wear and short life, necessitating frequent renewals. On the other hand, when the principal railroads in the country were using cast iron wheels, and brake shoes were of wrought iron strips riveted to cast iron blocks, it was found that the wrought iron flowed with the heat and, piling up metal on the shoe, gouged the wheel.

In an attempt to meet both of these objections, Isaac H. Congdon, Master Car Builder of the Union Pacific Railroad, patented a brake shoe in 1876 which provided for inserts of wrought iron to be embedded transversely in the cast iron. The advantages of the Congdon type of shoe were brought out at the Master Car Builders' convention in 1878, and it was gradually adopted by most of the railroads using cast iron wheels.

In the meantime, important developments were being made in wheels. The Pennsylvania Railroad had experimented with steel-tired wheels, but it was soon found that the Congdon shoe, with its inserts of wrought iron, scored the steel treads, and railroads using steel or steel-tired wheels were obliged to return to the use of the common cast iron brake shoe.

These experiences illustrate the "trial and error" methods which the railroads had to follow in order to determine whether or not certain designs and materials of brake shoes were adaptable for their use.

The gradual increase in train speeds and in the weight of equipment, together with the introduction of power brakes with quicker action, tended to put more work on the cast iron shoes and thus wear them out rapidly. On this account many ef-

Diamond-S and Streeter brake shoes, 1897 and 1899

FIG. 5

Showing construction of modern Diamond-S brake shoe

FIG. 6

forts were made to devise a brake shoe that would be strong and durable and yet not injure the wheels.

One such effort resulted in a patent, issued in 1884, to J. J. Lappin, a molder in a foundry located in Toronto. His idea was to make a brake shoe of hard iron cast against chill blocks spaced so as to form chilled areas at intervals in the wearing face. While this shoe to some extend accomplished the purposes aimed for, some of the frictional qualities of the unchilled shoes were sacrificed.

The Lappin shoe had sacrificed some frictional value while increasing durability. Could a shoe be made that would have long life and yet have proper frictional qualities? In 1896 Eugene W. Applegate was granted a patent on a brake shoe with inserts of very soft cast machinery iron set in a body of high chilling cast iron, with chilled ends. It was claimed that the softer inserts would compensate for the low friction of the chilled ends. This shoe was known as the Corning shoe, taking its name from Corning, New York, where it was made.

The Corning shoe was a real advance, but this improvement was destined to be overshadowed by an entirely different brake shoe which was developed the following year, for it was in 1897 that the Diamond-S type of construction appeared. In this construction hard iron was cast around a bundle of expanded mild sheet steel, producing a brake shoe with a wearing face of strong iron interlaced with strands of mild steel. This retarded the rapid grinding away of the cast iron, thereby obtaining a higher degree of durability without any sacrifice of friction. Due to the method of distribution of the mild steel in the shoe face, no injury was done to the steel tire. This development constitutes one of the most outstanding improvements that have been made in brake shoe construction.

Meanwhile, an effort was made to improve the Corning shoe. Realizing that a hard iron shoe, portions of which are chilled, is structurally weak, A. L. Streeter, a brake shoe manufacturer of Chicago, was issued a patent in 1899 which utilized a body of soft nonchilling iron surrounding a continuous spiral insert of very hard white iron disposed in the wearing face of the shoe. But the Streeter shoe met with little success as it did not equal the service given by the Diamond-S shoe.

However, the Diamond-S type of shoe, as first designed, caused a great deal of production trouble. The molten metal burned the insert as it was poured around it, resulting in voids or cavities which, in turn, frequently caused damage to the wheels. For this reason many railroad officials and brake shoe manufacturers opposed its use, even though its advantages were generally conceded. F. W. Sargent, and one or two of his associates, having a firm faith in this type of brake shoe construction and its superiority over any heretofore brought out, persisted in their experiments. Finally, through a rearrangement of the expanded metal, and the development of a protective coating, production troubles were overcome to a considerable extent. Research between 1930 and 1935 resulted in a further improvement in the coating which has reduced to a minimum the possibility of voids or cavities in the shoes.

Early examples of reinforced brake shoes, 1890

FIG. 7

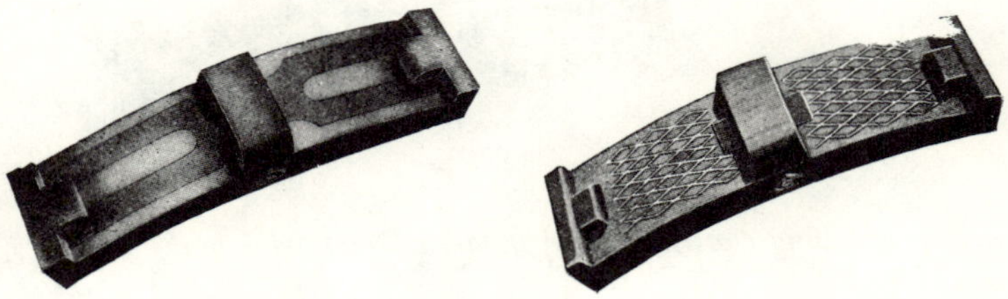

Steel plate and Diamond-S reinforcements, 1893 and 1897

FIG. 8

Duplane steel back reinforcement

FIG. 9

The Diamond-S shoe was adopted by all of the leading railroads, shortly after it was introduced, for use on passenger cars and locomotive tenders, and also by the Pullman Company for its equipment. In 1936 it was adopted as standard by the Association of American Railroads for interchange repairs on freight cars.

CAR SHOE REINFORCEMENT

Meanwhile the introduction of the Christie idea, which provided a separate head and shoe, had created a new problem. When the head and shoe were combined, the wearing metal was the main concern, but with a shoe separate from the head it became necessary, in addition, to develop some means of reinforcement to hold the shoe together.

In 1890, Charles R. Herron, a brake shoe manufacture of Cincinnati, Ohio, placed two rods of wrought iron longitudinally behind the inserts as used in the Congdon shoe. The purpose of these rods was to strengthen the shoe against fracture before and after going into service. However, it was found that these rods were likely to be burned by the molten metal cast around them, or to be worn through before the shoe was worn out, so a better means of strengthening the shoe was introduced in 1893. This important step in the improvement of brake shoes consisted of locating a plate of mild steel flush with the back of the shoe and firmly anchoring it by allowing the body metal to flow over the steel back at the ends and center. This enabled the shoe to be worn out completely without destroying the effectiveness of the reinforcement. Also, it was possible to continue the use of the brake shoe even though the body metal was cracked. The importance of this development is evidenced by the fact that this type of reinforcement, with modifications and improvements, has been carried through to this day.

While the standard steel back, which originated in 1893, proved satisfactory for passenger cars and locomotive tenders, it has been found that in freight service many shoes break in two, due to the

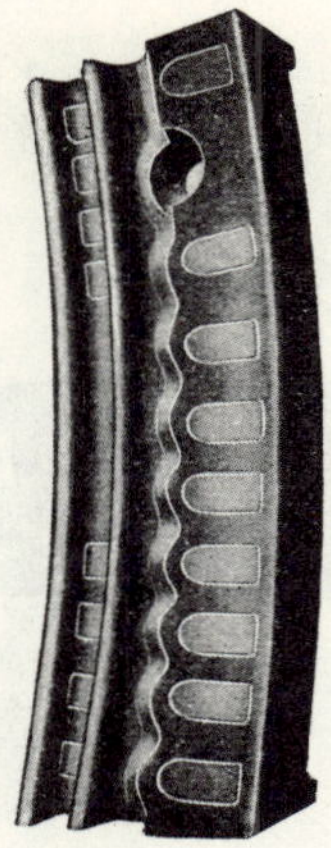

Ross-Meehan driver shoe

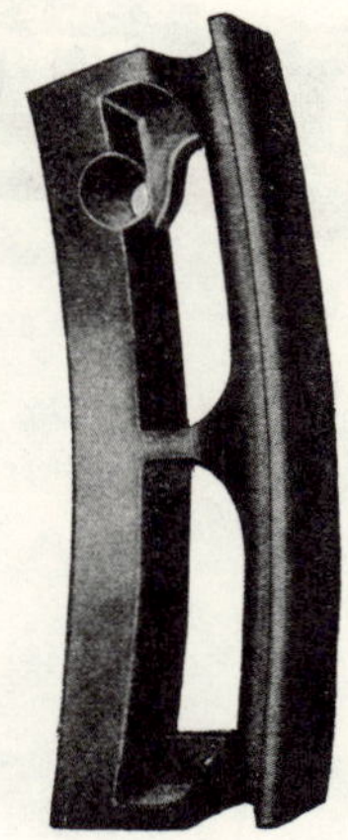

Skeleton-type driver shoe

FIG. 10

fracture of the steel back reinforcement. In 1930 the engineers of the American Brake Shoe and Foundry Company again undertook the task of developing a better type of steel back reinforcement. Approximately 80 designs of steel back reinforcements were put through severe tests in the laboratory. One by one they were eliminated, finally leaving the Duplane steel back as the one with the best record.

In this development a special hot rolled section is utilized, which is formed in two planes throughout its entire length. This design gives great stiffness and therefore resists vibration. It also makes it possible for the lug to be an integral part of the reinforcement.

These laboratory results were soon confirmed by road tests and brake shoes with the Duplane steel back reinforcement began to come into general use. This type of brake shoe was adopted by the Association of American Railroads in 1936 as an alternate standard for freight cars in interchange repairs.

BRAKE SHOES FOR LOCOMOTIVE DRIVER WHEELS

Soon after brakes were applied to locomotive driver wheels, the brake shoes were extended over the flange of the wheel, primarily to help keep the shoe in alignment with the wheel. The earlier driver shoes, being made of wrought metal, caused a certain amount of wheel wear and assisted in the grooving of the wheel tread by the rail. To overcome this condition various designs of brake shoes were employed. The first idea to be advanced was the recessing of the face of the brake shoe over the limits of the rail wear, but as this design permitted the shoe to move out of place on the wheel tread, it was soon discarded.

Later, in 1884, G. B. Ross of Buffalo, New York, received a patent which covered a shoe that was recessed over the limits of the rail wear on the wheel treads, but provided a lip on either side of the wheel flange to hold the shoe in place. This shoe, however, did not prove satisfactory. Made of cast steel, the inner flange lip wore the wheel flange at the throat. To eliminate this objection, while retaining the general idea of the Ross patent, a shoe design was brought out some time later in which the body metal was cut away, leaving only enough bearing to keep the shoe in position. This eliminated the objectionable cutting of the throat of the wheel flange.

The fact that locomotive driver wheels become unsafe for use when badly grooved, and have to be removed from service and turned on a lathe, led to the inspiration of making the brake shoes assist in keeping true the contour of the wheel tread. When cast iron shoes are used, no wear occurs on the wheel except that caused by the rail. The idea was conceived, therefore, of using inserts of hard cutting metal in the cast iron body metal of the shoe, and in 1885 J. B. Meehan of Buffalo, New York, patented such an idea. This patent covered inserts located in the shoe so as to act upon the outer tread and the top of the flange of the wheel where there is no wear by the rail. Later, the ideas of Ross and Meehan were combined by Charles R. Herron, who manufactured such a shoe, known as the Ross-Meehan shoe. It was claimed for this shoe

Perfecto Type driver shoe

Improved Perfecto driver shoe

FIG. 11

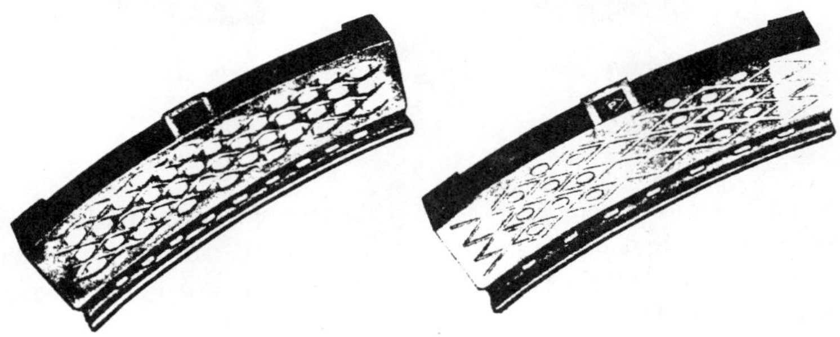

Diamond Perfecto driver shoes, 1911 and 1921

FIG. 12

that the cast steel inserts would cut the flange and outer tread, thereby dressing the tire, while the cast iron body of the shoe maintained the frictional requirements.

With the advent of the present century, the speed and weight of locomotives showed marked increases, resulting in proportional demands on the brake shoes. To meet these demands, a locomotive driver shoe was introduced in 1904 which increased the general efficiency of driver brake shoes. Known as the Perfecto Type, this shoe was made of strong cast iron and in the wearing face were embedded extremely hard non-annealing metal inserts, which mildly dressed the tire. The ends of the shoe were tapered to provide a greater area to radiate heat and were cast against chill blocks, which produced a hard surface to resist wear. This shoe not only partially dressed the tire of the wheel, but also provided sufficient frictional qualities. With improvements, it is used extensively today.

Because of the satisfactory results obtained by the use of the Diamond-S construction in car shoes, a locomotive driver shoe was developed in 1911 which likewise embodied the Diamond-S construction. This is known as the Diamond Perfecto Type and consists of adding a layer of expanded metal to both the face and the back of the body of the Perfecto Type. This produces a wearing face of strong iron interlaced with strands of mild steel and special metal non-annealing inserts.

REINFORCEMENT OF DRIVER SHOES

With the introduction of the Perfecto Type of driver shoe, a major improvement was embodied. This provided for the reinforcement of the shoe with a back made from a plate of steel, to which the body was firmly anchored and by means of which the shoe could be safely worn down to the steel back before replacement.

Also at this time, a steel back with separate hook attaching lug came into rather general use, but in 1908 the attaching hook was made integral with the steel back – a decided improvement in attaching arrangements over anything up to that time. This type remained in service for years despite two fundamental disadvantages. The shoe would wear down to the steel reinforcement on one end and be almost unworn at the other; and its design did not permit it to be reversed, so that a stock of right and left hand patterns was required.

To overcome these disadvantages a new center fastening attaching arrangement was designed for the driver shoe. Interchangeable from one side of

First hook and bolt type— cast iron hook

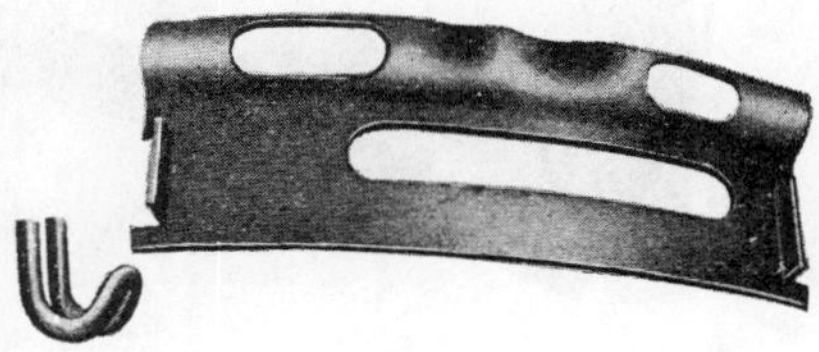

Hook and bolt type reinforcement — separate hook

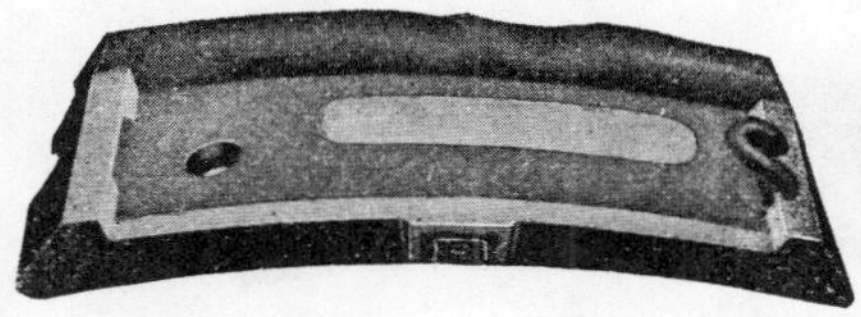

Driver shoe with above reinforcement

Reinforcement with integral attaching hook, 1908

Reinforcement with center fastening lug, 1907

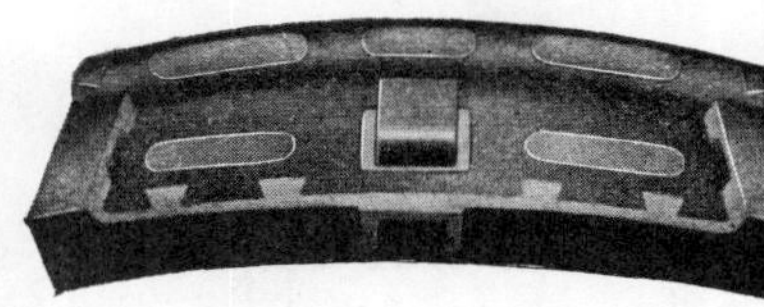

Driver shoe with above type of center fastening

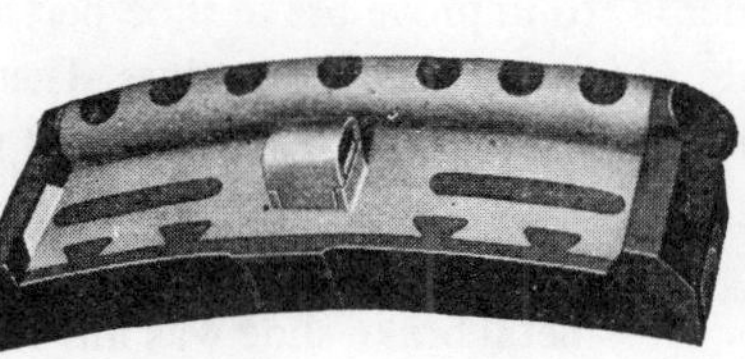

Improved center fastening driver shoe, 1922

FIG. 13

the locomotive to the other, the center fastening design also permitted the shoe to be reversed if worn unevenly from end to end. The advantages of this design were so apparent that brake head designs were modified to permit its use. Today the center fastening shoe is standard for locomotive driver wheels.

An improvement in the center fastening lug for driver shoes was introduced in 1922. This consisted of setting the shoulders of the lug on an upturned portion of the steel back and tightly clinching the lug toes under the steel back, thus altogether forming a lug construction that is very sturdy and one that will stand thrusts or blows in any direction, transmitting them to the steel back instead of to the cast iron body.

When modern operating conditions brought higher speeds and heavier loads, the burden on driver brake shoes was increased and pieces of the body metal dropped out due to heat cracks. To provide an improved method of body metal anchorage the Engineering Department of the American Brake Shoe and Foundry Company developed the Samson type of steel back for driver shoes,

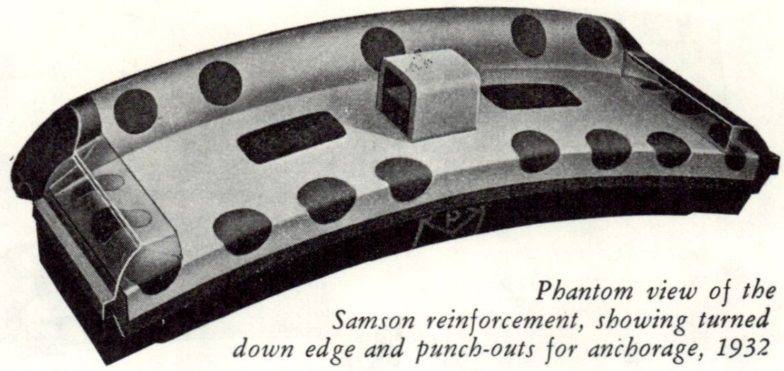

Phantom view of the Samson reinforcement, showing turned down edge and punch-outs for anchorage, 1932

FIG. 14

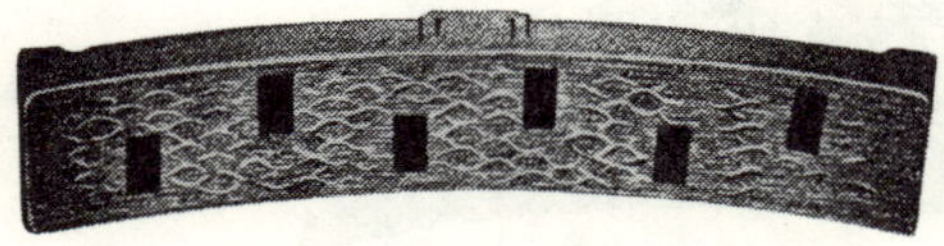

Carbon Insert Type

FIG. 15

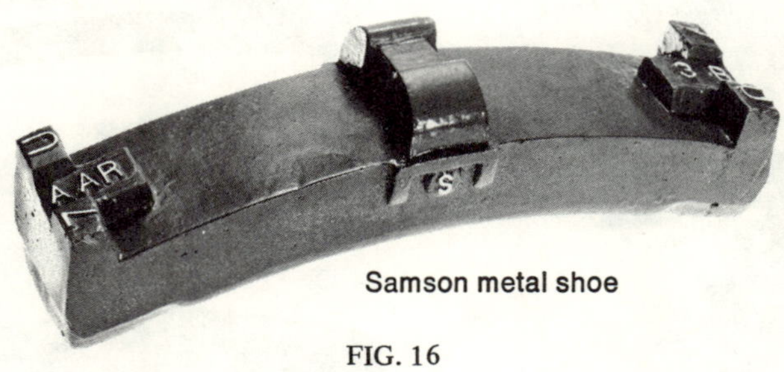

Samson metal shoe

FIG. 16

which began to come into use in 1932. The feature of this reinforcement is that the steel back is wider than the tire. It is turned down over the outside edge and thus encases the body of the shoe. Holes are punched in the steel back, into which the molten iron flows, securely anchoring the body metal to the reinforcement and thus preventing pieces of the body metal from falling out.

The Samson reinforcement was the outcome of extensive research and the results obtained in actual service have verified the experimental findings. Of the long and varied list of developments made in the design of reinforcements for driver brake shoes to bring them to their present high state of efficiency, the Samson type of reinforcement is most outstanding and is now in general use on the railroads of the United States and Canada.

In the 1930's the American railroads introduced streamlined passenger trains capable of 100 mph speeds. These high speeds increased the demand on the brakes that heavy sparking and short brake shoe life were sometime observed. The carbon insert brake shoe was developed beginning in 1935 to improve brake shoe performance on high speed trains and locomotives. The carbon inserts prevented rapid oxidation and wear of the cast iron body metal under the high energy conditions.

In 1970, the high phosphorus Samson cast metal brake shoe was introduced and has been widely applied to railroad equipment all over the world. This development began in England where it was found that cast iron containing higher than normal phosphorus content drastically reduced sparking, improved life and somewhat increased friction. The high phosphorus material is, however, more brittle than ordinary cast iron so that a steel bundle of expanded metal was needed to reinforce the high phosphorus brake shoe internally and to prevent a break-up of the shoe material.

Cast iron, malleable iron, and both hard and soft steel have been used in making brake shoes. Long experience has proved that wrought iron, when used in brake shoes, rolls and bunches up with the heat, while brake shoes of cast steel have a low friction value, require a large surface for heat dissipation, and when heated will flow and do extreme damage to the treads of steel wheels. At an early date, experience proved that malleable iron was not suitable for brake shoes as, after repeated heating and cooling, it returned to its original state of white iron.

Cast steel as a brake shoe body metal remained a rival of cast iron the longest, but about 1910 it went out of use on the last major railroad. Cast iron thus became the sole survivor and is today the best material that can be used successfully in cast metal shoes. By using steel back reinforcements, expanded steel bundles, metal inserts, carbon inserts and increased phosphorus content, the cast metal brake shoe has been developed to meet the demanding requirements of today's railroads.

21 Development of Brake Shoes: Part II—High Friction Composition

D. G. BLAINE

Continued attempts were made to improve brake shoe wear, reduce rubbing noise and obtain better, more uniform friction, at both high and lower speeds. Characteristic friction of cast iron shoes on cast iron or steel wheels was relatively low at high speeds, but rapidly increased as speed reduced, particularly from 20 mph down to the stop. In passenger services the train brake had to be graduated off or wheel sliding and rough stops would likely result. A notable development in the second decade of the 20th Century was the Pittsburgh Filled brake shoe. This consisted of a pressed steel cup-like container, the size of a typical shoe filled with composition material (Fig. 17). The filled shoe was rather widely used on some railroads and private car lines and worked well on

FIG. 17. PITTSBURGH FILLED BRAKE SHOE

chilled cast iron wheels. However, this composition, which included asbestos, did not give very good wheel wear on the steel wheels which came into general use in the late 1920's and early 1930's and the filled composition shoe was withdrawn from the market.

The great explosion of Light Weight High Speed Diesel Electric Streamlined Passenger Trains in the mid-to-late 1930's brought about another set of braking problems. The streamliners operated 20 to 40 mph faster than the older heavy-weight conventional passenger trains and often exceeded 100 mph. In order to reliably stop within existing signal spacings, considerably higher braking ratio and brake shoe forces were required with clasp brake (two shoes per wheel) and cast iron shoes. However, these very high shoe forces needed in the high speed zone to obtain the required retardation at high speed had to be reduced as speed reduced or severe wheel sliding was likely to occur. This gave rise to the HSC Speed Governor Control equipment which automatically reduced brake cylinder pressure and consequently shoe force as train speed reduced.

Truck-mounted brake cylinders and truck-contained brake leverage systems were commonly applied to passenger and diesel electric locomotives to improve overall efficiency and deliver maximum useable forces to the two brake shoes operating on each wheel. Brake shoe forces above 65 mph were from about 200 percent to 250 percent of the empty weight on each wheel. The braking ratios were approximately one-third greater than those employed on older equipment.

As mentioned previously, longer, larger, and more sophisticated cast iron shoes were developed for these exacting streamlined train services. Improvements included shoes with carbon block inserts to minimize the effects of sparking, rapid wear and very high heat input to the wheel tread during high energy dissipation, slow down, and stop braking.

A development of considerable importance in the late 1930's and 1940's was the North American Railway Disc Brake. The prime reason was to improve high speed braking by removing the heavy thermal load from the wheel tread and simplifying overall equipment by providing a higher

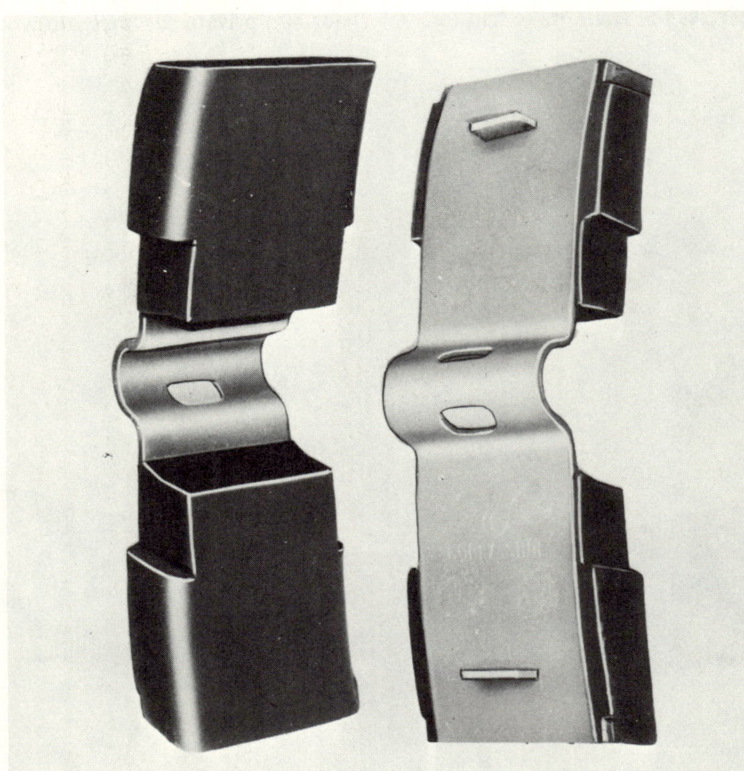

FIG. 18. MODERN HIGH FRICTION COMPOSITION TREAD BRAKE SHOE

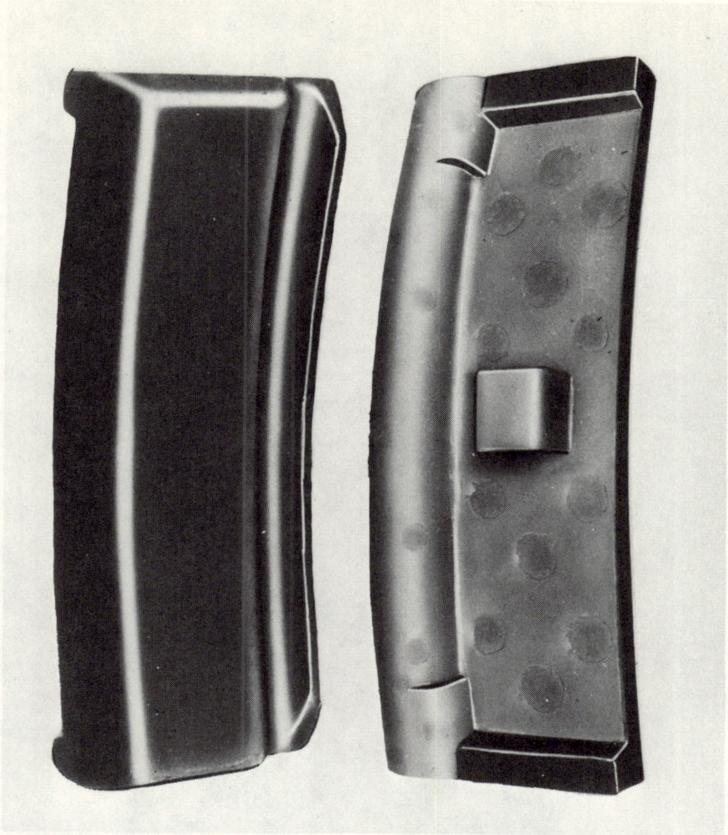

FIG. 19. FLANGED STYLE HIGH FRICTION COMPOSITION BRAKE SHOE

and virtually uniform friction level at high and low speeds. Attaching a two-faced disc to the inside hub of each wheel also reduced wheel wear effects from the brake shoes rubbing on the tread and contributed to longer disc shoe life. A combination of woven and molded composition friction materials was devised to provide better, more uniform friction, over a wide speed range and considerably longer wear life without the tendency to spark or flame sometimes encountered with iron shoe types. The disc brake eliminated the need for the complexities and weight of high force clasp brake rigging. This enabled considerable truck frame simplification and weight saving therefrom. In addition, the disc brake eliminated the need for the electro-pneumatic speed governor control.

However, automatic wheel slip control was found to be a necessity with high performance disc brakes because the retardation levels in higher speed ranges often exceeded practical day-to-day wheel rail adhesion under heavy service and emergency applications.

Intensive laboratory dynamometer investigations and road service tests in the early 1950's sought to adapt the advantages of the better friction and longer life characteristics of molded composition friction material to brake shoes operating on the wheel tread. In 1954 the first practical molded high friction composition (HFC) tread brake shoes were introduced for general service (Fig. 18). By 1955-1956 three (3) U. S. and Canadian companies were offering molded HFC shoes. These enabled major reductions in braking forces since essentially the same stopping and slow down performance could be obtained with from one-third to one-half the shoe force and with improved shoe wear. In most cases this has allowed one brake shoe per wheel to do the work of two (Fig. 19). The friction level of HFC shoes was adjusted to more closely follow the practical wheel rail adhesion curve vs speed than the original disc brake.

General acceptance of the high friction composition tread brake shoe occurred in the mid 1960's and virtually all new freight cars are so equipped.

FIG. 20. NORTH AMERICAN STYLE DISC BRAKE SHOE

In the passenger car field the high friction composition shoe made possible the development of the true package brake unit in which the force actuator (brake cylinder or air/hydraulic booster), leverage and automatic slack adjuster are combined in a compact housing with a brake head and brake shoe extending outward to contact the wheel tread. This can usually be located directly in line with the wheel between the wheel and truck frame transom or bolted to the truck side. Most of the modern transit and rail suburban cars employ some form of tread brake package unit and a single high friction composition shoe at each wheel for the friction brake.

Modern disc brake shoes or friction blocks using a single molded pad have been developed (Fig. 20). Newer main line passenger cars generally employ disc brakes utilizing these molded high friction type composition materials. Very recent developments include combination disc and tread brake systems with air or air/hydraulic actuation. This type of combination brake reduces peak thermal loads on both wheels and discs to optimize wheel and disc wear, shoe wear and retardation performance (Fig. 21).

FIG. 21. COMBINATION RAILWAY DISC BRAKE AND TREAD BRAKE

22 Rapid Transit Cars, Trucks, Propulsion 100 Years Development

J. P. VAN OVERVEEN

Where a rail transportation system is used in and around major population centers, it is usually identified as a suburban railroad, street car, or rapid transit system.

This mode of transportation started with the horse-drawn street car and has been more mechanized, electrified and automated ever since Thomas A. Edison built his first electric locomotive in 1880. Siemens in Germany built an electric locomotive pulling three cars (in 1879) for a Berlin exhibition. Current was collected from the running rails.

Leo Daft in 1882 experimented with a 450 lb locomotive and in 1883 built a 25 hp locomotive weighing 2 tons. It was powered by a "third rail". This engine was wrecked in an accident that same year. However, it was the forerunner of better equipment that ran as electric street-cars with overhead trolleys in Asbury Park, New Jersey in 1887. Other engineers that made a reputation for themselves in the early development of electric traction for street cars were Charles Van De Poele and Frank Sprague. Frank Sprague did much basic work. Many of his electrical patents form the foundation on which

were developed the first city-wide trolley systems, subways and elevated trains, and automatic train safety controls.

He sold his patents to such firms as General Electric and Westinghouse Electric. He designed the first multiple unit train controls and was actively engaged in these engineering endeavours until his death in 1934.

CAR DEVELOPMENT

The rail car and street car designs of the early days were based on the wooden, four wheel, horse drawn cars which soon gave way to the larger cars with eight wheels and two trucks around 1890.

Many electric street car companies started up around the turn of the century. Many built their own cars, which involved some engineering, a lot of craftsmanship and hand labor. Some of the cars, now being restored by Rail Historical Museums, show evidence of the level of expertise of the craftsman fitting all the wooden parts together for strength and the artistic quality of the interior finish.

Figure 1 shows an early type of car where the floor construction provided the basic strength of the car and the sides and roof are basically shelter enclosures. This car is a Brill Semiconvertible of the San Diego Electric RR. Figure 2 shows a two axle street car prior to its reconstruction. The side and center sills are heavy wooden beams, augmented by steel girders. This car, when restored, will look somewhat like the very early street car shown in Fig. 6. Figures 3 and 4 show a construction where the structural strength of a suburban type rail self-propelled car has been lodged in the wooden side wall. Note the heavy beam structure providing a bridge type beam support. Figure 4 shows a detail of a reconstructed portion of that same car, which illustrates the high quality of wood working craftsmanship that went into the construction. These rail cars operated at speeds in excess of 60 mph.

In the early nineteen hundreds, more and more cars were built of steel construction with heavy side and center sills. The U. S. Suburban Railway industry was in its heyday and in the forefront of the car construction industry for these types of vehicles on a world wide scale. The *Story of the London Underground* makes several references to cars built after American designs for their surface lines of that early period.

FIG. 1 CAR UNDER RESTORATION, RIO VISTA, CALIF. SAN DIEGO ELECTRIC CO. BRILL SEMICONVERTIBLE CIRCA 1900. NOTE ABSENCE OF TIMBER TRUSS IN CAR SIDE BENEATH WINDOW BELT RAIL. VERY HEAVY WHITE OAK FLOOR TIMBERS SUPPORT BODY SUPERSTRUCTURE AND CONTROL EQUIPMENT UNDERNEATH.

FIG. 2 BODY DETAIL BEFORE RESTORATION. SAN DIEGO ELECTRIC. BRILL TWO AXLE CAR. RIO VISTA, CALIF.

Suburban Electric Railway cars were adapted for all kinds of uses. There were special funeral cars (Fig. 17), parlor cars and even sleeping cars on the early lines. As suburban rail and street cars became less popular and were replaced by the automobile and bus, the U.S. rail car building industry played an ever decreasing role on the world market. Since the 1950's, the major technical developments in suburban rail car, subway and street car development have originated in Europe and Japan. One outstanding exception is the monocoque car design used on the San Francisco and Washington, D.C. rapid transit systems, which follows the principles of airplane construction (Fig. 25). This produces a light weight vehicle, less expensive to assemble and therefore at low initial cost. A possible drawback of this construction is the difficulty of repair of minor damage such as may occur during a side swipe at a yard switch, which is an occasional problem on any rail operated system. Since the sides of the car provide the structural strength, any damage to that structure is a serious matter.

In the fifties, stainless steel became a very popular material of construction for subway and suburban cars. In the last twenty years aluminum has found many applications. There is still a wide difference of opinion as to whether the light weight of aluminum outweighs its lower resistance to corrosion, mechanical fatigue and lower mechanical properties such as strength and yield characteristics, when compared with stainless steel.

Plastics are used extensively in the interior of the modern transit car. In addition, some front ends and operator cabs are enclosed in molded parts of epoxy fiberglass. Front windshields and side window panes are made of special composition such as acrylic and polycarbonate materials. These materials particularly are used to protect passengers and operators against acts of vandalism

FIG. 3 CAR SIDE WALL CONSTRUCTION BEFORE RESTORATION. NOTE THE HEAVY TIMBER BRIDGE TRUSS SIDEWALL CONSTRUCTION BETWEEN BELT RAIL AND SIDE SILL – TYPICAL OF WOODEN RAILROAD AND INTERURBAN CAR CONSTRUCTION FROM THE 1870's UNTIL STEEL FRAMING BECAME COMMON IN 1910-1920 PERIOD.

FIG. 4 CAR SIDE WALL CONSTRUCTION RESTORED.

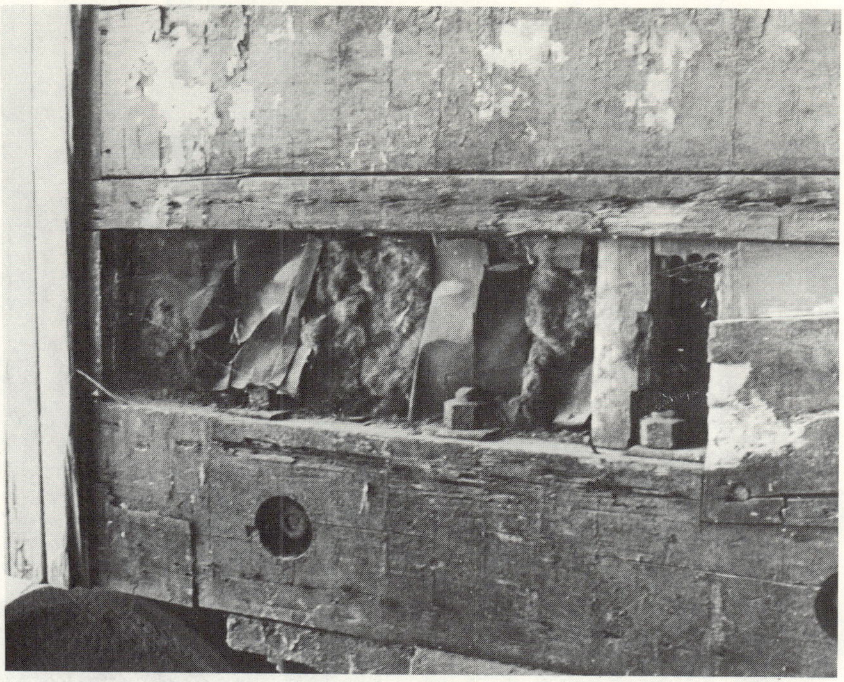

FIG. 5 HORSE HAIR SIDE WALL INSULATION.

that could destroy windshields or windows. Efforts are under way to use plastics of greater fire resistance than used heretofore.

Interiors of cars used in the mass transit service vary from the austere to very plush. Where short distances and large numbers of passengers are involved, the cars are designed for a large number of standees and only a small area of the available floor space is assigned to seats. Of late, more attention is given to adequate seating areas. Depending upon the possibility of damage by vandalism, these seats vary from plain fiberglass molded seats to foam rubber (or plastic) cushioned seats with vinyl or nylon upholstery (Fig. 26).

Lighting of the car interior has greatly improved. The noise attenuation has become a very important factor in the car insulation and design.

Today all types of fiberglass and foam materials are used for insulation. In the early days much horse hair was used, as shown in Fig. 5.

A most influential design of a Street Car or Light Rail Vehicle was the President's Conference Committee Car, more widely known as the PCC car. This design evolved around 1930 from the desire of the Rapid Transit Industry to arrive at a standard car design which could be used by all operating properties, manufactured in larger numbers and therefore be economically attractive.

These types of cars are still operating today in San Francisco, Boston, Philadelphia, Pittsburgh and elsewhere all over the world.

The design introduced the rubber sandwich wheel construction to reduce noise, the electromagnetic track brake to increase braking capability and an advanced electric traction controller for smooth and high rate acceleration.

Many recently built European light rail vehicles still have design features based on the PCC car.

Figures 6 thru 15 show a variety of street car designs. Figures 13 thru 25 show the trend of development of Rapid Transit or suburban vehicles, starting with the steam powered elevated railroad of New York City.

These illustrations do not intend to cover all designs that have operated or are still operating, but are shown as examples of the changes in concept and design.

This paper would not be complete without mention of the cable cars, which have become a National Monument in San Francisco. This operation was the recipient of the first Historic Engineering Achievement plaque issued by ASME.

Much has been done in the design of mass transit vehicles. The subject has been reviewed by a great number of books containing many photographs and illustrations. The purpose of this con-

FIG. 6 SAN FRANCISCO-MARKET STREET RAILWAY. CAR 578 – PECKHAM TRUCK – RIO VISTA MUSEUM, CALIF.

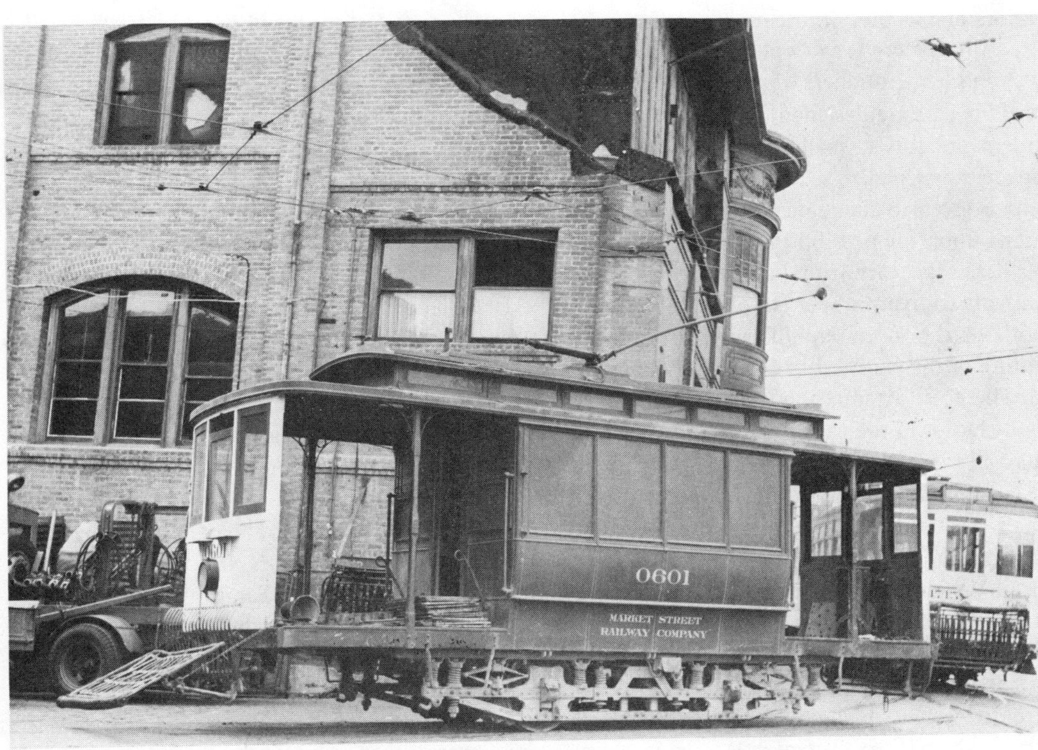

FIG. 7 SAN FRANCISCO-MARKET STREET RAILWAY – SAND CAR PECKHAM TRUCK–BUILT CIRCA 1895.

FIG. 8 WORK CAR BUILT BY OAKLAND TRACTION CO. 1906.

FIG. 9 CINCINNATI STREET RAILWAY – CURVED SIDE CAR – BUILT BY CINCINNATI CAR CO.

FIG. 10 SAN FRANCISCO MUNICIPAL RAILWAY CAR NO. 1 "ARNOLD CAR" – HOLMAN CAR CO.-1911 PCC CAR – ST. LOUIS CAR 1952

FIG. 11. SAN FRANCISCO MUNICIPAL RAILWAY. TYPE K. 47 FT LONG X 9 FT WIDE X 12 FT HIGH. BUILT BY BETHLEHEM SHIPBUILDING CO. DELIVERED TWENTY BETWEEN JULY 1 AND NOVEMBER 9, 1923. COST $16,500. EACH. STANDARD CAR TRUCKS WITH 34 INCH DIAMETER WHEELS. SUPPLY CURRENT AT 600 V DC, FOUR WESTINGHOUSE ELECTRIC MOTORS. CAR RATED AT 260 HP. WEMHL CONTROLS. AIR BRAKE WAB-SME.

FIG. 12 PITTSBURGH, PA. PITTSBURGH RAILWAYS – NOW PAT. PCC CAR ST. LOUIS CAR CO. 1949-B3 TRUCK.

FIG. 13 NEW YORK CITY. STEAM POWERED ELEVATED – PRIOR 1890.

FIG. 14 NEW YORK CITY – NINTH AVENUE ELEVATED 1936.

FIG. 15 NAPA VALLEY RR CALIFORNIA-1937. BUILT BY NILES 1908 – BALDWIN TRUCK.

FIG. 16 SACRAMENTO NORTHERN RR, BUILT BY HALL SCOTT MOTOR CAR CO. – BALDWIN TRUCKS

FIG. 17 CHICAGO TRANSIT AUTHORITY FUNERAL CAR #756, LATER #2756. BUILT 1898 BY BARNEY AND SMITH AS A MOTOR CAR. CONVERTED TO FUNERAL CAR BY PULLMAN IN 1907 AND OPERATED AS SUCH UNTIL 1926. BECAME MEDICAL EXAM CAR UNTIL 1952, WHEN USED AS TRAINING CAR. STEEL UNDERFRAME, WOOD BODY AND FINISH. ORIGINALLY 2-GE 55 MOTORS; NEW YORK STRAIGHT AIR BRAKE ORIGINALLY BUT CONVERTED TO WESTINGHOUSE AMR AUTOMATIC BRAKE IN 1904.

FIG. 18 CHICAGO TRANSIT AUTHORITY, STEEL FRAME, WOOD SHEATH CONSTRUCTION FOR THIRD RAIL AND TROLLEY OPERATION. BUILT IN 1907 FOR NORTHWESTERN ELEVATED RAILWAY CHICAGO BY ACF AS A NON-CONTROL TRAILER. CONVERTED IN 1950 TO A DOUBLE END CONTROL MOTOR CAR AND RENUMBERED 1811, USED AS WORK MOTOR CAR 1958 UNTIL SCRAPPED IN DEC. 1967. WABCO AMR AUTOMATIC BRAKE M-23 BRAKE VALVE.

FIG. 19 CHICAGO TRANSIT AUTHORITY – MIXED WOOD AND STEEL CAR M.U. CONSIST – CIRCA 1940. METROPOLITAN WEST SIDE EXPRESS (CHICAGO-MAYWOOD-WESTCHESTER) CROSSING 4 TRACK 3RD RAIL BRIDGE OVER CHICAGO RIVER AFTER STOPPING AT "CLINTON"-UNION STATION TRANSFER. NOTE CHICAGO UNION STATION BUILDING UPPER RIGHT AND CHICAGO, BURLINGTON AND QUINCY RAILROAD HEADQUARTERS BUILDING ABOVE TRAIN IN CENTER LEFT. NOTE ALSO 5 CAR CHICAGO AURORA & ELGIN RR TRAIN ON LEFT OUTSIDE PLATFORM AT CLINTON.

FIG. 20 CLEVELAND TRANSIT SYSTEM – AIRPORTER TYPE CARS – STAINLESS STEEL CAR CIRCA 1970 BUILT BY PULLMAN. STANDARD DOUBLE END CINESTON CONTROL, 4-100 HP MOTORS, TYPE RT-2 WABCO ELECTRO PNEUMATIC VARIABLE LOAD BRAKE COMBINING TREAD BRAKE AND DYNAMIC BRAKE. TYPE N-2 CAR AMR & ELECTRIC COUPLER. PHOTO TAKEN IN PULLMAN 111th ST. PLANT.

FIG. 21 NEW YORK METROPOLITAN TRANSIT AUTHORITY LONG ISLAND RAIL ROAD STAINLESS STEEL CARS – 1971 – BUDD BUILT. 4 160 HP MOTORS PER CAR, A-B TWO CAR SETS CABIN OUTER END. WABCO RT-5-C "P" WIRE BRAKE AND POWER CONTROL, AUTOMATICALLY BLENDED UNIT STYLE FRICTION TREAD AND FULL DYNAMIC BRAKE CONTROL. N-2-A CAR, AR, AND ELECTRIC COUPLER WITH AUTOMATIC CENTERING DEVICE.

FIG. 22 NEW YORK METROPOLITAN TRANSIT AUTHORITY NEW HAVEN LINE – 1976. G.E.-BUDD BUILT – "P" WIRE CONTROL. WESTCODE E.P. BRAKES WITH DYNAMIC BLENDING. UNIT STYLE TREAD BRAKES.

FIG. 23 PHILADELPHIA – PORT AUTHORITY TRANSIT COMPANY. STAINLESS STEEL – BUDD BUILT – CIRCA 1971. WABCO RT-5 BRAKE WITH "P" WIRE CONTROL OF POWER AND BRAKES, AUTOMATIC BLENDING OF UNIT STYLE TREAD BRAKE AND DYNAMIC.

FIG. 24 NEW YORK CITY TRANSIT AUTHORITY STAINLESS STEEL SUBWAY CARS – (R-42). A & B CAR SETS – WABCO SMEE (RT-2 TYPE) VARIABLE LOAD E.P. BRAKE EQUIPMENT WITH STAGE TYPE DYNAMIC BLEND.

FIG. 25 SAN FRANCISCO BAY AREA RAPID TRANSIT DISTRICT ALUMINUM MONOCOQUE CONSTRUCTION – ROHR CORP., 1971. P WIRE STYLE POWER AND BRAKE CONTROL. HURST ELECTRO HYDRAULIC FRICTION DISC BRAKE WITH AUTOMATIC BLENDING OF DYNAMIC MOTORS (4 PER CAR) AND CONTROLS BY WESTINGHOUSE ELECTRIC CORP.

FIG. 26 SAN FRANCISCO BAY AREA TRANSIT SYSTEM INTERIOR. SEATS 72. AIR CONDITIONED. RUG (WOOL) ON FLOOR. WALLS PLASTIC SHEET. SOFT SEAT CUSHIONS.

tribution is to merely indicate some of the engineering efforts that have been expended on this subject for the last hundred years.

TRUCK AND DRIVE TRAIN DEVELOPMENT

A variety of vehicles put on wheels were assembled onto supporting structures called trucks. Countless trucks have been built at one time or another to comply with the demands of some particular operating company. In a compilation, *Electric Railway Car Trucks,* E. Harper Charlton lists some 100 different truck manufacturers, many of which produced more than one model truck. Appended is a list of some of these manufacturers, showing the names of the truck designers, where known, and the year of the manufacture of the first truck.

Some examples of truck types which are not seen any more are:

1. The three axle radial truck, which had three single axle units with link coupling interconnection. The end units supported the car through king pin swivel supports. The center unit acted as a guide for the end units, and did not support the car body directly. It was used on the West End Street Railway of Boston (Fig. 27). The Pivotal Radial Trucks consisted of two single axle pivoting trucks interconnected to each other (Fig. 28).

2. The maximum traction truck in which the motor driven axle had larger diameter wheels than the guiding idler axle on the same truck (Fig. 29).

Spring supports consisted of the primary springs on the journal box pedestal, while in most instances the center support or truck bolster was suspended from the truck frame by a variety of spring support mechanisms consisting of either eliptical or coil springs or both, called secondary suspension. Some truck designs were very stiff frames, while others were constructed with side frames that had

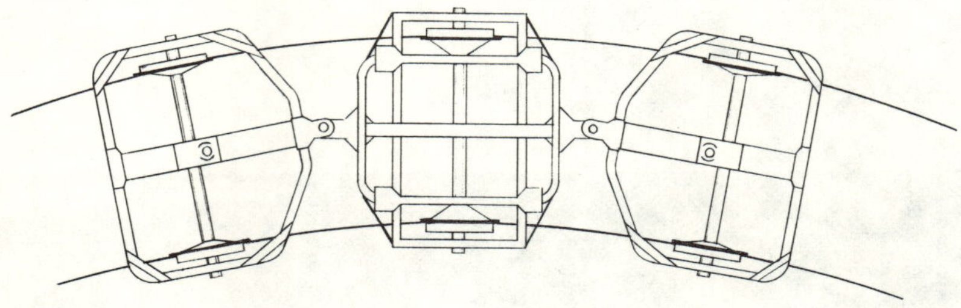

FIG. 27 SCHEMATIC OF THREE AXLE RADIAL TRUCK

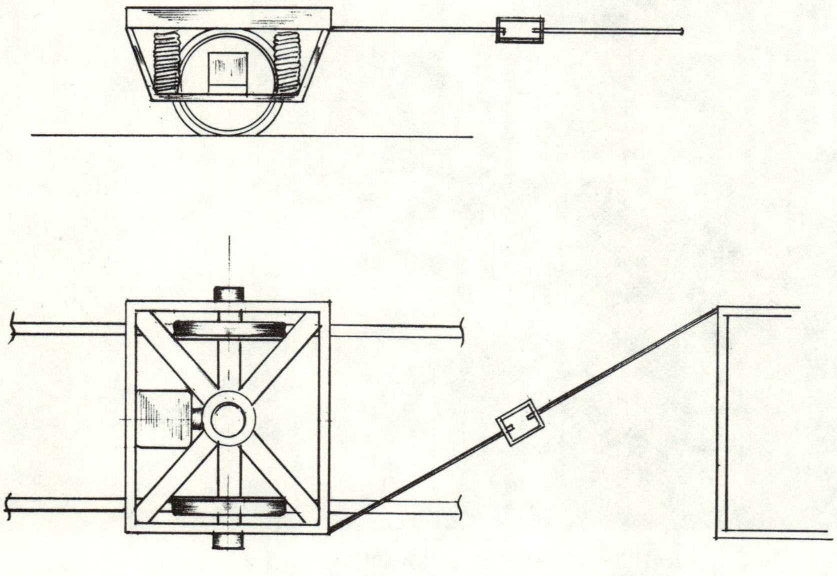

FIG. 28 SCHEMATIC OF PIVOTAL RADIAL TRUCK

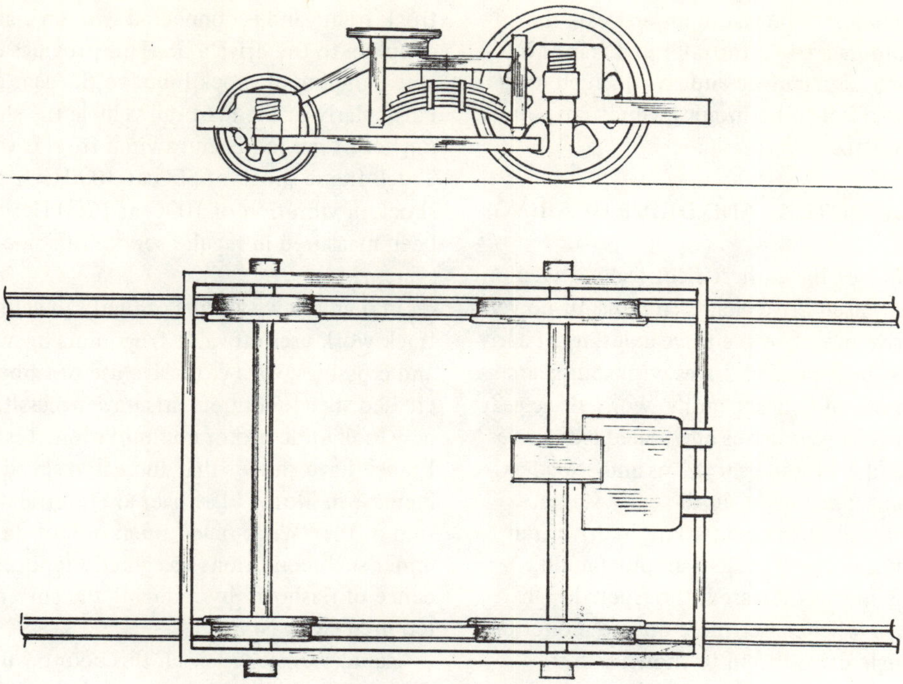

FIG. 29 SCHEMATIC OF MAXIMUM TRACTION TRUCK

the capability of some independent motion in the vertical plane.

It is interesting to note that as the riding comfort became more important, increased attention to compliant suspension and reduction of noise is noticeable. Particularly with the introduction of the air spring or air support bellows, the secondary suspension became very important.

3. The "rubber chevron" design, consisting of alternate rubber and steel plate layers, vulcanized together, provided novel methods for primary and secondary spring suspension.

In some designs, primary springing is nearly nonexistent and the journal bearings (roller bearings) are contained in a rubber cylindrical sleeve which is solidly mounted in the side frame.

Where the track alignment is not as good as preferred, the air spring suspension contributes greatly to riding comfort, and also provides greater compliance and better wheel contact with the track.

In addition to ride quality, the air spring provides a means to retain the car floor level at the same height above the top of the rail, regardless of the number of passengers carried, which allows the station platform and car floor level to be in the same horizontal plane at all times.

The increased or decreased air pressure in the air spring is also used to provide a so-called load weigh signal to the traction control, thereby enabling the proper propulsion or braking effort to be delivered, so that acceleration or deceleration is the same for empty or loaded trains. This can be very important for rail transit systems operating at a high frequency of train schedules during commute hours.

The latest available truck designs as used in New York City, Chicago, and San Francisco in rapid transit and light rail vehicle applications are shown in the appended illustrations. To the casual observer, some of these trucks may not look much different from those used 50 years ago. However, the details are greatly improved.

In present technology we can design for known shock and vibration input while in the past: "If it breaks, make it heavier." was the usual approach.

Rail transit and light rail designers now are looking at the wheel-rail contact as part of a system. The complexity of a truck can be reduced greatly if the track alignment is of high quality and can be maintained as such. Wheel tread profile and the relation between track gauge and wheel base of a truck are becoming more important as operating speeds increase.

On the whole, in the last hundred years the "art" of building a truck for rail rapid transit has become a precise scientific endeavor, which must consider the effect of harmonics, vibrations, shock and noise control.

ELECTRIC MOTORS AND DRIVE GEARING

While some of the earliest electrically driven on track vehicles used drive chains to transmit power from the drive motor to the drive axles, most early applications show parallel drives with spur gears.

With the use of helical gearing, worm drive gear and hypoid type gear drives and sealed axle drives increased, and enclosed gear drives and sealed-axle drives become a necessity. In the early stages, where speeds did not exceed 40 mph, drive train design did not seem to present an outstanding problem. As speeds increased, the truck designs show a longer wheel base where double reduction and right angle drives begin to show up in the designs.

The revival of rail rapid transit in the post-World War II era demonstrates designs that show increasing need for protection of the gear drives from shock, both torque and other, originating at the wheel-rail interface.

The Westinghouse Electric Company designed the so-called Trak Pak (Fig. 30), where the gear case is suspended by elastomer mounts from the truck frame and is connected with an elastomeric coupling to the drive axle. This provides considerable damping of shock input to the gear drives. Particularly in the light rail vehicle the shock impact of rail joints and switch frogs is very severe. Truck frame shock of 50G at 300 Hertz and wheel shock or vibration of 100G at 1200 Hertz have been measured in regular service of some transit systems.

In many new systems of light rail vehicles the track work uses movable frog joints in switches and crossings, with extensive use of ribbon rail. Increased speed requirements have necessitated a new look at dc motor commutation. Tests in France have shown that increased speed and increase in Motor Flashover are interrelated. Omission of the "Weak field" operation of dc motors under such conditions has greatly reduced the incidence of flashover by reducing the current level required (Ref. Nouvion).

Rapid Transit Trucks in this country use parallel drives with double or single reduction helical gear drives, or perpendicular drives through hypoid gearing. The new light rail vehicles manufactured by Boeing use monomotor trucks of German design (Figs. 31 and 32).

A very similar looking Monomotor Truck, (Fig. 33) using the motor frame as part of the truck structure, was designed in 1888 and used on the Elkhart, Indiana, street railway. Apparently some

FIG. 30 TRAK-PAK PARALLEL DRIVE (WESTINGHOUSE ELECTRIC)

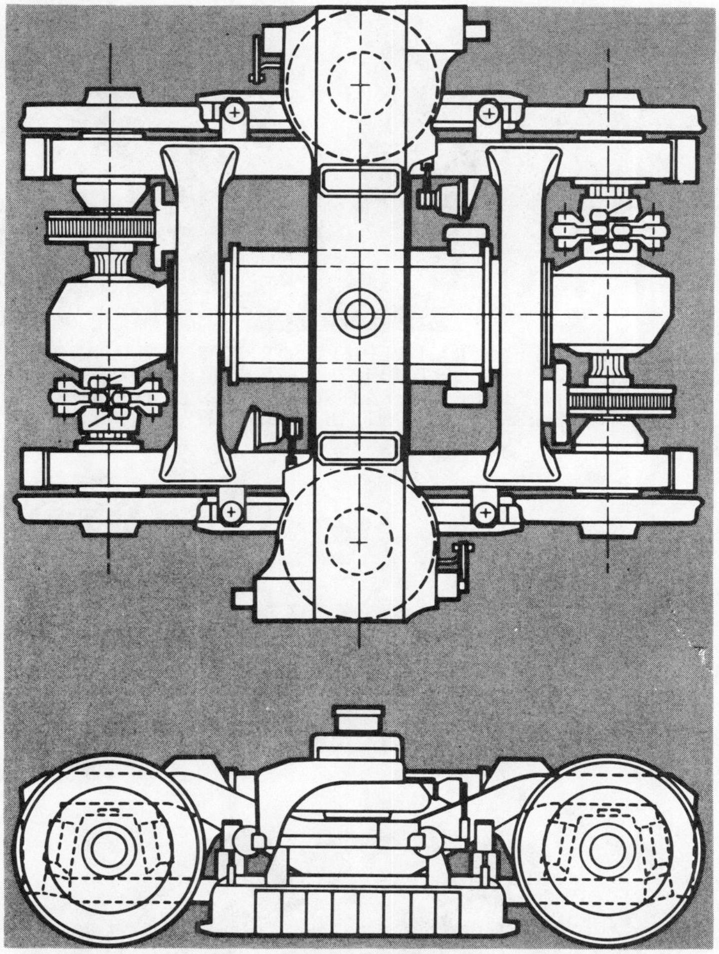

FIG. 31 SCHEMATIC OF MONOMOTOR TRUCK FOR LIGHT RAIL VEHICLE (BOEING-VERTOL) 1975.

FIG. 32 TRUCK FOR LIGHT RAIL VEHICLE, STREET RAILWAY AND ELEVATED SERVICE (BOEING-VERTOL) 1975. TRUCKS EQUIPPED WITH MAGNETIC TRACK BRAKE SHOES AND A DISC FRICTION BRAKE ON EACH AXLE. NEW YORK AIR BRAKE CONTROL EQUIPMENT.

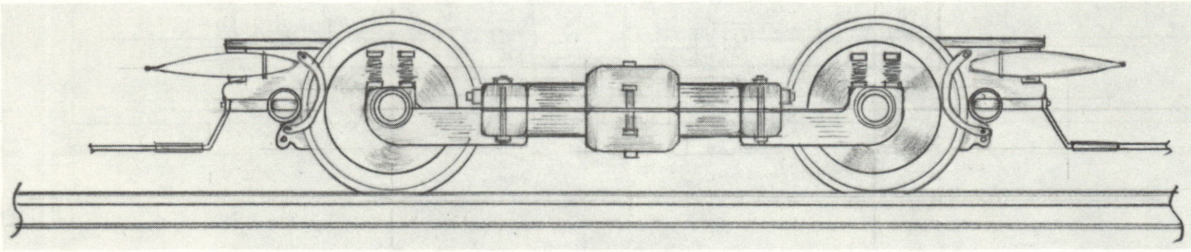

FIG. 33 SCHEMATIC OF MONOMOTOR TRUCK USED IN ELKHART, INDIANA 1888

FIG. 34 TRUCK FOR NEW YORK SUBWAY R-46 CARS. ROCKWELL ATCHISON DIV. 1976. AUTOMATIC BLENDING OF UNIT STYLE FRICTION TREAD AND DYNAMIC BRAKE BY WESTCODE.

of the new designs have been reinvented as has happened in other branches of engineering design.

This part of the 100 years of progress in railway mechanical engineering has only touched on the actual progress of the work horse beneath the vehicle, the truck and the drive train.

Attached hereto are additional illustrations and descriptions of trucks.

Figure 34 shows the latest truck design used by the New York Subway's R-46 Cars. The construction allows flexing of the side frames (Rockwell).

Figure 35 shows the truck assembly used by the San Francisco Bay Area Rapid Transit. It is a flexible frame design (Rockwell).

Figure 36 shows the side view of an early vintage truck, probably a Baldwin Motor Truck Class K.

Figure 37 shows a modified St. Louis MCB-23 truck. The connecting rod was applied to provide greater traction to a work car. The truck is equipped with one parallel drive motor and used under the work car shown in Fig. 8.

Appendix C shows a series of trucks used by the London Transport Subway, starting from the year 1906 till 1973. The designs show typical changes with the times. Until the 1940's these designs were based on the most successful U.S. designs, modified for the lower car profiles needed for that system.

The rapid transit suppliers and users have an increasing exchange of technical information the world over. The present day advancement of the technology is pursued intensively in Japan, Germany, Switzerland, France and in this country, with much interchange of information between individuals engaged in this endeavor. The next one hundred years will see further and rapid advancement in the state of this art.

FIG. 35 TRUCK FOR SAN FRANCISCO BART CARS. ROCKWELL ATCHISON DIV. 1970.

FIG. 36 BALDWIN TWO MOTOR TRUCK CLASS K. RIO VISTA RAILROAD MUSEUM

FIG. 37 ST. LOUIS MCB-23 TRUCK, MODIFIED WITH CONNECTING ROD FOR MAXIMUM TRACTION. RIO VISTA RAILROAD MUSEUM.

APPENDIX A

Below is a list of truck manufacturers showing the year of the first truck built.

American Car Company
Designer William Sutton, Pres., 1891
Maximum Traction Type Truck, 1895

American Car and Foundry – 1905
Anger Trucks named for John E. Anger, a native of England, associated for a time with Gilbert Car Manufacturing Company. First built 1891.

Baker Trucks named for W. E. Baker, master mechanic of the West End Street Railway of Boston, Mass. First built by Loconia and Bemis in 1895.

Baldwin Locomotive Co.
First date shown – 1915

Baltimore Car Wheel Co. – 1891

Barber Trucks designed by Edward A. Barber Fabricated by Peckham – 1907

Barney and Smith Car Co. – 1891

Bay State Street Railway Co. (Boston) design built by St. Louis Car Co. – 1909

Bemis Car Box Company, Designer Summer A. Bemis – 1891

Bentley Knight Double Motor Truck – 1884. First true frame electric truck for Cleveland Transit.

J. R. Blackwell of Chicago and Joliet Electric Railway in 1927 and 1930 designed two different experimental trucks, together with Timken, Westinghouse Electric, Westinghouse Air Brake and International Motors Company, using automotive suspension and worm drive.

J. G. Brill and Company, Philadelphia, the largest builder of electric railway car trucks, started in 1887. Manufactured some 95 major types of trucks (last year 1929).

Brownell Car Co. – 1892

Budd Co. – First Truck 1960.

Chicago Electric Truck Company
Designed by George H. Graham – 1894

Cincinnati Car Company
Elliott Truck – 1902

Commonwealth Steel Co. – 1921

Dorner and Dutton Manufacturing Co. – 1890

General Steel Castings (later General Steel Industries)

Took over Commonwealth Steel Co. – 1930.

Charles E. Keevil designed truck for Chicago Transit Authority
This principle still used by Rockwell LFM Division.

Mc Manufacturing Co.
Moses Hubbard Designer – 1890

Peckham Manufacturing Co.
Edgar Peckham Designer – 1889
These designs also used in England.

Transit Research Corporation for P.C.C. type cars
Build by Clark Equipment Co.

REA Motor Truck
Original design of Mono Motor Truck – 1888
Used Motor Housing as frame of truck
Elkhart, Indiana Street Railway.

St. Louis Car Co.
Truck Dates from 1894 extensively used in Berlin

Standard Steel Truck Co.
W. G. Price Designer – 1904

Taylor Electric Truck Co.
John Taylor – 1890

Timken – Detroit Axle Co.
Automotive Design Principles – 1927
No frame, but Elliptic Springs to connect axles – 1928

Tripp Manufacturing Co.
Earliest user of roller bearings – 1888

APPENDIX B

BIBLIOGRAPHY

Flanagen, R. C., and Fraelich Jr., K. H., *A New Transit Car Propulsion System for PATH,* Transportation Equipment Div., Westinghouse.

"Les Materials Moderne De La Regia Autonome," *Des Transports Parisiens,* October 1970.

"Air Suspension Bogies for Train Set," *Elektrische Bahnen,* Series 420, Issue 11, 1969, pp. 263-269.

Rashin, Donald, and Chergwin, Keith M., *Energy Storage Cars.*

"Energy Storage Propulsion Systems for Rapid Transit Cars," UMTA Report No. UMPTA-NY-06-0006-75-1.

Nouvion, F., *The Evaluation of Traction Motors on the SNCF.*

Nouvion, F., "Diesel and Electric Railroad Operation in France," ASME/IEEE Railroad Conf. 1972, IEEE Paper 72-936-8-IA.

Appleby, E. C., "A Space Age Drive for Rapid Transit Cars," ASME Paper 63-RR-5.

"Power Energy Drives – The State of the Art," *Metropolitan,* Jan.-Feb. 1976.

"Chicago Rapid Transit Rolling Stock 1947-1976," Bulletin 115 Central Electric Railfan Association.

Rowsome Jr., Frank, *Trolley Car Treasury,* Bonanza Books, Crown Publishers, 1956.

Carlton, E. Harper, *Electric Railway Car Trucks,* Harold E. Cox, 1967.

Day, John R., *The Story of London's Underground,* London Transport Publicity, 1974.

Thorn, L. C., "London Transport Traction Motors," *Railway Engineering Journal,* Sept. 1972.

Graff-Baker, W. S., "Consideration of Bogie Design," with particular reference to "Electric Railways," *Institution of Mechanical Engineers,* January 1952.

ASSISTANCE IS ACKNOWLEDGED FROM:

K. L. Jackson GSI – St. Louis, Mo.

F. Nouvion – Paris, France

T. J. Lowe, London Transport

R. F. Townley, Bechtel Corp.

G. Krambles, CTA, Chicago, Ill.

W. Vigrass, PATCO, Philadelphia, Pa.

A. R. Vollmecke, Boeing Vertol Co., Philadelphia

E. K. Arrelly, PATH, Jersey City, New Jersey

E. C. Appleby, Westinghouse Electric Corp., Philadelphia, Pa.

Franck Bonneres, Metro Cammell Limited, Birmingham, England

J. Grosser, Westinghouse Electric, Sao Paulo, Brazil

R. J. Bruggen, Rockwell Transportation Division, Atchison, Kansas

D. G. Blaine – Westinghouse Air Brake, Chicago, Ill.

S. A. Driver, London Transport

Harre W. Demoro – Oakland, California

Rio-Vista Railroad Museum – Rio Vista, California

APPENDIX C

LONDON TRANSPORT TRUCKS

FIG. C-1 1906 – STANDARD TYPE O TRAILER

FIG. C-2 1921 – TUBE MOTOR TRUCK

FIG. C-3 1922 – TYPE H MOTOR TRUCK

FIG. C-4 1923 – BRUSH MOTOR TRAILER

FIG. C-5 1923 – TYPE Y MOTOR TRUCK

FIG. C-6 1926 – TYPE W TRAILER

FIG. C-7 1938 – TUBE MOTOR TRUCK

FIG. C-8 1947 – PART RIVETED TRUCK

FIG. C-9 1967 – VICTORIA LINE TRUCK (AUTOMATED LINE)

FIG. C-10 1973 – TUBE MOTOR TRUCK